Lecture Notes in Civil Engineering

Volume 112

Lecture Notes in Civil Engineering (LNCE) publishes the latest developments in Civil Engineering - quickly, informally and in top quality. Though original research reported in proceedings and post-proceedings represents the core of LNCE, edited volumes of exceptionally high quality and interest may also be considered for publication. Volumes published in LNCE embrace all aspects and subfields of, as well as new challenges in, Civil Engineering. Topics in the series include:

- Construction and Structural Mechanics
- Building Materials
- Concrete, Steel and Timber Structures
- Geotechnical Engineering
- Earthquake Engineering
- Coastal Engineering
- Ocean and Offshore Engineering; Ships and Floating Structures
- Hydraulics, Hydrology and Water Resources Engineering
- Environmental Engineering and Sustainability
- Structural Health and Monitoring
- Surveying and Geographical Information Systems
- Indoor Environments
- Transportation and Traffic
- Risk Analysis
- Safety and Security

To submit a proposal or request further information, please contact the appropriate Springer Editor:

– Mr. Pierpaolo Riva at pierpaolo.riva@springer.com (Europe and Americas);
– Ms. Swati Meherishi at swati.meherishi@springer.com (Asia - except China, and Australia, New Zealand);
– Dr. Mengchu Huang at mengchu.huang@springer.com (China).

All books in the series now indexed by Scopus and EI Compendex database!

More information about this series at http://www.springer.com/series/15087

Mahdi O. Karkush · Deepankar Choudhury
Editors

Modern Applications of Geotechnical Engineering and Construction

Geotechnical Engineering and Construction

Springer

Editors
Mahdi O. Karkush
Department of Civil Engineering
University of Baghdad
Baghdad, Iraq

Deepankar Choudhury
Department of Civil Engineering
Indian Institute of Technology Bombay
Mumbai, Maharashtra, India

ISSN 2366-2557 ISSN 2366-2565 (electronic)
Lecture Notes in Civil Engineering
ISBN 978-981-15-9398-7 ISBN 978-981-15-9399-4 (eBook)
https://doi.org/10.1007/978-981-15-9399-4

This Springer imprint is published by the registered company Springer Nature Singapore Pte Ltd.
The registered company address is: 152 Beach Road, #21-01/04 Gateway East, Singapore 189721, Singapore

Contents

About the Editors

Prof. Mahdi O. Karkush is the President of Iraqi Scientific Society for Soil Mechanics and Foundation Engineering, and Professor of Civil Engineering at the Department of Civil Engineering at the University of Baghdad, Iraq. He received his Ph.D. in Civil Geotechnical Engineering from University of Baghdad, Iraq in 2008. He pursued his B.Sc. in Civil Engineering from University of Babylon, Iraq in 1995 and M.Sc. in Civil Structural Engineering from University of Babylon, Iraq in 1998. Prof. Karkush has authored over 65 technical publications including 50 peer reviewed journal papers. He has supervised over 20 Ph.D. and M.Sc. students. He is a member of the editorial board of several scientific journals. He is a Fellow of The Iraqi Engineers Association (IEA), International Society for Soil Mechanics and Geotechnical Engineering (ISSMGE), UK, and Fulbright scholar, USA.

Prof. Deepankar Choudhury is Institute Chair Professor of Civil Engineering at Indian Institute of Technology Bombay, Mumbai, India and Adjunct Professor ofAcademy of Scientific and Innovative Research (AcSIR) at CSIR laboratories, India. Prof. Choudharyreceived his Ph.D. in Civil Geotechnical Earthquake Engineering from IISc, India in 2002. He served as BOYSCAST Fellow at UC Berkeley USA, Alexander von Humboldt Fellow at TU Darmstadt Germany, JSPS Fellow at Kagoshima University Japan, and as Visiting Fellow/Professor at NUS Singapore, UoW Australia, INU South Korea. Prof. Choudhury has authored over 300 technicalpublications including over 160 peer reviewed journal papers. He has supervised 25 Ph.D. students. He isa Fellow of The National Academy of Sciences, India (FNASc), Fellow of Institutionof Engineers India (FIE), Indian Geotechnical Society (FIGS), Indian Society of Earthquake Technology (FISET). His research contributions in the domain of geotechnical earthquake engineering, foundation engineering and computational geomechanics have been included in several text books, design manual, codes and implemented in field projects.

Evaluating the Performance of Flexible Passive Pile Group in Cohesionless Soil Under the Effect of a Nearby Embankment

Mahdi O. Karkush, Ghofran S. Jaffar, and Omar K. Al-Kubaisi

Abstract Constructing embankments or multi-story structures can induce lateral soil movement especially for cohesionless soil due to its weak structure. In case of having these constructions near an existing pile group, damage can occur to these piles due to the passive loads induced by the soil movement especially when these loads exceed the capacity of the piles. In the current study, the performance of the existing flexible (2 × 1) pile group installed in cohesionless soil has been investigated experimentally under the effect of lateral soil movement induced by a nearby embankment. To simulate the cohesionless soil, a sandy soil having a dry unit weight of 13.5 kN/m^3 has been used in this study. While an aluminum pipe with an outer diameter of 10 mm and a length of 500 mm has been used to simulate the piles. An embedded length (L_e) of 420 mm has been adopted in this study to ensure the flexible behavior for the piles. The embankment loads have been applied at distances of 2.5D, 5D, and 10D from the edge of the pile. The results show that the soil reaction has decreased with increasing the distance between the edge of the piles and the embankment due to the reduction of soil movement pressure. The maximum soil reaction has decreased by (25–33)% in the loaded pile group (LG) and by (50–60)% in the unloaded pile group (UG) with increasing the distance from 2.5D to 10D, respectively.

Keywords Passive pile group · Flexible pile group · Nearby embankment · Cohesionless soil · Experimental study

M. O. Karkush (✉) · G. S. Jaffar
Department of Civil Engineering, University of Baghdad, Baghdad, Iraq
e-mail: mahdi_karkush@coeng.uobaghdad.edu.iq

G. S. Jaffar
e-mail: ghofran_civi@yahoo.com

O. K. Al-Kubaisi
The Faculty of Engineering, The School of Civil Engineering, The University of Sydney, Sydney, Darlington, NSW 2006, Australia
e-mail: omar.ismael@sydney.edu.au

M. O. Karkush and D. Choudhury (eds), *Modern Applications of Geotechnical Engineering and Construction*, Lecture Notes in Civil Engineering 112,
https://doi.org/10.1007/978-981-15-9399-4_2

1 Introduction

In the case of lateral loads, direct or passive, piles behave like transversely loaded beams and transfer loads to the surrounding soil by using the lateral resistance of the soil. Application of lateral loading on piles, a part or whole of the pile tries to move horizontally in the direction of loading which causes bending moment, rotation, and movement of the pile [1, 2]. The laterally loaded pile may be classified as an active pile or passive pile, regarding the method of transferring the loading direction between the pile and the surrounding soil [3–5]. An active pile is principally loaded at its top, with the lateral load being transferred to the soil, such as piles acting as foundations for transmission towers and offshore structures. A passive pile usually sustains lateral thrusts along its shaft arising from horizontal movement of the surrounding soil, such as piles in a moving slope [6]. Passive piles are extensively used next to the embankments and in moving slopes. In these systems, passive lateral pressure exerted to the pile by the moving soil causes displacements, moments and shear forces. The correct prediction of these values is a key element in the design, construction, and serviceability of the piles.

The influence of a passive pile on three-dimensional soil deformation around a laterally loaded pile and on the response of an active pile's in sandy soil has been investigated by conducting a series of model tests using a newly developed technique named Stereo-PIV [7]. A simple analytical method has been developed by Zhao et al. [8] and Zhang et al. [9] for simulating the effects of soil movement resulted from soil excavation on the behavior of pile groups. The lateral response of passive pile groups obtained from the proposed method has found to be well agreed with those obtained from centrifuge model tests. Furthermore, the behavior of piles subjected to lateral soil movement resulted from slopes has been investigated by conducting large-scale shear box [10]. The results of tests have been compared with those found in the literature to contribute to understanding the behavior of passive piles. Large geotechnical centrifuge model tests of two design schemes have been conducted to simulate the sheet-pile wharves with a load-relief platform in fine sand to investigate the distribution of lateral pile-soil pressure and to distinguish the pile's passive part from the active part [11]. The behavior of passive piles in fine-grained textured soil contaminated with two ratios of petroleum products (MFO) has been investigated under the effects of lateral soil movement [12, 13]. The obtained results of tests have shown an increase in the percentage of soil contamination has caused an increase in the impacts of embankment on the response of passive piles. A three-dimensional finite element model was established and the influence of pile spacing and pile-cap spacing on the displacement of soil around the pile has been analyzed. The results have shown that when the pile-cap spacing was less than 50 m, the soil settlement coefficient has almost changed linearly [14]. In the present study, the response of the existing flexible pile group of 2×1 to the lateral soil movement resulted from constructing a nearby embankment has been investigated.

2 Properties of the Materials

In this study, the soil, which was used, has been classified as (SP-SM) river sand according to the unified soil classification system (USCS). The geotechnical properties of the soil sample, which have been listed in Table 1, have been tested according to both the American (ASTM) standards and the British (BS) specifications [15, 16]. An aluminum pipe with a total length of 500 mm and a circular cross-section has been used to simulate the pile in this study by closing its both ends to prevent the soil to go inside the pile during the installation process. The mechanical properties of the aluminum pile have been listed in Table 2. A ratio of the pile length to its diameter (L/D) of 50 and an embedded depth (L_e) of 420 mm have been used in this study to ensure the flexible behavior of the pile group according to the flexibility factor (K_R) shown in Eq. 1 [17]. Where: E_p is the pile modulus of elasticity; I_p is the pile second moment of inertia; E_s is the soil secant modulus of elasticity and L_e is the pile embedded length.

$$K_R = \frac{E_p I_p}{E_s Le^4} < 10^{-5} \tag{1}$$

Four pairs of strain gauges, denoted as SG1 to SG4, with a gauge resistance of 120 ohms and a gauge factor of 2.12 ± 1% have been used to measure the bending strain along the model piles. The strain gauges have been fixed in two vertical lines with a 120 mm spacing, with the first pair at the soil surface, on both the near and the

Table 1 The geotechnical properties of the soil sample

Property	Value	Property	Value
Specific gravity (Gs)	2.67	Min dry unit weight (γ_{dmin}) [kN/m^3]	11.87
Coefficient of uniformity (Cu)	2.934	Max. dry unit weight (γ_{dmax}) [kN/m^3]	15.14
Coefficient of curvature (Cc)	1.188	Dry unit weight (γ_d) [kN/m^3] at Dr = 56%	13.5
Percentage of fines [%]	9.8	Friction angle (φ) [°] at Dr = 56%	35°
Percentage of sand [%]	90.2	Cohesion (c) [kPa]	9
Relative density (Dr) [%]	56	Confined elasticity modulus (E_{oed}) [kPa]	65770

Table 2 The mechanical properties of the model pile

Property	Value
Pile outer diameter (D)	10 [mm]
Thickness of pile wall	1 [mm]
Pile Length (L)	500 [mm]
Pile weight	42 [gm]
Density of aluminum	2.97 [gm/cm^3]
Pile modulus of elasticity (Ep)	69.871 [GPa]

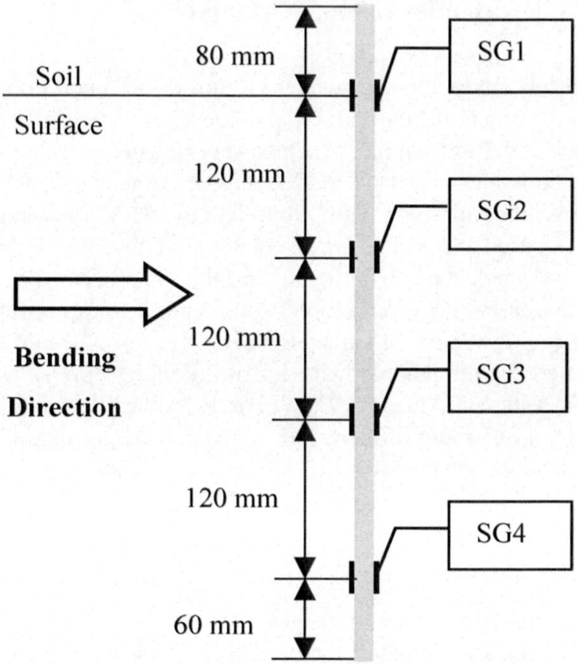

Fig. 1 Stain gauges arrangement along the pile in a half-bridge configuration

rear sides of the model pile with respect to the location of the embankment as shown in Fig. 1. Each pair of strain gages has been connected to a half-bridge configuration by Wheatstone bridge.

3 The Experimental Work

A steel box with a length of 800 mm, a width of 800 mm, and a height of 800 mm has been used to conduct the experimental work of this study. Sand has been poured in the box and densified to the required dry unit weight using the raining technique which required dropping the soil from a specific height. To calculate the required height, four dropping height ranged from 100 mm up to 400 mm with an increment of 100 mm have been adopted to study the effect of the dropping height on the dry unit weight of the soil as shown in Fig. 2. Based on Fig. 2 and to achieve a dry unit weight of 13.5 kN/m^3, a 240 mm dropping height has been adopted in this study. To maintain the dropping height, the dropping cone has been lifted a distance that equals the poured soil layer thickness.

A loading frame has been used to support the dropping cone system that used for the sand raining technique as well as a hydraulic jack of 10 tons capacity which is used for the piles installation process and the application of the embankment loads. Two (2 × 1) pile groups have been installed. One group, labeled as (LG) due to the

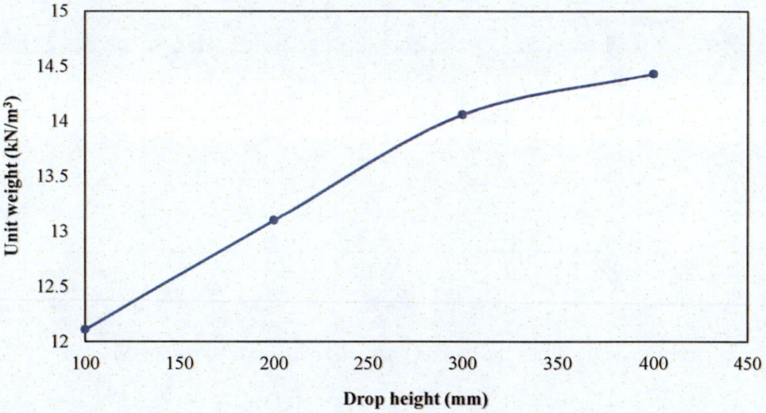

Fig. 2 The relation between the soil dry unit weight and the dropping height

application of axial loads, has been installed in the soil at a distance that exceeded 10 times the diameter of the pile from the steel box walls to ensure the elimination of any tip resistance effect [18]. While to ensure the elimination of rigid boundary effects [19], the other group, labeled as (UG) for the unloaded group, has been installed in the soil at a distance that exceeded 15 times the diameter of the pile from the loaded group (LG). Four dial gauges, two for each group, have been installed, one at the soil surface and the other one at the pile cap, to measure the horizontal displacement of each group. Surcharge loads of 10, 20, 30, 40, 50, and 60 kPa have been used to simulate the embankment loads at distances of 2.5, 5, and 10 times the diameter of the pile. A load cell mounted at the end of the hydraulic shaft has been used to monitor the magnitude of the applied surcharge loads. Each applied surcharge load has been maintained for 2 min as recommended by the literature [20, 21]. At each surcharge load, the strain gages readings as well as the dial gauge readings have been recorded with time.

4 Analysis of the Experimental Data

The flexural stress (σ_z) at the strain gauge locations has been calculated based on the measured bending strain (ε_z) by applying the elastic stress–strain relationship. Then, the discrete bending moment (Mz) has been calculated based on the calculated flexural stress (σ_z) using Eq. 2.

$$\mathrm{M_z} = \frac{2\sigma_z \mathrm{I_p}}{\mathrm{D}} = \frac{2\mathrm{E_p}\varepsilon_z \mathrm{I_p}}{D} \tag{2}$$

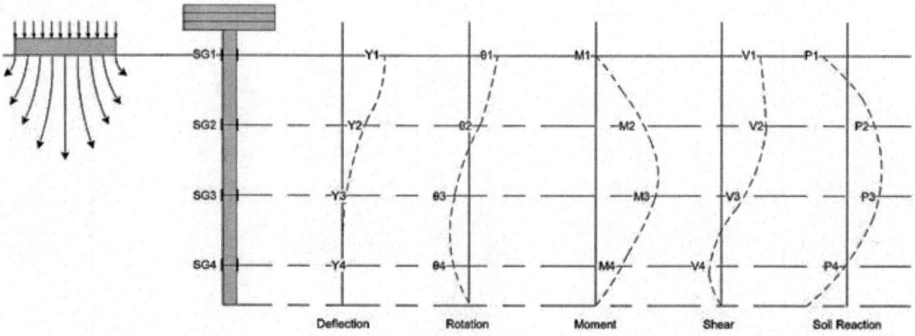

Fig. 3 Strain gages locations along the pile and patterns of calculated parameters

To convert the discrete bending moment to a continuous moment diagram along the pile, the literature has reported two approaches. One approach had been developed using the best-fit polynomial curve of the 4th to the 7th order. However, some drawbacks have been reported in the literature regarding this approach such as the inconsistency in selecting the best-fit curve as well as, in some cases, the sudden jump in the magnitude of the soil reaction at the pile tip [20, 22]. The other approach, [23–25], had been developed using the beam theory in which the responses of the pile and the reaction of the soil, see Fig. 3, could be derived from the bending moment.

Numerical integration following the trapezoidal rule has been adopted in this study to generate the profile of the bending moment along the pile. After this, the finite difference method [26], has been used to generate both the rotation and the deflection profiles along the pile as expressed in Eqs. 3 and 4. Where (θ_o and Y_o): are the rotation integration constant and the displacement integration constant, respectively, measured directly from the dial gauges at the head of the pile and Δ_z: is the distance between any two successive strain gauges as shown in Fig. 1a.

$$\theta_i = \sum_{i=0}^{n} \frac{M_i + M_{i+1}}{2} \nu z - \theta_o \tag{3}$$

$$Y_i = \sum_{i=0}^{n} \frac{\theta_i + \theta_{i+1}}{2} - \Delta z - n\Delta z\theta_o + Y_o \tag{4}$$

Finally, both the shear force profile (V_i) and the soil reaction profile (P_i) have been obtained by applying Eqs. 5 and 6, respectively, using five bending moments [27]. Two imaginary moments, labeled as M_5 and M_6 which assumed to be equal to M_4 and M_3, respectively, have been adopted to find the soil reaction (P4) based on the method described by Scott [28].

$$V_i = \frac{1}{2} \frac{M_{i-1} - M_{i+1}}{\nu z} \tag{5}$$

$$P_i = \frac{1}{7}\frac{2M_{i-2} - M_{i-1} - 2M_i - M_{i+1} + 2M_{i+2}}{\nu z^2} \quad (6)$$

5 Results and Discussion

The effects of constructed embankment adjacent to the pile group have been studied at three different distances from the edge of the model pile group equal to 2.5, 5 and 10 times the outer diameter of the model pile. Two loading conditions, axially loaded (LG) and unloaded (UG) pile groups, have been adopted in this study. The displacement of UG has been more than that of LG by 5% and 24% when the embankment constructed at distances of 5D and 10D, respectively, as shown in Fig. 4. While at a distance of 2.5D, LG has displaced more than UG at the soil surface by 21% due to the high pressure induced by the soil movement. Based on the results, the maximum displacement at the soil surface for the pile group has decreased by (60–64)% for LG and by (48–55)% for UG. The results obtained from the vertical loading tests and the tests of simulated embankment nearby axially loaded piles group are presented and discussed in this section.

The maximum rotation at the soil surface of LG increased by 415% with increasing the distance of the embankment from 2.5D to 5D because LG was rotated toward the embankment at distance 2.5D due to the high soil movement pressure as shown in Fig. 5. With increasing the distance from 5 to 10D, the maximum rotation at the soil surface decreased by 42% because the pressure of soil movement reduced as the distance increased. The rotation at the soil surface of UG decreased by (43–82)% as the distance increased from 2.5D to 10D as shown in Fig. 6. The UG group was not restrained with axial load and the rotation has decreased with increasing the distance due to the reduction in the pressure of soil movement.

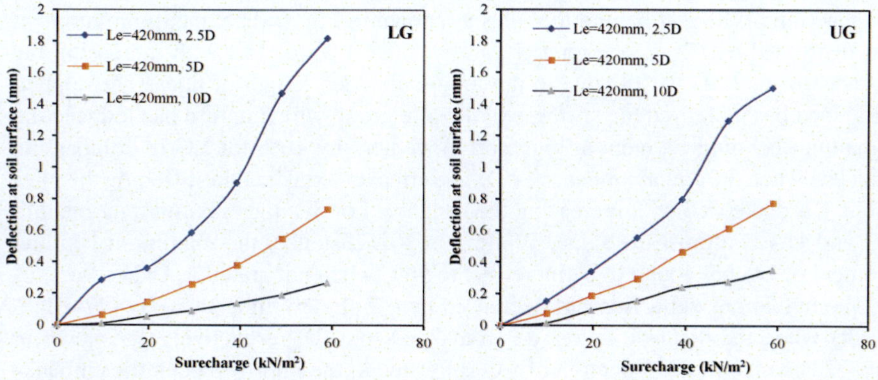

Fig. 4 Displacement of the pile group at the soil surface

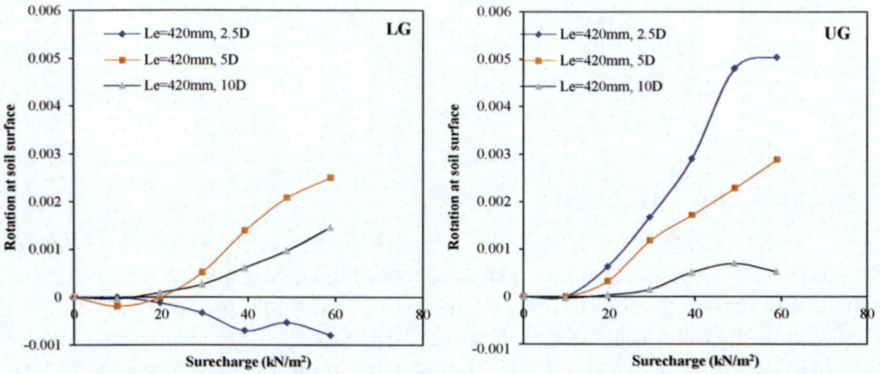

Fig. 5 Rotation of the pile group at the soil surface

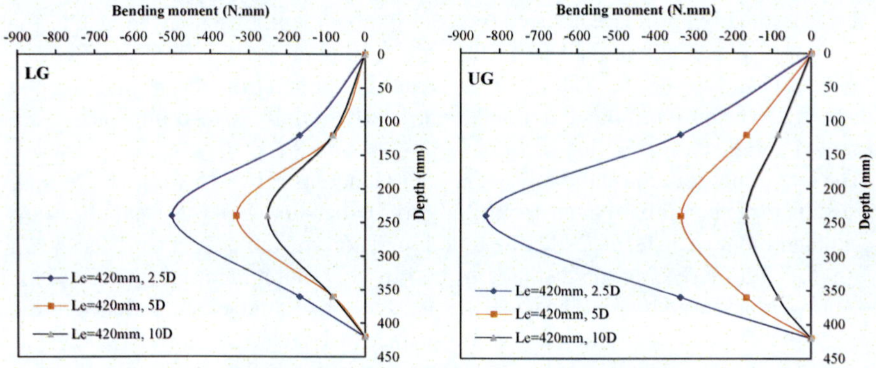

Fig. 6 Bending moment profiles of the pile group

The maximum moment is negative because the piles in the groups had bent under the reaction of the soil behind the piles as shown in Fig. 6. The maximum moment has decreased by (25–33)% for LG and by (50–60)% for UG with increasing the distance from 2.5D to 10D due to the long-embedded length of piles. The applied axial load has increased the stiffness of the pile group which in turn has reduced the maximum bending moment at high soil movement by 40% for LG in comparison with UG when the embankment at 2.5D. The effect of axial load in stiffening the piles group has decreased with increasing the distance. For 5D, the maximum moment in LG and UG is almost the same. While at a 10D distance, the maximum bending moment of LG has found to be increased to 50% when compared to UG.

The maximum deflection has decreased by (72–154)% in LG and by (47–61)% in UG when the distance increased from 2.5D to 10D, respectively, as shown in Fig. 7. The maximum deflection of LG is located at the piles tip when the embankment constructed at distance 2.5D due to the high soil movement and application of axial load which in turn has caused rotation toward the embankment. The maximum

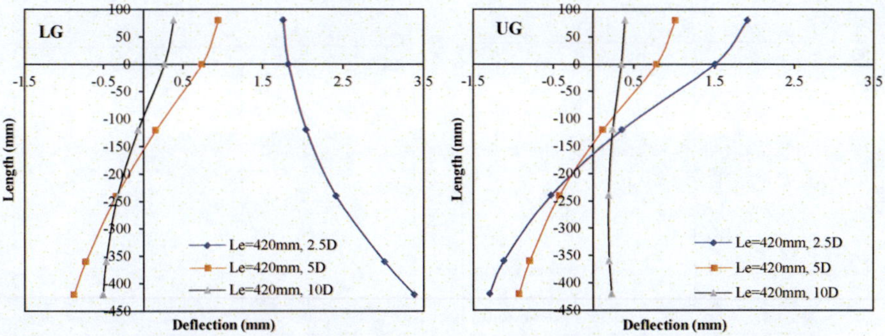

Fig. 7 Deflection profiles of the pile group

deflection of LG (at piles tips) has found to be more than that of UG (at piles heads) by 78% when the embankment constructed at a distance of 2.5D because LG has rotated toward the embankment due to the high soil movement and by 229% at distance 10D. While the maximum deflection of UG has found to be more than that of LG (at piles heads) by 7% when the embankment constructed at a distance of 5D.

The maximum rotation of LG is more than that of UG by 209% as shown in Fig. 8 when the embankment at 2.5D and 10D. While, the maximum rotation of UG is more than that of LG by 13% at 5D. The maximum rotation has increased by 146% then has decreased by 42% in LG, while it has decreased by 43% and 146% in UG with increasing the distance from 2.5D to 5D and from 5 to 10D, respectively. The maximum shear force has decreased by (25–33)% in LG and by (50–60)% in UG as the distance has increased from 2.5D to 10D, respectively, as shown in Fig. 9. This reduction has found to be related to the reduction in the bending moment for both LG and UG as described earlier.

The maximum soil reaction of negative sign has been defined as soil movement, while the soil reaction of positive sign has been defined as soil reaction. The soil reaction has decreased with increasing the distance between the edge of the pile

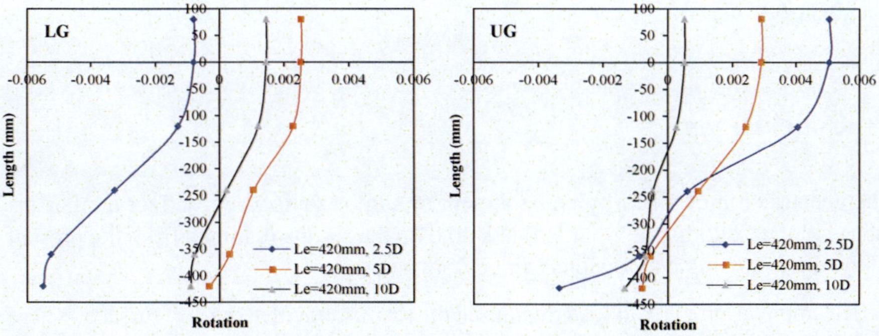

Fig. 8 Rotation profiles of the pile group

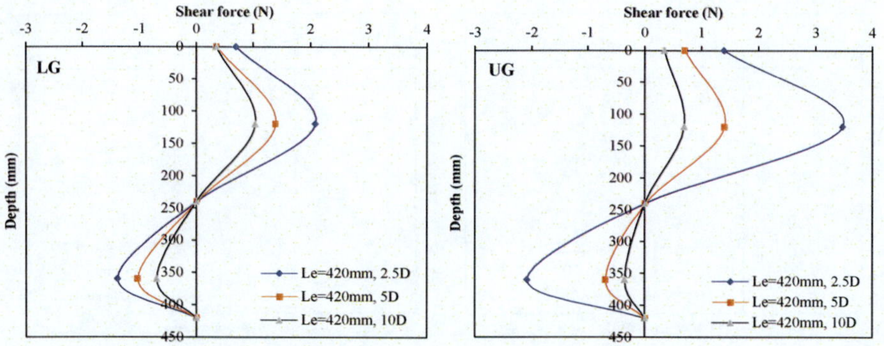

Fig. 9 Shear force profiles of the pile group

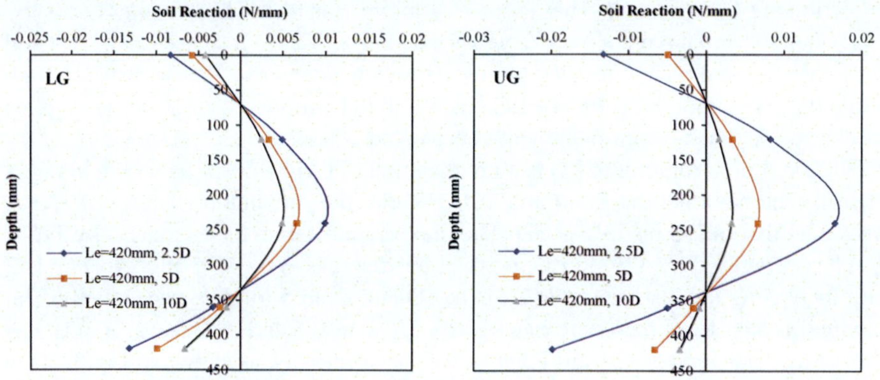

Fig. 10 Soil reaction profiles of the pile group

group and the embankment due to the reduction of soil movement pressure as shown in Fig. 10. The maximum soil reaction has decreased by 33% and 25% in LG and by 60% and 50% in UG with increasing the distance from 2.5D to 5D and from 5 to 10D, respectively.

6 Conclusions

The construction of a nearby embankment or surcharge load will affect the performance of the existing pile foundation. Based on the results obtained from the present study, the following conclusions can be drawn out:

- The application of axial load decreased the displacement at the soil surface, where UG displaced more than LG at 5D and 10D, respectively. However, at 2.5D, LG has displaced more than UG.

- The maximum deflection of LG has found to be more than that of UG when the embankment constructed at distances of 2.5D and 10D, respectively. However, the maximum deflection of UG has found to be more than that of LG by 7% at 5D.
- The maximum shear force in UG is more than that in LG by 40% when the embankment at distances of 2.5D and 5D. At 10D, the maximum shear force in LG is more than that in UG by 50%.
- The maximum displacement at the soil surface has decreased with increasing the distance between the pile and the edge of embankment from 2.5D to 10D in LG and UG, respectively.
- The soil reaction has decreased with increasing the distance between the edge of the piles and the embankment due to the reduction of soil movement pressure.

References

1. Salgado R, Basu D, Prezzi M (2008) Analysis of laterally loaded piles in multilayered soil deposits. Joint Transportation Research Program, Paper 330
2. Fleming WGK (1992) A new method for single pile settlement for prediction and analysis. Geotechnique 42(3):411–425
3. Ercan A (2010) Behaviour of pile groups under lateral loads. M.Sc. Thesis, Civil Engineering Department, Middle East Technical University
4. Fleming WGK, Weltman AJ, Randolph MF, Elson WK (2008) Piling engineering. Taylor and Francis, London and New York
5. Beer D (1977) Piles Subjected to static lateral loads. Doctoral dissertation, Gent Royal Institute for Soil Mechanics
6. Qin H (20101) Response of pile foundations due to lateral force and soil movements. Ph.D. Thesis, Griffith University, School of Engineering
7. Yuan B, Chen R, Teng J, Peng T, Feng Z (2014) Effect of passive pile on 3D ground deformation and on active pile response. Sci World J
8. Zhao M, Liu D, Zhang L, Jiang C (2008) 3D finite element analysis on pile-soil interaction of passive pile group. J Cent South Univ Technol 15:75–80
9. Zhang CR, Huang MS, Liang FY (2009) Lateral responses of piles due to excavation-induced soil movements. In: Ng, Huang, Liu (eds) Geotechnical aspects of underground construction in soft ground 2009. Taylor & Francis Group, London
10. Ersoy CO, Yildirim S (2014) Experimental investigation of piles behavior subjected to lateral soil movement. Teknik Dergi 25(4):6867–6888
11. Ren GF, Xu GM, Gu XW, Cai ZY, Shi BX, Chen AZ (2018) Centrifuge modelling for lateral pile-soil pressure on passive part of pile group with platform. In: Physical modelling in geotechnics vol 1. CRC Press, pp 583–587
12. Karkush MO, Kareem ZA (2018) Investigation the impacts of fuel oil contamination on the behaviour of passive piles group in clayey soils. Eur J Environ Civil Eng 1–17
13. Karkush MO, Kareem ZA (2019) Behavior of passive pile foundation in clayey soil contaminated with fuel oil. KSCE J Civil Eng 23(1):110–119
14. Liu N, Yu J, Gong X, Chen Y (2018) Analysis of soil movement around a loaded pile induced by deep excavation. In: IOP conference series: earth and environmental science, vol 189(3). IOP Publishing, p 032053
15. American Society for Testing and Materials (2003) Annual book of ASTM standards: soil and rock. ASTM, Philadelphia, PA

16. British Standards BS 1377 (1976) Methods of testing for civil engineering purpose. British Standards Institution, London
17. Poulos HG, Davies EH (1980) Pile foundation analysis and design. Wiley, New York, NY
18. Bolton MD, Gui MW, Garnerier J, Corte JF, Bagge G, Laue J, Renzi R (1999) Centrifuge cone penetration tests in sand. ASCE 49(4):543–552
19. Kishida H (1967) The ultimate bearing capacity of pipe piles in sand. In: Proceedings of the 3rd asian regional conference of soil mechanics and foundation engineering, vol 1. pp 196–199
20. Springman SM (1989) Lateral loading on piles due to simulated embankment construction. Unpublished doctoral dissertation, Cambridge University, England
21. Bransby MF, Springman SM (1997) Centrifuge modeling of pile groups adjacent to surcharge loads. Soils Found Jpn Geotechn Soc 37(2):39–49
22. Stewart DP (1992) Lateral loading of piled bridge abutments due to embankment construction. Unpublished doctoral dissertation, University of Western Australia, Australia
23. Ismael OK, Han J (2015) Model tests of laterally loaded piles under a horizontally scoured condition. In: IFCEE 2015, pp 802–808
24. Ismael OK (2014) Evaluating the behavior of laterally loaded piles under a scoured condition by model tests. Master Thesis, University of Kansas
25. Muthukkumaran K (2013) Effect of slope and loading direction on laterally loaded piles in cohesionless soil. Int J Geomech 14(1):1–7
26. Chen F, Yang M (2005) Numerical analysis of piles influenced by lateral soil movement due to surcharge loads. Chin J Geotechn Eng 27(11):1286–1291
27. Levachev SN, Fedorovsky VG, Kurillo SV, Kolesnikov YM (2002) Piles in hydrotechnical engineering. Taylor and Francis
28. Scott RF (1981) Foundation analysis. Prentice-Hall, Englewood Cliffs, NJ

Evaluating End-Bearing and Skin-Friction Resistance of Test Pipe Pile in Sand Soil

Wissam H. S. Al-Soudani and Bushra S. Albusoda

Abstract The present study explained a technique to separate the bearing capacity of pipe pile into skin friction and end bearing by adopting an experimental testing program. Close-end and open-end test pipe pile have considered. The technique depends on the manufactured movable steel container for a mini circular load cell (Dm = 15 mm) to measure the end bearing resistance of the pipe pile with a diameter 40 mm. The competence of this technique was studied by manufacturing and modeling the two pipe piles types with different lengths according to ratios of the lengths to a diameter that was considered in this study (L/D = 15 and L/D = 20). The model piles were tested in loose dry sand. The soil plugging phenomenon and the Incremental Filing Ratio IFR were studied. The complete setup is manufactured for installing and loading the piles at a constant rate of penetration. The obtained results show that the skin-friction less than the end bearing resistance for all pipe pile tests. The distribution of skin-friction tends to become constant below the critical depth of the pile into the sand container. Also, the distribution of skin friction will depend on the lengths to diameter ratios. Also, the end bearing resistance is more than the exterior shaft friction resistance for both pile length 600 and 800 mm.

Keywords Open-ended (OE) pipe pile · Closed-ended (CE) pipe pile · End-bearing capacity · Skin friction capacity · Plug capacity

1 Introduction

The foundation of the building structures in the offshore and land are still nowadays used the deep foundation type hollow pipe pile OE and CE. The OE is classified as a low displacement pile and has a low effect on the surrounded soil during installation.

W. H. S. Al-Soudani (✉) · B. S. Albusoda
Department of Civil Engineering, University of Baghdad, Baghdad, Iraq
e-mail: wissam.hadi@gmail.com

B. S. Albusoda
e-mail: albusoda@yahoo.com

M. O. Karkush and D. Choudhury (eds), *Modern Applications of Geotechnical Engineering and Construction*, Lecture Notes in Civil Engineering 112,
https://doi.org/10.1007/978-981-15-9399-4_3

During the pile driving into the ground, the soil will move up inside the pile and formed a soil plug. The entrance of soil will continue until the inner soil cylinder mode develops sufficient resistance to prevent further soil intrusion. If the stresses in the inner soil are sufficient to resist and prevent further soil intrusion inside the pipe pile, the pile will act as a full plugged CE. Thus, if the shear resistance developed along the length of the column of the restricted soil more than the end bearing capacity at the base of the soil plug, the pile will fail in the plugged mode. On the other hand, the pipe pile will fail in unplugged mode when the shear failure occurs between the soil plug and the pile shaft. The analytical methods and some field tests like chamber tests are still considered as design criteria for OE, suggested from [1–4].

American Petroleum Institute [5] used to evaluate the bearing capacity for OE for the different modes of the OE like unplugged and full plugged mode. The OE when installed in sands behave in partially plugged mode. Stefanoff and Boshinov [6], referred possibility the soil plug can be modeling by one-dimensional analysis, by taking a series of horizontal thin discs to simulate the length of the soil plug and these discs are connected by equilibrium condition represented by a force applied to each disc in final these forces used to measure the plug capacity of the OE. Fattah and Al-Soudani [7] explained that when soil entering the pipe pile during installation, the pile can be fully plugged or partially plugged where the external skin friction and internal skin friction will be mobilized during driving the pile into the ground [8]. Tested a special type of full-scale OE in the field (northern Indiana), that is represented by the double-walled OE that is constituted of two hollow pipes with different diameters as shown in Fig. 1a. Figure 1b explained the curve of separation total pile capacity to major components (shaft friction capacity and end bearing capacity) during the applied loading test. The figure explained a constant behavior resistance of the shaft

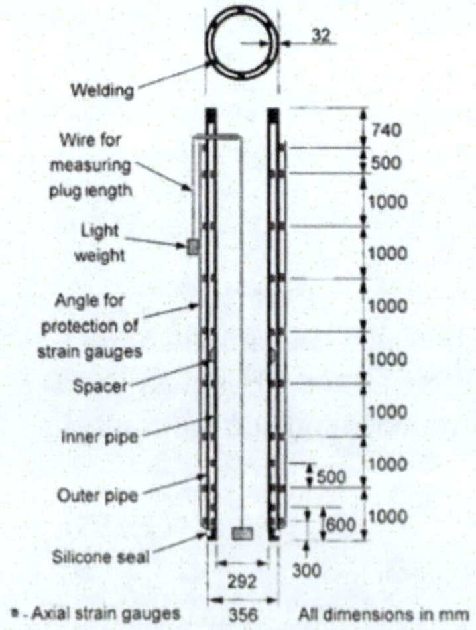

Schematic double-walled (OE)

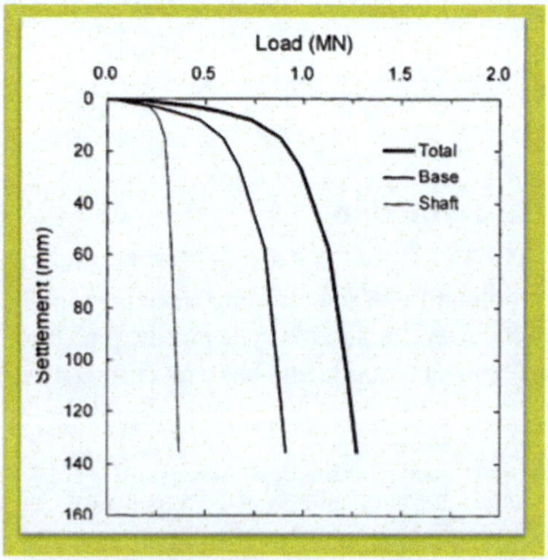

Separation total pile capacity to major components

Fig. 1 Schematic and Load settlement curves of full-scale test pile by [9]

friction capacity at the end of the loading test, while the end bearing capacity can be considered at the settlement of 35.6 mm, corresponding to 10% of the pile diameter.

2 Experimental Work

Four steel traditional pipe piles (open-ended and closed-ended) are used for testing (static axial compression loading) within loose sandy soil.

Soil Properties. The soil used in the study was clean sand from Karbala city, with poorly graded, the properties of the sand (physical and mechanical) were evaluated according to the ASTM standards, that explained in Table 1.

The loose density of the soil sample was prepared by letting the sands falling from a constant height like a raining by depending on the raining method. And to decreases the speed of the falling soil particles, by sand diffuser that includes sieves No. 10 to get sand with a relative density of (25%).

Model Setup Formulation. The loading system is manufactured to simulate that of the field-test, it comprises Steel Structure of frame model; steel tank container; steel moveable girder loading system; steel moveable axial loading system; compression load cell type(s); mechanical jack; gearbox; weighing indicator; speed regulator; raining apparatus; dial gauge indicator; pile driving installation system and reading board. The moveable loading system allows testing the pile models at any location in the steel container and can apply different types of loading in a different direction, as shown in Fig. 2.

Table 1 Properties of the sand (physical and mechanical) considered are in the study

Index	Value	Standard
Coefficient of uniformity (Cu)	2.63	–
Coefficient of curvature (Cc)	1.42	–
Grain size analysis	–	ASTM D 422-2007
Specific gravity (Gs)	2.63	ASTM D 854-2014
D10 (mm)	0.16	–
D30 (mm)	0.28	–
D50 (mm)	0.35	–
Soil classification (USCS)	SP	–
Maximum dry unit weight (kN/m^3)	17.4	ASTM D 4253-2014
Minimum dry unit weight (kN/m^3)	15.46	ASTM D 4254-2000
Maximum void ratio	0.667	–
Minimum void ratio	0.482	–

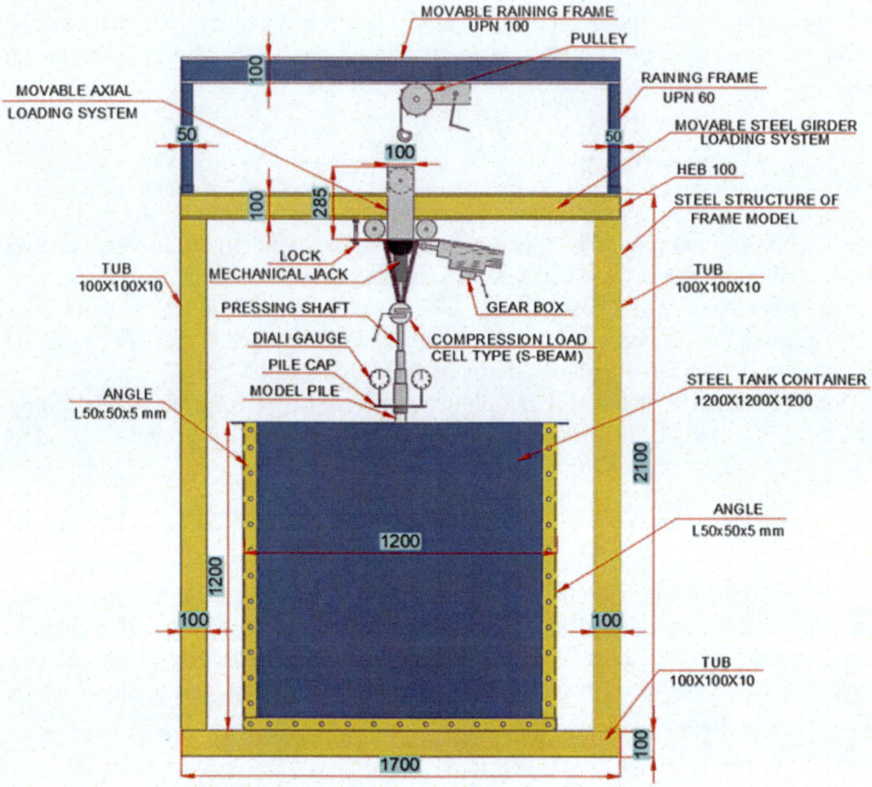

Fig. 2 Complete set up for model pile testing

Model Pile. Four models of the pile (steel pipe) were used and investigated in the study. These piles executed in the loose sand by hammering (free drop hammer). The piles model was illustrated by a diagram in Fig. 3. The lengths of the piles model (600 and 800 mm) were considered in the experimental testing program, with an outside diameter of 40 mm and thickness of the pile wall (2 mm). These models have named as Open-End Pipe Piles (OE) and Closed-End Pipe Piles (CE). An OE piles commonly was behaved in a partially plugged mode when driven into sands. Then the total capacity of OE composed of the annular load capacity, plug load capacity, and shaft load capacity.

3 Testing Program

Four models (OE and CE) tests have executed in dry sand soil samples, with loose relative densities. These piles were driven by a drop hammer (40 N) with a free-drop from a height of 200 mm. When pile driving, the pile penetration depth, and soil plug

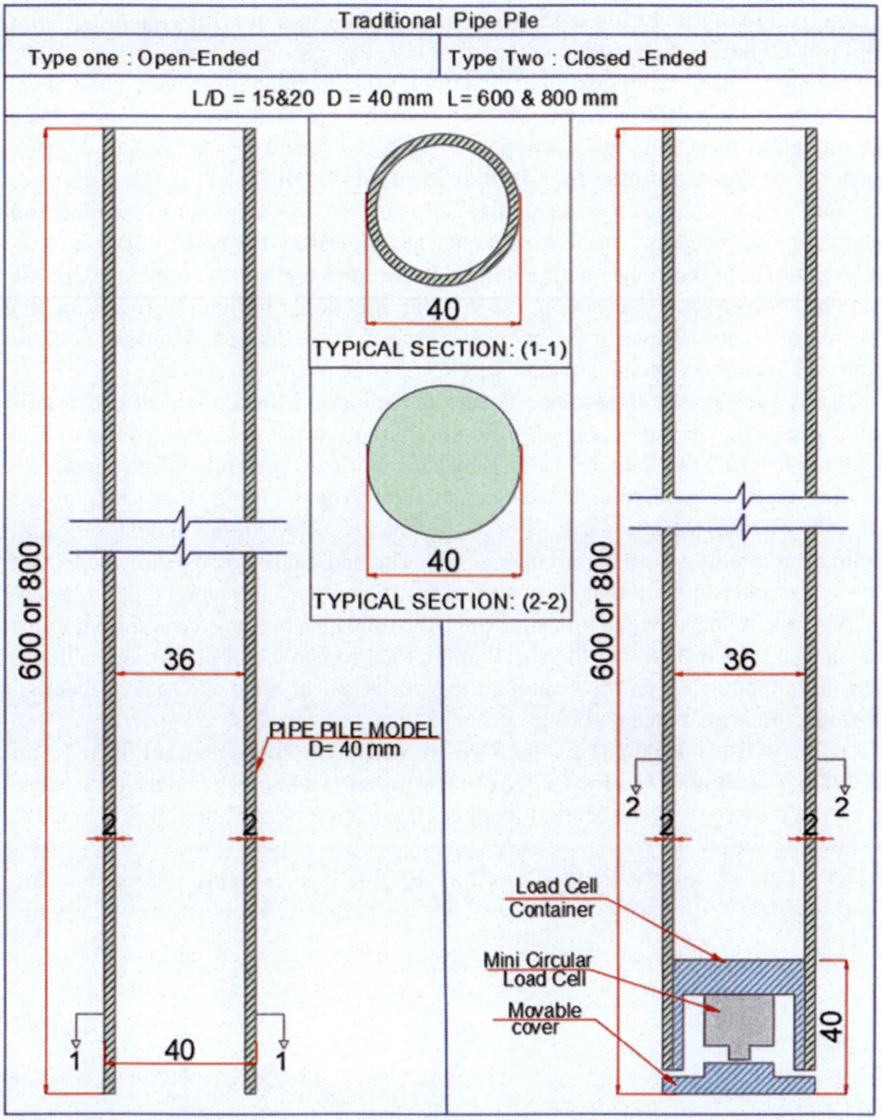

Fig. 3 Traditional OE and CE were piles used in the study

length were measured for every five blows intervals to calculate the incremental filling ratio (IFR). The variation in the soil plug length when pile driving was measured using a ruler entered inside the pipe pile. Axial static load compression has performed at a CRP by using a mechanical jack. The total applied load on the pile has evaluated by a load cell, and the settlement of the pile has evaluated by dial gages. The load tests have continued until the plunging failure.

Separation of the shaft friction resistance and the end bearing resistance from total pile capacity. To parting the total pile load capacity into shaft friction and the end bearing resistance, an instrumented circular steel container includes a mini-load cell inside it was used to measure the end bearing resistance of the closed-end pipe pile model, as shown in Figs. 4 and 5.

Steel Circular Container for Circular Mini Load Cell. A steel circular container was manufactured, to put a mini circular cell inside it, with a diameter (39 mm), and height 40 mm. The steel circular container has locked by the cover. It contains an apex aciculate in the center with a height of 2 mm to connect with the apex of the load cell. The cover is movable in the vertical direction, through a vertical interval distance of (1 mm) between the apex as shown in Figs. 4 and 5. The load cell has stabilized inside the steel container.

Distribution of the Resistance Forces of Soil along the Length of the Traditional Pipe Piles. Figure 6 explains the distribution of the resistance forces of soil along the external piles' body. Three types of traditional pipe pile have considered. The first type represents the distribution of the resistance forces for an unplugged pipe pile, whereas the total capacity of the pile includes the exterior and interior shaft friction forces along the external piles' body. The end-bearing resistance forces that represented only under the annular wall of the pipe pile. The second type for open-end pipe pile with a plugged mode in which the total resistance force is similar to the behavior of the third type of the CE, that the total resistance of these piles includes the exterior shaft friction forces along the external piles' body and the end-bearing resistance for a total area end bearing.

Length of the Soil Plug (Lp) and Measurements the Incremental Filling Ratio (IFR) for Traditional OE Pipe Pile. The (Lp) as shown in Fig. 7 and the (IFR) have measured for two piles of the traditional open-end pipe piles with different lengths, as shown in Fig. 8.

Definition of failure load: According to [8], that explains two criteria are commonly used to define the failure axial pile load capacity under static compression load.

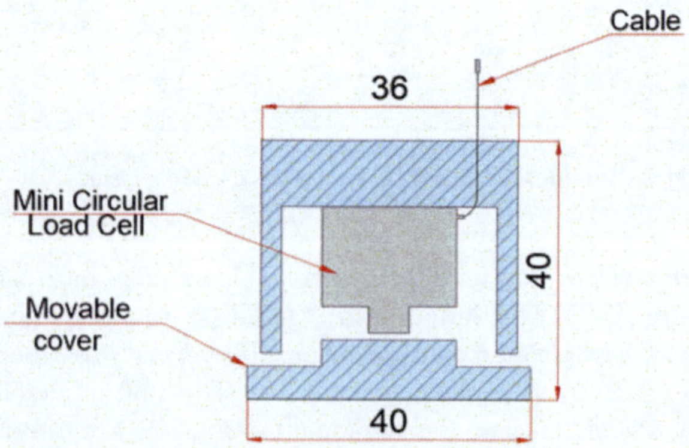

Fig. 4 Details of the steel container of load cells

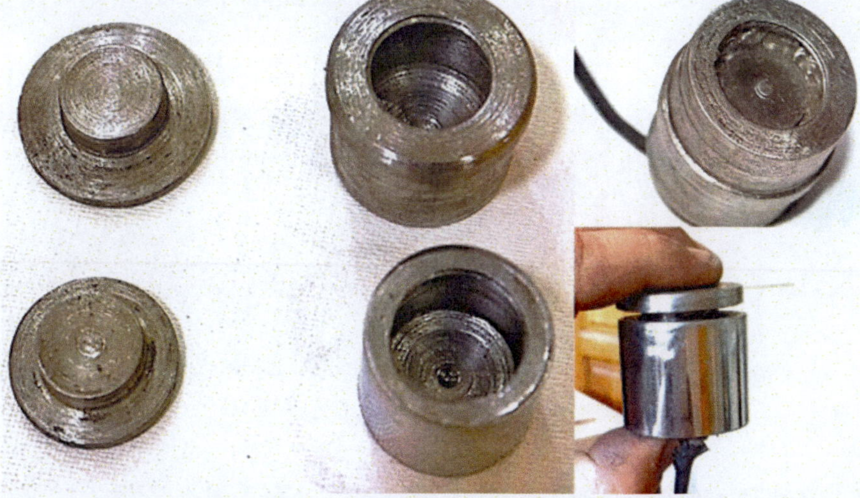

Fig. 5 Types of the steel container of load cells to measure end bearing resistance

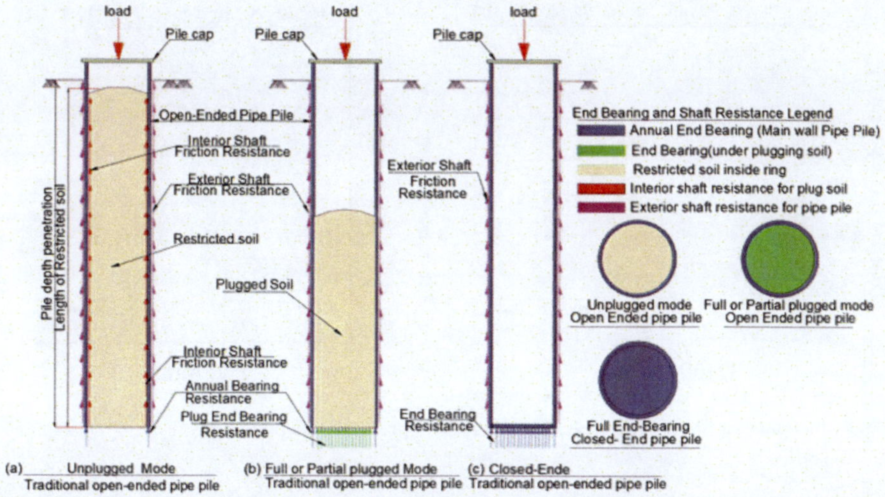

Fig. 6 Distribution of the resistance forces of soil along the external piles' body

Ultimate capacity (plunging failure). The failure load represents the load at which settlement continues indefinitely, as shown in Fig. 8. At field testing, pile load testing cannot use sufficient loads to reach this type of failure. Plunging failure has been reached when using a constant rate of penetration test (CRP).

Deformation controlled capacity. This concept is based on the failure load. It is the load at which the settlement of pile reaches 10% of the diameter of the pile

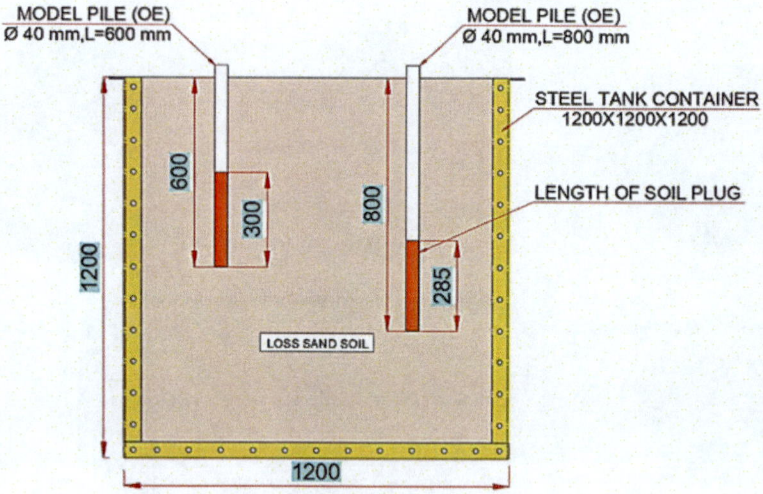

Fig. 7 Length of soil plug for traditional open-end pipe pile

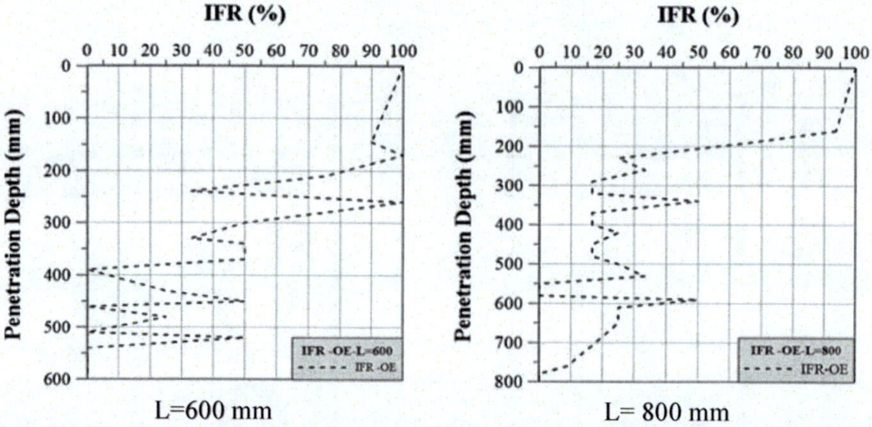

Fig. 8 Incremental filling ratio (IFR) for traditional open-end pipe pile (OE)

base, as shown in Fig. 9. In the present study, the failure load (ultimate capacity) has considered as the load that causes plunging failure, due to the type of loading test is the constant rate of penetration (CRP).

Interpretation of ultimate pile load capacity (failure load). A 30 mm displacement of the pile into the sand container has considered ensuring the mobilization of the end bearing and pile reaches the failure condition. Fleming et al. [10] showed that the true ultimate capacity of the tested pile can be evaluated when the shaft friction essentially is fully mobilized. It is recommended to mobilize the conceivable rate of end bearing. This means that for good analysis, it needs a settlement over around

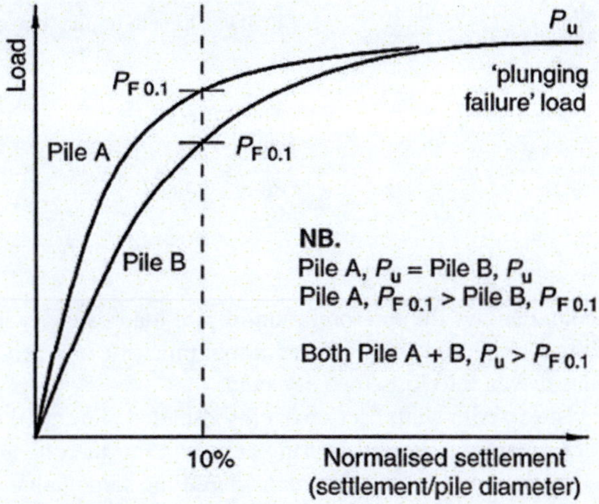

Fig. 9 Definitions of pile failure

25 mm for the tested pile. The [11] and [8] defined the load at failure as the load at which the rate of settlements continues lowering without further increment of the load.

Presentation Result for Traditional Pipe Pile (Open-End and Closed-End). Four traditional pipe piles with two types, open-end (OE) and closed-end (CE), with different lengths to the diameter ratios (L/D = 15 and 20). Figure 10 presents the curves of the relation between the applied load and pile head settlement for traditional OE with partial plugged and CE. It can be noticed that all curves of the relation between applied load and pile head settlement exhibit punching shear failure.

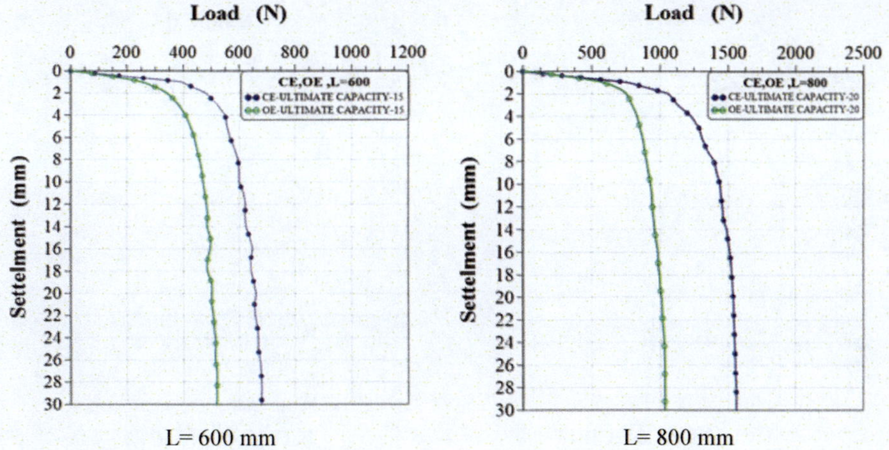

Fig. 10 Curves of the relation between applied load and pile head settlement (OE and CE) performed in loose sand

Table 2 Interpretation of ultimate pile load capacity

Pile ID	Length of pile (mm)	Ultimate pile load capacity according to [11] and [8]
OE-15	600	520
CE-15		680
OE-20	800	1030
CE-20		1555

Table 2 shows the interpretation of pile load capacity according to this method for 600 and 800 mm long for traditional pipe pile and pre-settlement open-ended pipe piles driven into loose density sand.

The exterior shaft friction of closed-ended pile can be calculated by subtracting the total capacity from end-bearing capacity. For the open-ended pipe pile, the exterior shaft friction resistance has considered the same value of the exterior shaft friction resistance of the closed pipe pile, due to the same size and material type of the piles considered in this study. The interior shaft friction resistance and the end bearing capacity under the soil plug for open-end can be calculated by subtracting the total capacity from exterior shaft friction capacity. Figure 11a shows the load settlement curve for separating the end bearing resistance and the exterior shaft friction resistance from the load capacity of closed-ended pipe pile length 600 mm. Both the end bearing resistance and the exterior shaft friction resistance will be increased with the first stage of the applied load. The increase of the end bearing resistance will be continuous until the settlement of the pile head reaches a value of (4 mm) corresponding to 10% of the pile diameter. After that, the increase in bearing capacity continuous in decreasing rate. While the increases of the exterior shaft resistance

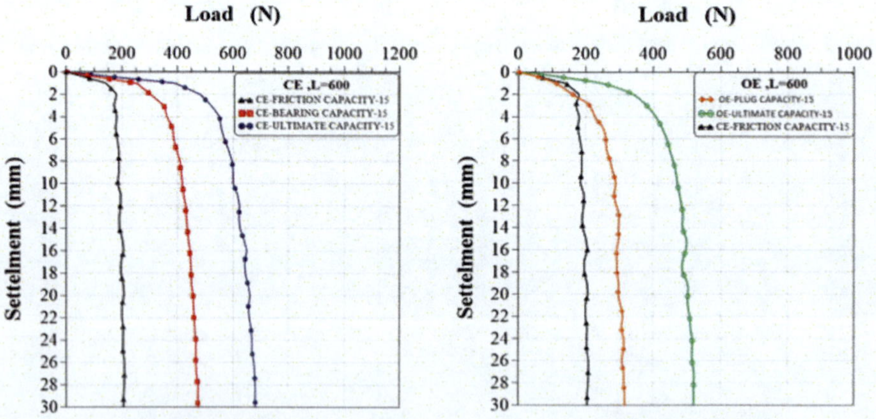

a. Closed-Ended (CE) b. Open-Ended (OE)

Fig. 11 Load-settlement curves for friction resistance and end bearing resistance for performed in loose sand, L = 600 mm will be continuous until the settlement of the pile head reaches a value of (2 mm) corresponding to 5% of the pile diameter. Then, the increase in bearing capacity continuous at a slow rate, and became constant when the settlement of the pile head reaches a value of (8 mm) corresponding to 20%.

Figure 11b shows the load settlement curve for soil plug resistance and exterior shaft friction resistance for open-ended pipe pile length 600 mm. The plug capacity resistance increased with the first stage of loading. This increase will be continuous until the settlement of the pile head reaches a value of (8 mm) corresponding to 20% of the pile diameter. The increase in bearing capacity continued in decreasing rate and became constant at a settlement of (10 mm) corresponding to 25% of the pile diameter. These results agree with the results of [12]. Their results showed that the settlement about (2.5–10 mm) was enough to mobilize full shaft friction. The mobilization of end bearing resistance needs a large settlement about (10%) of the base diameter.

Figure 12a explained the load settlement curve for separating the end bearing resistance and the exterior shaft friction resistance from load capacity for closed-ended pipe pile length 800 mm, the end bearing will be increased with the first stage of the applied load, and the end bearing will be continuous until the settlement of the pile head reaches a value corresponding to 10% (4 mm) of the pile diameter, and after that the increase in end bearing capacity continuous with a little rate. While the increases of the exterior shaft resistance will be continuous until the settlement of the pile head reaches a value corresponding to 5% (2 mm) of the pile diameter and after that the increase in bearing capacity continuous with a little rate, and became constant when the settlement of the pile head reaches a value corresponding to 20% (8 mm). Figure 12b explained the load settlement curve for soil plug resistance and exterior shaft friction resistance for open-ended pipe pile length 800 mm. The plug capacity resistance will be increased with the first stage of the applied load; this

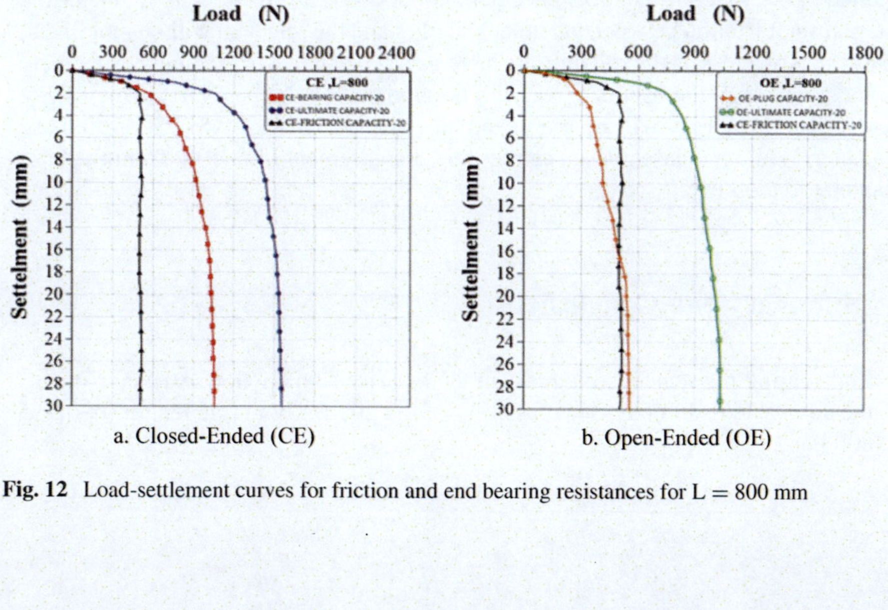

Fig. 12 Load-settlement curves for friction and end bearing resistances for L = 800 mm

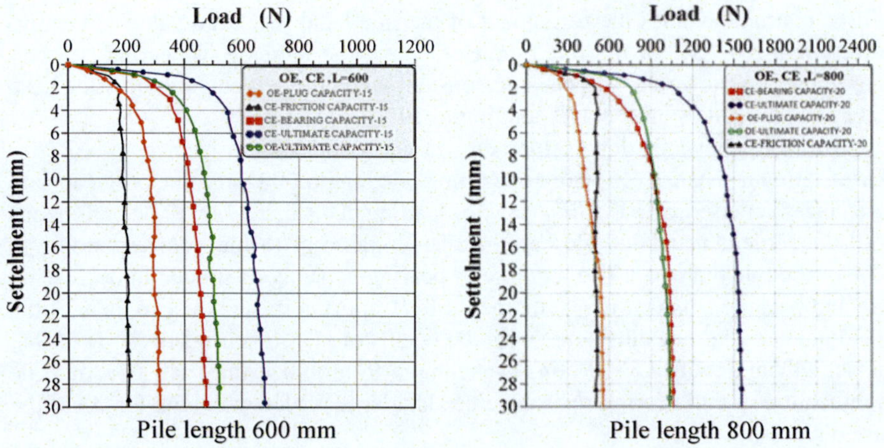

Fig. 13 Load-settlement relations separate for friction resistance and end bearing resistance for traditional pipe pile OE and CE driven piles in loose sand

increase will be continuous with increases of the settlement of the pile head due to the development of the soil plug phenomenon until the failure of the pile.

Figure 13 explained the End bearing resistance is more than the exterior shaft friction resistance due to the compacted soil under the closed tip of the pile for both pile length (600 and 800 mm). These results agree with the result of the study by [13] which investigated the effect of different density of soil on the load-transfer mechanism of soil plug when testing steel (OE) with an external diameter 42.7 mm, the internal diameter of 36.5 mm and length 908 mm and was explained the soil plug resistance is more than the external shaft resistance. The mobilizing shaft friction and end bearing capacity for both piles length (600 and 800 mm) agree with the result obtained [14], that shows at displacement about 0.2–0.5% of the pile diameter the internal shaft friction between the inner soil plug and the pile wall will be mobilized. The effects of stress on the soil plug are increased at the base will gradually exacerbate up the plug. Also, the results agree with the findings of [10] and [15], which show that the shaft capacity will be mobilized at a displacement of 0.5–2% of the base diameter. While the base resistance required displacement of 5–10% from the base diameter to mobilize.

4 Summary and Conclusions

- End bearing resistance is bigger than the exterior skin friction resistance due to the compacted soil under the closed tip of the pile for both pile length (600 and 800 mm).

- Mobilizing the end bearing capacity corresponding to a settlement of 8 mm, 10% of the pile base agrees with the available results in the literature.
- Mobilizing of the friction capacity corresponding to settlement 5% of the pile base (2 mm) agree with the results available in the literature.

References

1. Klos J, Tejchman A (1977) Analysis of behavior of tubular piles in subsoil. In: Proceedings of the 9th Internationl Conference on Soil Mechanics and Foundation Engineering, Tokyo, pp 605–608
2. Nishida H, Ohta H, Matsumoto T, Kurihara K (1985) Bearing capacity due to plugged soil on open-ended pipe pile. Proc Jpn Soc Civ Eng 364, 219–227
3. Randolph MF, Leong EC, Houlsby GT (1991) Onedimensional analysis of soil plugs in pipe piles. Geotechnique 41, 587–598
4. Jardine RJ, Overy RF, Chow FC (1998) Axial capacity of offshore piles in dense North Sea sands. J.Geotech Geoenviron Eng 124:171–178
5. American Petroleum Institute (API) (2014) RP2A: Recommended practice of planning, designing and constructing fixed offshore platforms-working stress design. America Petroleum Institute, API RP2A, Dallas
6. Stefanoff G, Boshinov B (1977) Bearing capacity of hollow piles driven by vibration. In: Proceedings of the 9th International Conference on Soil Mechanics and Foundation Engineering, Tokyo, pp 753–758
7. Fattah MY, Al-Soudani WHS (2016) Bearing capacity of open-ended pipe piles with restricted soil plug. Ships Offshore Struct 11:501–516
8. British Institution of Civil Engineers (2013) ICE manual of geotechnical engineering, Volume II, Published by Thomas Telford Publishing, Thomas Telford Ltd, 1 Heron Quay, London
9. Kyuho P, Salgado R (2003) Determination of bearing capacity of open-ended piles in sand. J Geotechn Geoenviron Eng ASCE 129:46–57
10. Fleming WGK, Weltman AJ, Randolph MD, Elson WK (2008) Piling Engineering. CRC Press
11. British Code of Practice for Foundations (2017) This Code of Practice is available for viewing in the Buildings Department. https://www.bd.gov.hk
12. Kouretzis G (2018) Fundamentals of Geotechnical Engineering and their Applications. The University of Newcastle, Australia Edition
13. Paik KH, Lee SR (1993) Behavior of soil plugs in open-ended model piles driven into sands. J Mar Georesour Geotechnol 11:353–373
14. Randolph MF, Wroth CP (1978) Analysis of deformation of vertically loaded piles. J Geo Eng Divn ASCE 104:465–1488
15. Smith GN (1993) Elements of soil mechanics for mining and civil engineers. Crosby Lockwood Staples, London

Case Study of Retaining Wall Supporting Swelling Soil in Mosul City

Khawla A. Al-Juari, Mohammed Y. Fattah, Suhail I. A. Khattab, and Mohammed K. Al-Shamam

Abstract The main objective of the study is to investigate the effect of expansive soil on the resulting horizontal stresses and deformations inflicted on a retaining wall. Real concrete cantilever retaining wall resting on and supporting backfill of expansive soil was selected from Al-Mosul city. The depth of the water table plays an important role in the expansion of swelling soil. The major variable considered in the study is the depth of the water table and its effect on the lateral swelling, vertical swelling, and the lateral deflection of the wall is studied. The results show that the lateral stresses increased with the depth of the wall and reached the maximum value then decreased. The lateral deflection of the wall increased with the height of the wall and raising the level of the water table. The vertical swelling decreased with raising the water table. Generally, the vertical swelling along the horizontal distance behind the wall increased with raising the level of the water table.

Keywords Retaining wall · Swelling soil · Unsaturated soil · Case study · Numerical analysis

1 Introduction

The seasonal climatic change considers the main reason for the volume change of the soil. So, it swells and shrinks when it is exposed to any source of water, such as rains,

K. A. Al-Juari (✉) · S. I. A. Khattab · M. K. Al-Shamam
Civil Engineering Department, University of Mosul, Mosul, Iraq
e-mail: khawlaaljuari@gmail.com

S. I. A. Khattab
e-mail: suhailkhattab@yahoo.com

M. K. Al-Shamam
e-mail: drmohammedalshmam@gmail.com

M. Y. Fattah
Civil Engineering Department, University of Technology, Baghdad, Iraq
e-mail: myf_1968@yahoo.com

M. O. Karkush and D. Choudhury (eds), *Modern Applications of Geotechnical Engineering and Construction*, Lecture Notes in Civil Engineering 112,
https://doi.org/10.1007/978-981-15-9399-4_4

water pipes leakage, flooding, water table rising, and garden watering. All lightweight structures constructed on/in this type of problematic soil, such as residential building, pavement road, airport, basement wall and foundations frequently exhibit total or differential heave due to vertical swelling. Therefore, the great destruction of these structures is associated with the foundation cracks [1–7]. The retaining wall is a structure constructed to withstand the lateral stresses of the backfill material. The design of these types of retaining walls depends mainly on the basic principles of Rankine and Coulomb theories. These theories did not consider the influence of the lateral swelling pressure on the lateral earth pressure calculation, when the backfill material is clayey soil. Sometimes plastic soil expands in horizontal more than vertical directions [8, 9], it causes substantial distress to landslide and slope failure as well as civil infrastructure involving piles and retaining walls as a result of lateral swelling [3, 10, 11].

Marsh and Walsh [12] and Thomas [13] stated that the common causes of large damages and distresses observed in the retaining wall due to the creep of clayey backfill, expansive surcharge, and the inadequate surface and wall drainage. Also, Ozer et al. [14] illustrated the destruction of the building wall because it sustains clayey soil. The lateral swelling pressure exceeds and reaches seven times the combined pressures exerted by the active soil and the hydrostatic pressure. This condition worsens with increasing the depth and surcharge pressures [15, 16]. Al-Juari et al. [17] tried to establish experimental and numerical modeling for the wall movement in expansive soil. Achieving this was made by manufacturing a laboratory model of a steel box having 950, 900, and 600 mm dimensions for height, length, and width, respectively, and the soil was laid in the box using a specified compaction technique by layering. Specified condition of saturation and normal stress was applied. Certain wall movements were allowed and the effect of those movements on the settlement of retained soil both vertically and horizontally were studied and evaluated. Also, during the unloading stage, the upheave of expansive soil behind the wall was studied. The results showed that the vertical settlement of the soil increased linearly with wall movements. The settlement of the soil behind the wall was more than that away from the wall. The objective of this work is to investigate the effect of the depth of water table on the distribution of lateral swelling pressure along the depth of the field cantilever retaining wall, the resulted deflection of the wall, and the vertical swelling of the soil surface.

2 Description of the Project

The concrete cantilever retaining wall lies in Hai Al-Arabie district on the east bank of Tigris river in Mosul city. The wall was constructed recently in 2013 to retain large and open areas of natural expansive soil. This area is designed for the construction of residential buildings on it. The wall is composed of 25 parts, the total length of the wall is 315 m. The stem height is 7.5 m, and the width of the stem is 0.3 and 1.0 m at the top and bottom of the stem, respectively. The height of the wall footing

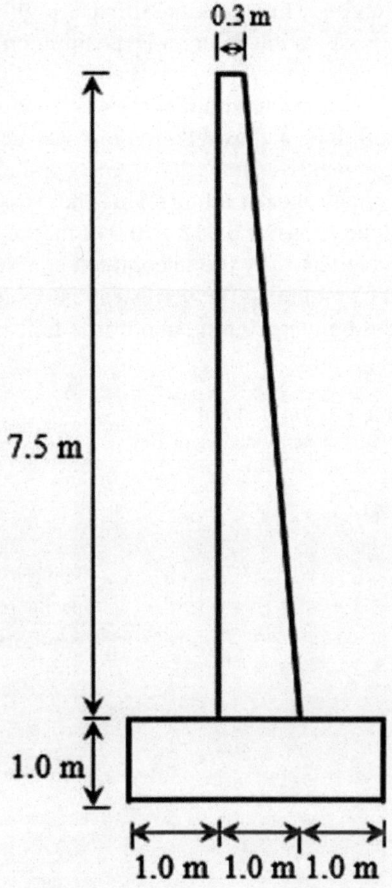

Fig. 1 Schematic diagram of concrete cantilever retaining wall constructed in Hai Al-Arabi district

is 1.0 m with a width of 3.0 m. Figure 1 shows a schematic diagram of the wall with the dimensions of its parts.

3 Experimental Work

Disturbed and undisturbed soil samples were collected from the project site, the samples were placed in plastic bags and tightly closed to prevent loss of moisture content of the samples. Then, the samples were brought to the scientific research laboratory at the University of Mosul for testing the samples. Disturbed samples were utilized for conducting initial moisture content test, Atterberg limits tests, grain size distribution test and chemical tests, whereas undisturbed samples were used for finding the initial dry unit weight of the soil, consolidation, swelling, direct shear and unconfined compression tests. The engineering properties are summarized in

Table 1. The soil is classified as highly plastic silty soil according to the Unified Soil Classification System. Depending on the percent of clay size and plasticity index of the soil, the soil is classified as medium swelling soil according to Williams [18].

A constant volume test was conducted to determine the swelling pressure of the soil, then a consolidation test was completed on the same sample. The unconfined compressive strength test was conducted on the samples with the same initial water content and dry unit weight. The cross-sectional area of the soil sample is square with dimensions of 0.042 × 0.042 m, the height of the sample is 0.095 m. The samples were tested by the unconfined compression device for rock with a constant rate of 0.02 mm/min. The vertical stresses were applied to the sample until failure achieved throughout 2.0 min, and brittle failure occurred. The failure plane is approximately

Table 1 Physical and engineering properties of Hai Al-Arabi soil

Properties of the soil		Value
Atterberg's limits	Liquid limit (%)	62
	Plastic limit (%)	38
	Plasticity index (%)	24
Soil classification		MH
Specific gravity (Gs)		2.72
Chemical tests	Total soluble salts, TSS (%)	1.4
	Organic matter (%)	3
	Sulfate salts (SO_3)	0.005
Grain size analysis	Sand (%)	9
	Silt (%)	44
	Clay (%)	47
Activity = P.I/C		0.523
Natural dry unit weight (kN/m^3)		17.81
Natural moisture content (%)		19.0
Swelling pressure (kN/m^2)		50
Unconfined compressive strength	θ (degrees)	82
	ø (degrees)	74
	σ_1 (kN/m^2)	3495
	c (kN/m^2)	245.5
Modulus of elasticity (kPa)		122.75
Poisson's ratio		0.49
Hydraulic conductivity (m/day)		6.2×10^{-7}
C_c		0.039
C_s		0.024
c_v (m^2/day)		2.35×10^{-3}

vertical and the angle of failure (θ) is about 82°. The cohesion of the soil is about 245.5 kN/m^2 as listed in Table 1.

4 Numerical Analysis

The cantilever retaining wall is simulated in the GeoStudio program to study the behavior of the retaining wall and soil when they are exposed to seasonal variation. The material category used in the simulation of the concrete retaining wall is total stress parameters with a linear elastic model. The soil is simulated by effective parameters with pore water pressure change as a material category, and linear elastic model. The geometry of the project is shown in Fig. 2, the soil is extended to a horizontal distance of 15.0 m behind the wall and 7.5 m below the base of the wall footing. The location of the water table is assumed at different depths as follows:

- At depth 7.5 m which is equal to the height of the wall,
- At the base of the wall footing, and
- At a half depth of the wall.

The material properties of the used soil are given in Table 1. The SWCC of the Hai Al-Arabi was obtained from the Geostudio program, depending on the grain size

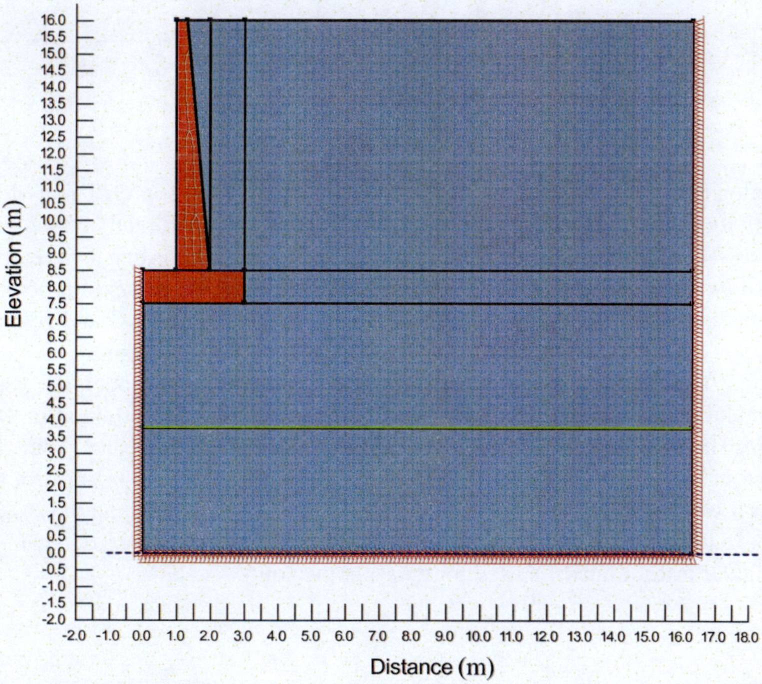

Fig. 2 Geometry and boundary conditions of the problem

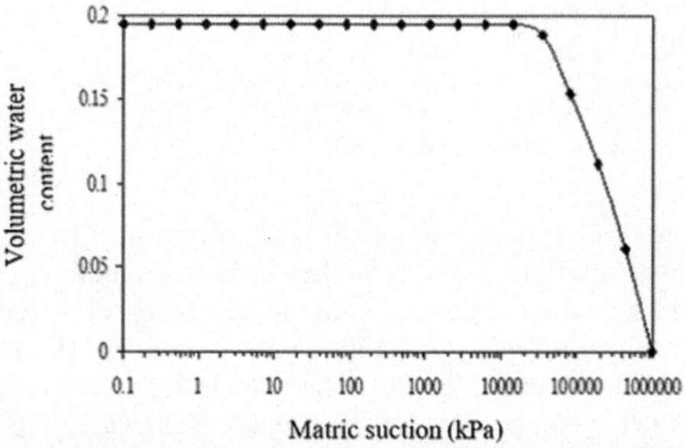

Fig. 3 SWCC of Hai Al-Arabi swelling soil

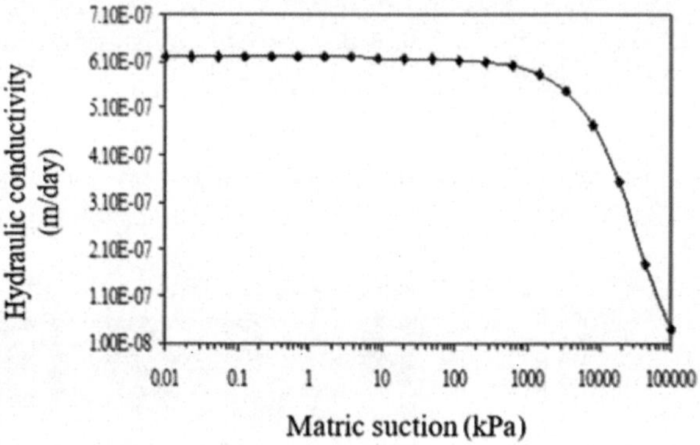

Fig. 4 Hydraulic conductivity curve of Hai Al-Arabi swelling soil

distribution curve and liquid limit of the soil, it is depicted in Fig. 3. The hydraulic conductivity curve for Hai Al-Arabi soil, which was obtained indirectly with parameters of the SWCC and the saturated hydraulic conductivities, according to the equation proposed by Fredlund and Xing [19] is presented in Fig. 4. Stress/strain boundary conditions are used for the finite element method stress analysis, and hydraulic boundary conditions are used for finite element method flow analysis.

In stress/strain analysis, and at the right and left sides of the model, the vertical soil boundaries are allowed to move in the Y-direction but are restrained in the X-direction. The bottom of the soil geometry is restrained on both the X and Y-directions as shown in Fig. 2. The mesh used in the simulation of the project is uniform mesh size was used for the whole domain of the model. The size of the element is (0.05 × 0.05) m. In the finite element model, 8 nodes quadrilateral isoparametric element of a rectangular grid of quads is used to represent the soil.

5 Results and Discussion

Effect of Rising of Water Table on Lateral Deflection of the Wall. The lateral deflection of the wall increased with raising the level of the water table, where the maximum deflection occurred when the water table is located at a half depth of the wall as depicted in Fig. 5. No wall deflection occurred when the water table is located at depth H below the base of wall footing. On the other hand, Fig. 5 shows that the lateral deflection of the wall increased when the water table is located at the base of the wall footing. The lateral deflection increased with the depth of the wall until reaching the maximum value of 7.4 mm at 0.87 m, which is approximately equal to the field observation, then decreased. The deflection of the wall increased also when the water table is raised to the half depth of the wall. The wall deflected away from the soil. The deflection increased with the height of the wall until reaching the maximum deflection which is about 19.5 mm at the top of the wall. The percent of increase in lateral deflection is about 100% at the top of the wall when the water table raised from depth H below the base of the wall footing to the half depth of the wall.

Effect of Rising of Water Table on the Lateral Stresses. The lateral stresses increased with the depth of the wall and reached the maximum value at 0.625 m then decreased as shown in Fig. 6. The lateral stresses remain constant throughout the height in the range between 2.75 and 7.5 m then started to increase until reaching the maximum lateral stresses which are about 128.4 and 416.8 kPa when the water table is located at depth H below the base of wall footing and the base of wall footing, respectively.

Effect of Rising of Water Table on the Vertical Swelling. The vertical swelling of the soil increased continuously during the time one year as shown in Fig. 7. The

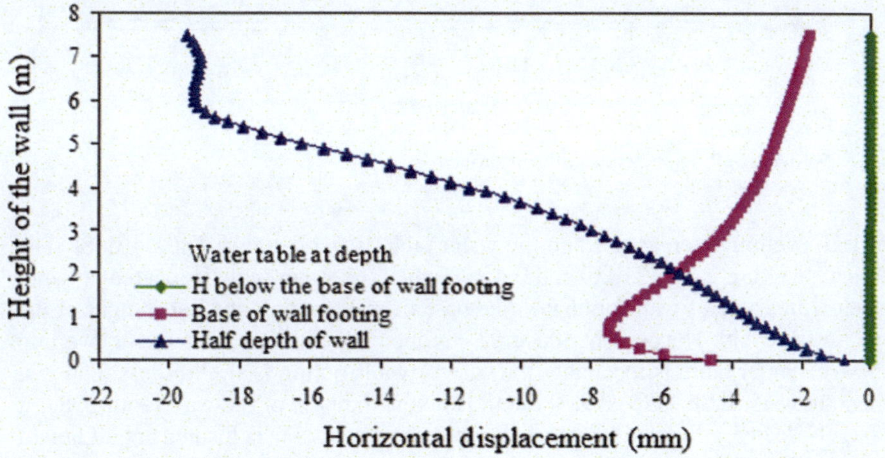

Fig. 5 Lateral deformation of the wall of height 7.5 m

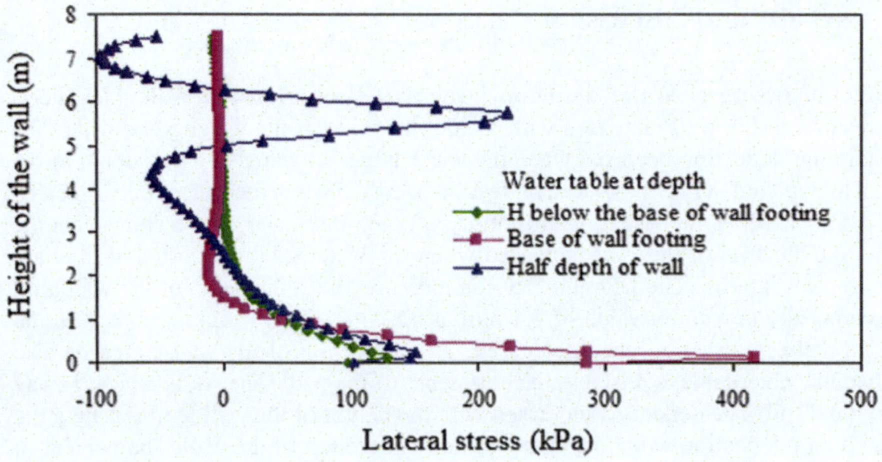

Fig. 6 Lateral stresses on the wall of height 7.5 m

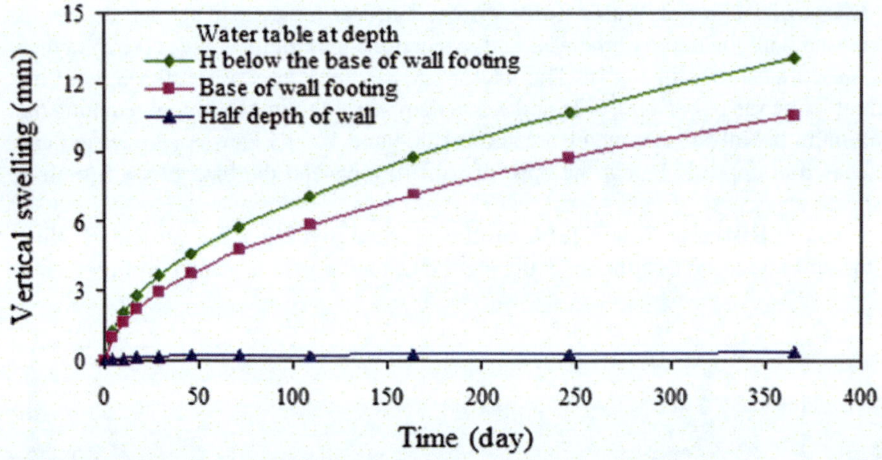

Fig. 7 Vertical swelling of the backfill soil surface

vertical swelling decreased when the water table is rising upward at all times. The vertical swelling is 13.1, 10.6, and 0.4 mm when the water table is located at depth H below the base of wall footing, at the base of wall footing and half-height of the wall, respectively. The percent of decrease is about 19 and 97%. The rate of swelling when the water table is located at the base of wall footing is slightly less than that when the water table is located at depth H below the base of the wall footing, but it is largely decreased with time intervals when the water table is located at half height of the wall.

Effect of Rising of Water Table on the Vertical Swelling of The Soil Surface. Figure 8a presents the vertical swelling of the soil surface when the water table is located at depth H below the base of wall footing. The soil surface uniformly swelled along the horizontal distance for all times, which is related to the large depth of water table and high suction value which caused uniform vertical swelling to the soil surface without wall deflection occurrence. The increase in vertical swelling at short times is less than that at longer times. Figure 8b shows the vertical swelling of the soil surface with horizontal distance behind the wall, when the water table is located at the base of the wall footing.

The vertical swelling slightly decreased with horizontal distance behind the wall up to 6.0 m then remained approximately constant with further horizontal distance. When the water table is raised upward to the half depth of the wall, the vertical swelling of the soil surface remained constant with the horizontal distance up to 2.5 m as shown in Fig. 8C, then started to increase at a low rate when the time interval is little but it increased with a high rate when the time interval is large. So, the vertical swelling reached an equilibrium state beyond a horizontal distance of 7.5 m.

6 Conclusions

Studying of the natural problem of cantilever concrete retaining wall in the field and analyzing it numerically, the following conclusions can be listed down.

- The lateral deflection of the wall increased with the height of the wall and raising the level of the water table. However, the percent of increase in lateral deflection is about 100 and 91% at the top of the wall when the water table is raised to the base of the wall footing and the half depth of the wall, respectively.
- The lateral stresses increased with the depth of the wall and reached the maximum value at 0.625 m then decreased. The maximum lateral stress is 416.8 kPa when the water table is located at the base of wall footing.
- The vertical swelling of the soil increased continuously with time. The vertical swelling decreased with raising the water table. The percent of decrease is about 19 and 97% when the water table is raised to the base of wall footing and at half the height of the wall, respectively.
- The vertical swelling decreased slightly with horizontal distance behind the wall up to 6.0 m, then it remained approximately constant when the water table is located at the base of the wall footing.
- The vertical swelling of the soil surface increases with the horizontal distance until it reached an equilibrium state beyond a horizontal distance of 7.5 m, when the water table is raised to half depth of the wall.
- The study shows a good agreement between the results obtained from the numerical analysis and the recorded observation on the constructed concrete

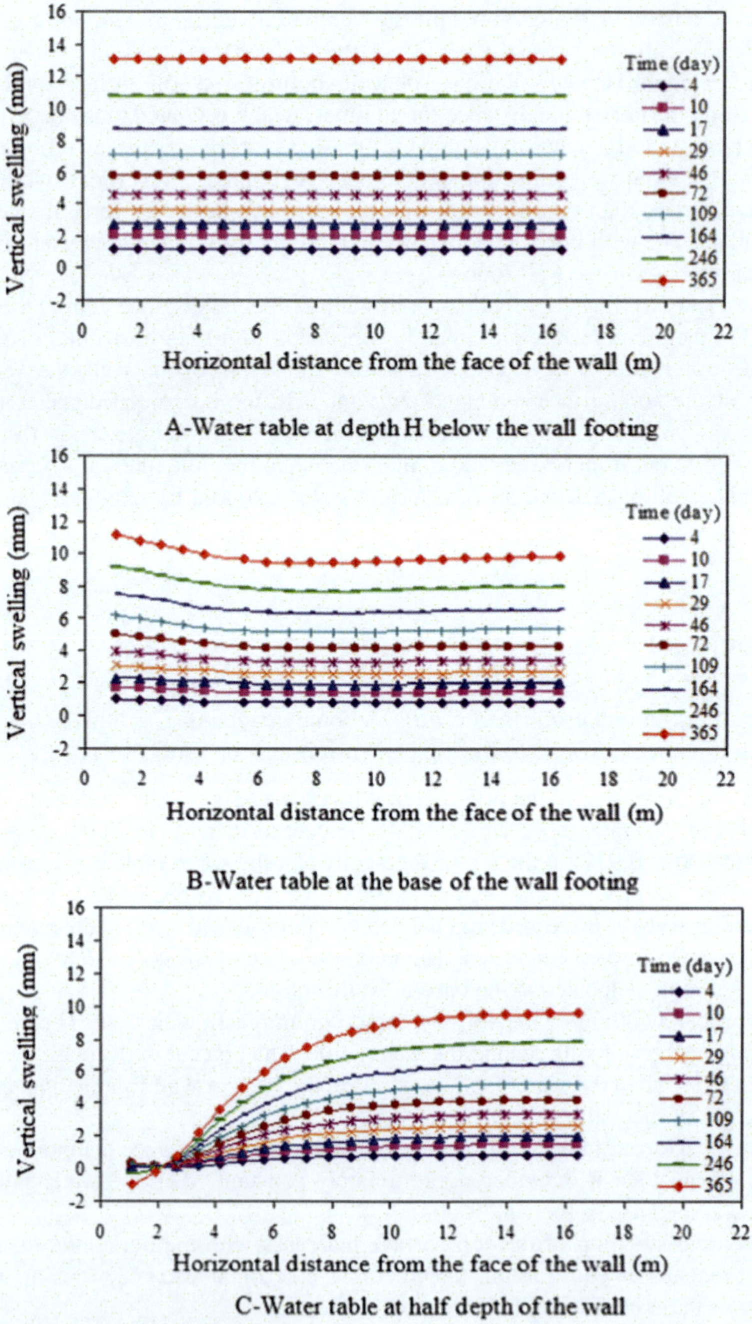

Fig. 8 Vertical swelling of the soil surface behind the wall of height 7.5 m

retaining wall in Mosul city. This reflects the successful modeling of the swelling phenomenon and soil behavior upon wetting.

References

1. Zhang X (2004) Consolidation theories for saturated-unsaturated soils and numerical simulation of residential buildings on expansive soils. Ph.D. Dissertation in Civil Engineering. Texas A&M University
2. Erguler ZA, Ulusay R (2006) Swelling behavior of Ankara clay predictive techniques, damage details, and swelling maps, expansive soils: recent advances in characterization and treatment. Taylor & Francis Group, London, UK
3. Hong GT (2008) Earth pressures and deformations in civil infrastructure in expansive soils. Ph.D. Dissertation, Texas A&M University
4. Dafalla MA, Shamrani MA (2012) Expansive soil properties in a semi-arid region. Res J Environ Earth Sci 4(11):930–938
5. Karunarathne AMAN, Sivanerupan S, Gad EF, Disfani MM, Rajeev P, Wilson JL, Li J (2014) Field and laboratory investigation of an expansive soil site in Melbourne. Aust Geomech 49(2):85–94
6. Li J, Cameronand DA, Ren G (2014) Case study and back analysis of a residential building damaged by expansive soils. Comput Geotech 56:89–99
7. Dafalla MA, Al-Shamrani MA (2014) Swelling characteristics of Saudi Tayma shale and consequential impact on light structures. J Civil Eng Archit 8(5):613–623
8. Andy BF (1989) Laboratory evaluation of lateral swelling pressure. J Geotechn Eng ASCE 115(10):1481–1486
9. Erguler ZA, Ulusay R (2003) A simple test and predictive models for assessing swell potential of Ankara (Turkey) clay. Eng Geol 67:331–352
10. Jiang H, Wang B, Inyang HI, Liu J, Gu K, Shi B (2013) Role of expansive soil and topography on slope failure and its countermeasures, Yun County, China. Eng Geol 152:155–161
11. Hou TS, Xu GL, Shen YJ, Wu ZZ, Zhang NN, Wang R (2013) Formation mechanism and stability analysis of the Houba expansive soil landslide. Eng Geol 161:34–43
12. Marsh ET, Walsh RK (1996) Common causes of retaining-wall distress: case study. J Performance Construct Facil 10(1):35–38
13. Thomas MG, Puppala AJ, Hoyos LR (2009) Influence of swell pressure from expansive fill on retaining wall stability. In: International Foundation Congress and Equipment Expo. American Society of Civil Engineering
14. Ozer M, Ulusay R, Isik NS (2012) Evaluation of damage to light structures erected on a fill material rich in expansive soil. Bull Eng Geolog Environ 71:21–36
15. Mohamed, OZ, Taha YK, Abd El-Aziz SM (2014) Experimental study on the effect of lateral swelling pressure of expansive soil on retaining structure. J Eng Sci 42(1):84–92
16. Nelson JD, Chao KC, Overton DD, Nelson EJ (2015) Foundation engineering for expansive soils. Wiley, Hoboken, New Jersey
17. Al-Juari KAK, Fattah MY, Khattab SIA, Al-Shamam MK (2019) Experimental and numerical modeling of moving retaining wall in expansive soil. Geomechan Geoeng
18. Williams AAB (1957) Discussion on the prediction of total heave from the double oedometer test. In: Symposium on Expansive Clays, South African Institution of Civil Engineer, 57
19. Fredlund DG, Xing A (1994) Equations for the soil-water characteristic curve. Can Geotech J 31(4):521–532

Influence of Magnetized Water on the Geotechnical Properties of Expansive Soil

Salem M. A. AL-Ani, Mahdi O. Karkush, Askar Zhussupbekov, and Ahmed A. Al-Hity

Abstract The expansive soil is characterized by increasing its volume when subjected to water. The expansion of soil mostly resulted from the presence of clay minerals such as montmorillonite. The technology of magnetized water was used in many fields such as agriculture, industry, and other sciences. In the present study, the influence of potable water treated by reverse osmosis and passing from magnetic fields of different intensities (500, 1000, 1500, and 2000 Gauss) on the chemical and swelling properties of expansive soil have been investigated in details. Based on the results of experimental tests, the magnetized water helps in mitigating the expansion potential of expansive soil. The liquid limit was reduced from 113 to 72% and the plastic limit was reduced from 55 to 22%, free swelling of the soil was reduced from 33.7 to 25.4%, and the uplift movement of soil was reduced from 21.66 to 14.25 mm. Also, the unconfined compressive strength of soil was increased from 26.36 to 49.49 kPa at a water content of 32% and from 120 to 224 kPa at a water content of 22% with increasing the intensity of the magnetic field with a small increase in the strain corresponding to the unconfined compressive strength at failure was noticed. The technology of magnetized water is one of the methods that can be resigned within the sustainable definition to be used in geotechnical engineering.

Keywords Magnetic field · Swelling · Expansive soil · Plasticity index · And unconfined compressive strength

S. M. A. AL-Ani · M. O. Karkush (✉)
Department of Civil Engineering, University of Baghdad, Baghdad, Iraq
e-mail: mahdi_karkush@coeng.uobaghdad.edu.iq

S. M. A. AL-Ani
e-mail: salimalani39@yahoo.com

A. Zhussupbekov
Civil Engineering Department, Eurasian National University, Astana, Kazakhstan
e-mail: astana-geostroi@mail.ru

A. A. Al-Hity
Department of Water Resources and Dams, University of Anbar, Anbar, Iraq
e-mail: jubair3a@gmail.com

M. O. Karkush and D. Choudhury (eds), *Modern Applications of Geotechnical Engineering and Construction*, Lecture Notes in Civil Engineering 112,
https://doi.org/10.1007/978-981-15-9399-4_5

1 Introduction

Expansive soils showed significant volume changes by variations in soil moisture content, so it is considered one of the problematic soils. Wetting and drying are caused by volume changes in soil and huge damages for the structures resting on such soils. Generally, it is found in arid and semiarid regions that cover huge areas in the world. The polar molecule of water (H_2O) is composed of a V-shaped, it contains two hydrogen atoms of positive charge and the oxygen atom of negative charge. It has polar molecules partially negative dipole ends are attracted to the ions of clay minerals with unsatisfied forces. The saline water has a significant influence on the soil texture properties such as decreasing cracks (depth and width) and form of shrinkage cracks, where the saline water causes increasing the concentration of salts in soil [1, 2]. Also, increasing the salinity of pore fluid causes decreasing the liquid and plastic limits of soil [3–5].

Ajalloeian et al. (2013) explained the effect of water salinity on geotechnical properties of fine-grained textured soil such as Atterberg's limits, compression index, and swelling index. The results of tests showed decreasing of studied properties with the concentration of salinity of pore water in soil [6]. Vanda (2014) used NaCl and $CaCl_2$ in concentrations of 0.5 and 1 M to change the pore fluid chemistry and compared with the behavior of soil with distilled water and saline water. Increasing the concentration of salts which causes increasing the unconfined compressive strength is higher than that of the soil samples prepared with distilled water [7]. The factors affecting the behavior of expansive soil are the relative density, water content, compaction, permeability, density, the groundwater table, and the presence of trees and vegetation [8].

The particles of clay are very small and arranged in thin crystal frame layers. Many elements can become incorporated into the clay mineral structure (H^{+1}, Ca^{+2}, K^{+1}, Mg^{+2}), these are dissolved ions and influenced the behavior of the clay minerals and engineering properties of soil [9]. The circulation of water in the magnetic field causes increasing the positive ions in general except Ca^{+2} and the value of sulfur ions is decreased. The carbonates did not appear in magnetized water. Also, the magnetization of distilled water makes the precipitation of salts more than non-magnetized distilled water [10]. In the present study, the effects of several intensities of magnetic field (500, 1000, 1500, and 2000) Gauss on the Atterberg's limits, swelling potential, unconfined compression strength, and free swelling of expansive soil have been investigated experimentally in the laboratory.

2 Soil Sampling and Water Treatment

The expansive soil was obtained from Al-Anbar Governorate/Al-Rahhalyia which is located to the south of Al-Ramadi city and west of Al-Razzaza Lake. The soil samples were obtained from a 1 m depth below the existing ground level. Some of the geotechnical properties of used soil are given in Table 1 and Fig. 1.

In this investigation, the distilled water is used in typical traditional geotechnical tests and potable water treated with reverse osmosis (RO) is used in the treatment of swelling soil and used in the treatment of soil then it passing from magnetic fields of different intensities (500, 1000, 1500, and 2000) Gauss. In the water magnetization technique, nothing is added or extracted from water but the potable water is passing in the magnetic field which causes the rearrangement of water molecules and changes the content of minerals (Na, Ca, SiO_2, Mg) and dissolved salts (TDS). In the present study, the water container is made from plastic fiber (acrylic) to avoid the effects of chemical reactions. The container also contains an electrical pump (25 w, head 18 m, flow rate 1000 l/h, and AC220 V/50 Hz) and devices of magnetic fields (500, 1000, 1500, and 2000) Gauss with a plastic tube of 12 mm diameter, see Fig. 2. All the mentioned parts (pump and devices of magnetization) are connected by plastic tubes and installed in the water tank of capacity of 100 L. The magnetization of potable water treated by (RO) will cause changing the chemical and electrical properties of water, these properties will become approximately constant after 24 h of magnetization [11].

Table 1 Properties of the expansive soil

Properties of Clay	Value
Optimum moisture content (%)	22.6
Maximum dry density (g/cm^3)	1.377
Specific gravity	2.79
Void ratio	1.012

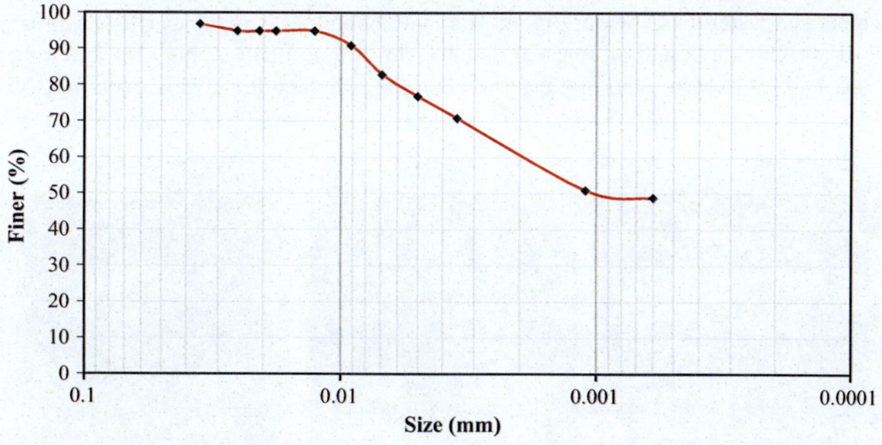

Fig. 1 Particle-size distribution curve of expansive soil

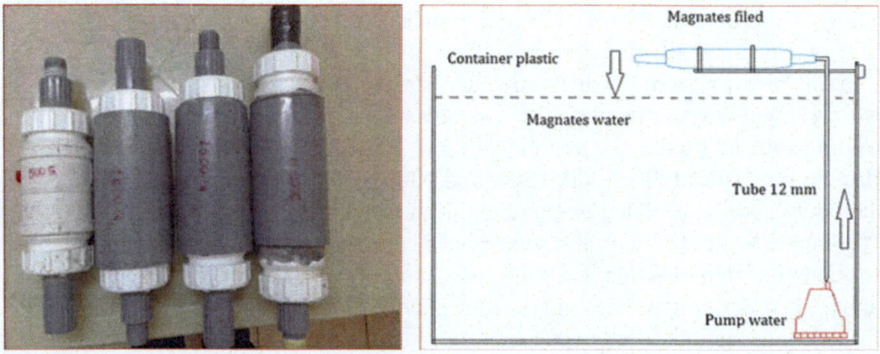

Fig. 2 Devices of magnetic fields and preparation of magnetic water

3 Laboratory Experimental Work

The laboratory experimental work includes measuring the effects of several intensities of magnetic fields on the key properties of swelling soil. The procedure of soil treatment by magnetized water and conducted tests are described in the following sections.

3.1 Preparation of Soil

A plastic pipe container of 5 mm wall thickness and an internal diameter of 580 mm and height of 740 mm is used as an outer container, this container used to keep the magnetized water as shown in Fig. 3. The inner container is made of CPVC plastic of 10 mm thickness and an internal diameter of 400 mm and a height of 700 mm, this container is perforated and used to keep the soil. The inner pipe contains 300

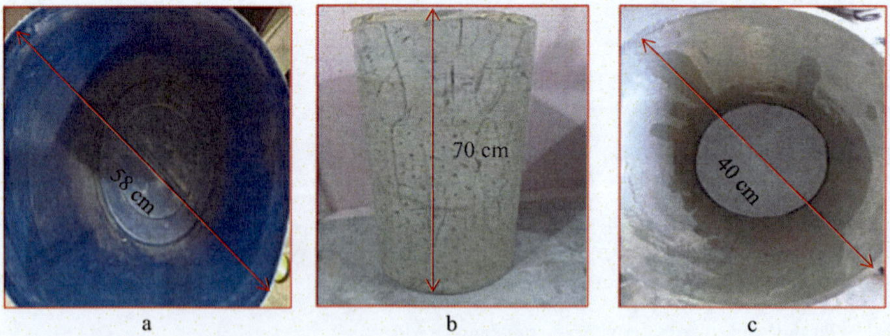

Fig. 3 **a** Water container, **b** Soil container, **c** Base of soil container

holes of 3 mm diameter distributed arbitrarily over 500 mm from the base of the inner pipe. The end of the outer container is closed by tightening CPVC plastic cap and the end of the inner pipe is closed with a perforated CPVC plastic cap as shown in Fig. 3.

3.2 *Soil Tests*

The key properties of expansive soils are measured in the present work to evaluate the effects of water magnetization on these properties. These properties are Atterberg's limits (ASTM D4318), free swelling test (IS 2720-40) [12], total dissolved salts (ASTM D5907), salinity concentration depends on NaCl concentration (Digital Salt Refractometer), and unconfined compression strength (ASTM D2166-00) [13]. In general, the soil may be rich in salts because the parent rock from which it was formed contains salts. The depth of influence for the swelling soil on a near-surface zone usually occurs to about 3 m depth from the surface and depending on the environment state and the properties of soil. The swelling potential of expansive soil depends on the initial void ratio, water content, soil structure, and stress condition [14]. In this test, using 10 g of soil (after dried in the oven) passing through sieve number 4, then put the soil in a glass tube of 100 mm^3 volume and added kerosene to reach 100 mm height in the tube. After 24 hrs., the free swelling is recorded and the same soil samples are prepared but treated with several types of water (RO and magnetized water 500, 1000, 1500, and 2000 Gauss). The free swelling is recorded after 24 hrs. as shown in Fig. 4. The value of free swelling is calculated using the following equation (IS 2720-40) [12].

$$\text{Free Swelling}(\%) = \frac{V_2 - V_1}{V_1} \tag{1}$$

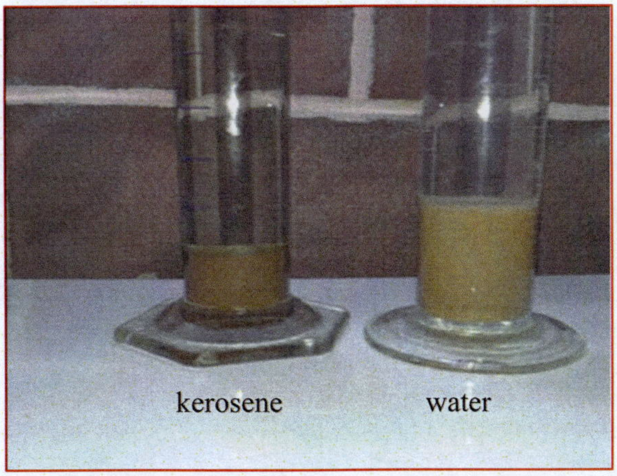

Fig. 4 Comparison of free swelling between kerosene and water

where V_1 is the volume of soil in kerosene after 24 h and V_2 is the volume of soil in water after 24 h.

3.3 Swelling Test of Expansive Soil

The inside face of the soil container was coated with lubricating oil to minimize the friction effect (interaction) between the soil and the wall of the container. A particularly designed hammer (8.5 kg in weight) was used to compact the soil in the container to reach the wet density of 1.68 g/cm^3, the density of each layer was checked by using an odometer ring which has constant dimensions of 50 mm diameter and 19 mm height. In this test, the height of the soil container was divided into nine layers of 70 mm thickness. Each layer was filled with 14.5 kg of wet soil in total (12 kg of dry soil and 2.5 kg of distilled water). Also, each layer was compacted by about 148 $\pm$ 2 blows of a hammer, and the falling height of the hammer was 400 mm to achieve compaction energy of 4429.52 kg cm/1000 cc. After completing the compaction of soil layers to achieve the wet density, the soil container is put in the water container. The water in this container was circulating in a magnetic field.

The uplift movement test (swelling test) was conducted on the soil surface by placing a circular steel plate of 250 mm diameter and 3 mm thickness as shown in Fig. 5. Two dial gauges were fixed on the steel plate surface to record the swelling of soil resulting from the wetting of soil and the effects of water magnetization on the swelling amount. Several types of water were used in this test (RO and magnetized water under several intensities of magnetization). During the test, the water supplied in the outer container is under continuous circulation through the magnetic field, where the electric conductivity (EC) of water was 145, 197, 226, and 265 μS/cm for water passing through magnetic fields of intensities 500, 1000, 1500, and 2000 Gauss, respectively [11].

4 Results and Discussion

The different magnetic fields have significant effects on the liquid limit (LL) of swelling soil and slight effects on the plastic limit (PL) as shown in Fig. 6. The liquid limit decreased by 17.16, 25.65, 32.32, and 35.8% for soil samples treated with magnetic fields of 500, 1000, 1500, and 2000 Gauss, respectively. The plastic limit suffering from a slight decrease due to the treatment of soil with magnetic water, where the plastic limit decreased by 0, 5.45, 7.27, and 9.09% for soil samples treated with magnetic fields of 500, 1000, 1500, and 2000 Gauss, respectively. Generally, the plasticity index (PI) decreased significantly as shown in Fig. 6.

Figure 7 showed increasing of TDS value by 186, 236, 268, and 300% for soil samples treated with tap water passing through the magnetic field of intensities 500, 1000, 1500, and 2000 Gauss, respectively. Also, the content of NaCl in soil samples

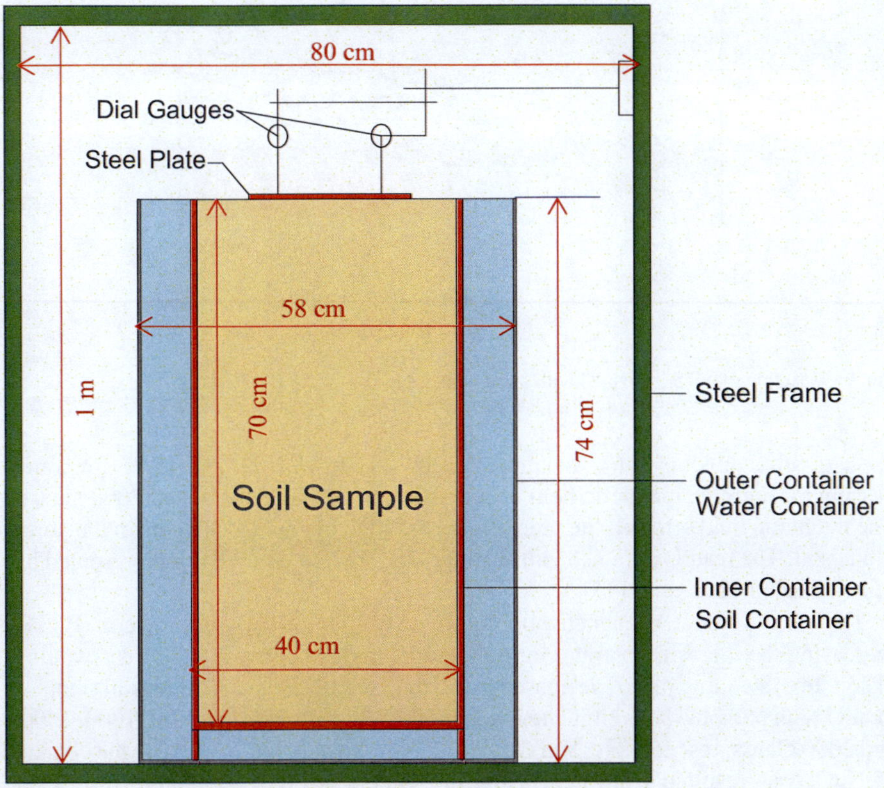

Fig. 5 Set up used to test the uplift movement of soil

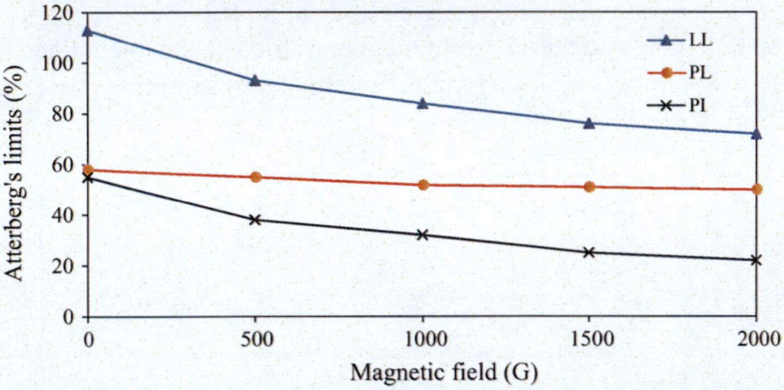

Fig. 6 Variation of Atterberg's limits and plasticity index with the magnetic field

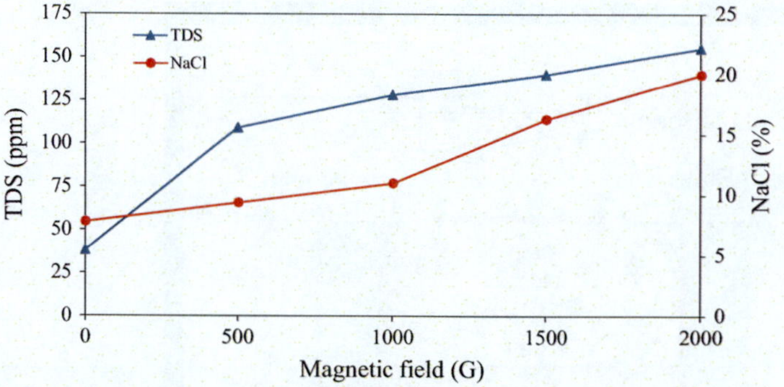

Fig. 7 Variation of TDS and NaCl with a magnetic field

treated with magnetic water was increased by 20.51, 41, 109, and 156% for water passing through the magnetic field of intensities 500, 1000, 1500, and 2000 Gauss. The magnetic field affected the precipitation of TDS and NaCl contents in the pores of the soil. The results obtained in this study are well agreed with those presented by Mostafazadeh-Fard et al. [15].

The effects of water magnetization on free swelling of soil are shown in Fig. 8. Increasing the intensity of water magnetization causes decreasing the free swelling of soil. Free swelling of soil decreased by 8.16, 11.92, 16.5, and 22% for soil samples treated with potable water circulate in magnetic fields of intensities 500, 1000, 1500, and 2000 Gauss, respectively. Van Olphen (1997) showed that increasing the content of NaCl was resulted from decreasing the thickness of the double-diffuse layer between quasicrystals of soil which consequently causes decreasing the free swelling of soil duo to liquid (contain salt) dielectric constant reductions the double-layer depth and permits to clay particles to style closer in it [16].

Figures 9 and 10 showed the effects of magnetics filed on the unconfined compression strength of swelling soil. Two types of swelling soil samples were tested by an

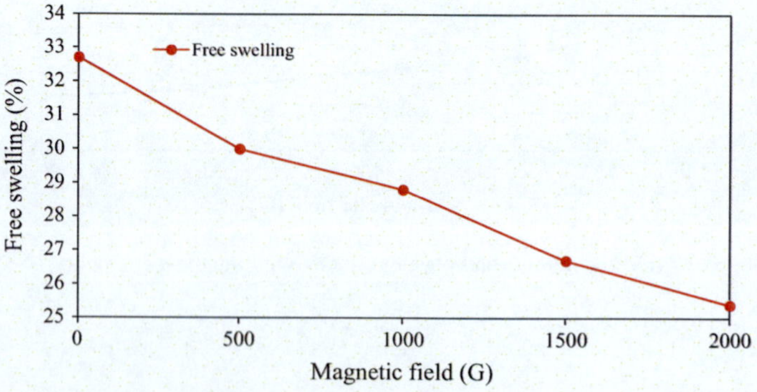

Fig. 8 Relationship between free swelling and magnetic field

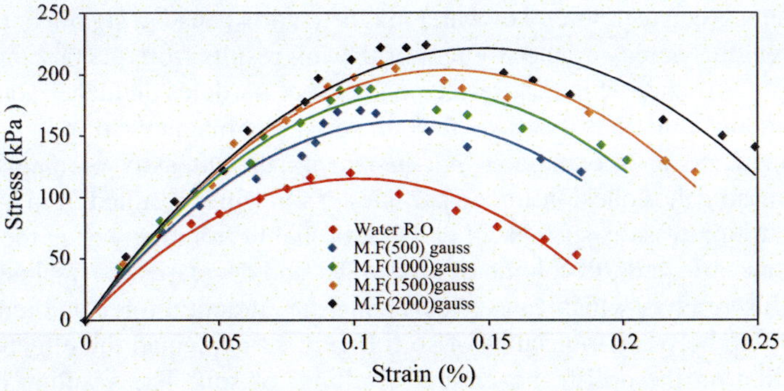

Fig. 9 Unconfined compression test results at a water content of 22%

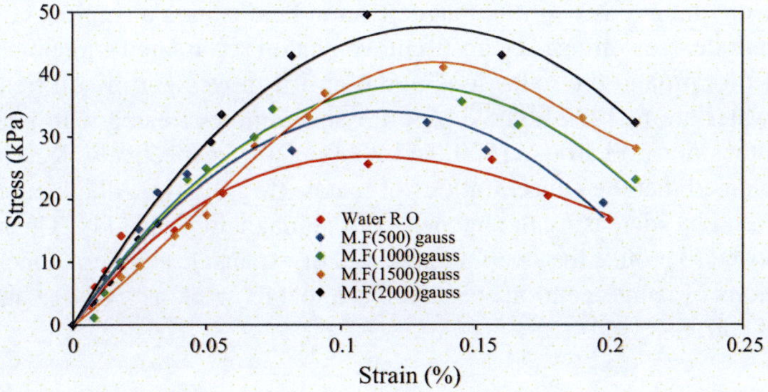

Fig. 10 Unconfined compression test results at a water content of 32%

unconfined compression test, the first one was wetted with 22% of distilled water and the second soil sample was wetted with 32% of distilled water. The results of tests showed decreasing the unconfined compressive strength of soil from 120.12 kPa to 26.36 kPa with increasing the water content from 22 to 32%. Generally, the unconfined compressive strength (q_u) of soil was increased significantly by increasing the intensity of the magnetic field that potable water circulated in it. The unconfined compressive strength of soil (q_u) samples prepared with an initial water content of 22% reached 170.52, 188, 208.56, and 224 kPa when treated with several intensities of magnetized water 500, 1000, 1500, and 2000 Gauss, respectively. The unconfined compressive strength of swelling soil is increased by 73.8% when treated by magnetized water of intensity 2000 Gauss. Also, the unconfined compressive strength of swelling soil of initial water content 32% was increased to be 32.27, 35.59, 41.06, and 49.49 kPa when treated with several intensities of magnetized water 500, 1000, 1500, and 2000 Gauss, respectively. The unconfined compressive strength of swelling soil is increased by 87.75% when treated by magnetic water of intensity 2000 Gauss. The initial water content affected negatively the unconfined compressive strength of the soil and using green technology of water magnetization causes a significant increase in such strength. Increasing the strength of the soil results from the precipitation of calcite in pores of soil which causes cementation of particles and takes some of the water for chemical actions that take place in soil under magnetization.

The conducted tests explained that increasing the intensity of magnetization causes increasing the cohesion among particles of soil which resulted from increasing the concentration of salts such NaCl or TDS, probably from increasing the concentration of salts affected the volume of pore water and decreases the thickness of the double-diffusive layer which causes increasing the attraction force and reduces the repulsive force between molecules of soil. Figure 11 explained the effect of water magnetization on the up-lift movement (swelling) of soil. The swelling of soil is increased gradually with increasing time and reached equilibrium after 110 hrs. test, after that the swelling of soil increased steeply for the period between 110 and 140 hrs.

Generally, using magnetized water reduces the swelling potential of swelling soil, and increasing the intensity of the magnetization field causes a significant decrease in the magnitude of swelling. The up-lift movement of soil treated with potable water (RO) is 21.66 mm and the reductions in the up-lift movement of soil treated with magnetic water are 6, 14, 24, and 34.2% for soil samples treated with magnetized water of intensities 500, 1000, 1500, and 2000 Gauss, respectively. Karkush et al. (2019) explained that the concentrations of metals (Mg, K, Na, Alkaline, and SiO_2) were increased by increasing the intensity of the magnetic field [11]. The magnetic water is proved to reduce the swelling potential of expansive soil duo to increase the concentrations of some metal in soil and these metals work as a bridge among the particles of soil.

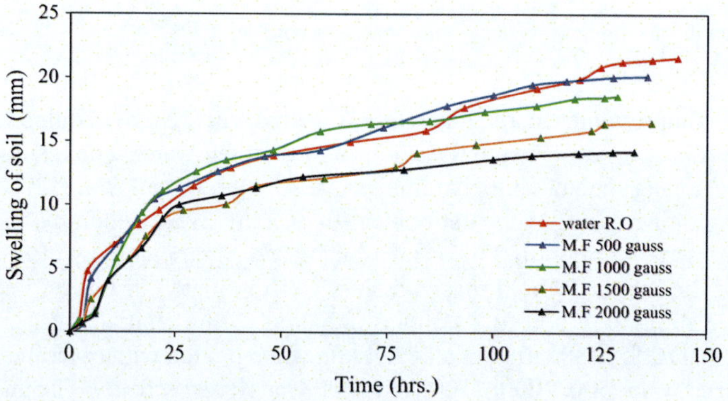

Fig. 11 Relationship between uplift movement and time for expansive soil

5 Conclusions

The different intensities of magnetic fields have significant effects on the key properties of swelling soil. The main conclusions collected from this study are:

- The plasticity properties (LL and PL) were decreased significantly with increasing the intensity of the magnetic field from 500 to 2000 Gauss but the change of PI is slow, which reduces the potential of soil swelling and change classification.
- The unconfined compressive strength at the failure of swelling soil increased with increasing the intensity of the magnetic field.
- The uplift movement (swelling) of soil is decreased significantly by increasing the intensity of the magnetic field.
- The strength of swelling soil was improved by reducing the collapse potential of such soil the water magnetization technique and without using any chemical or solution additives, probably this new technique is considered one of the sustainable solutions in geotechnical engineering.

References

1. Peron H, Laloui L, Hueckel T, Hu LB (2009) Desiccation cracking of soils. Eur J Civil Environ Eng 13(7–8):869–888
2. Musielak G, Śliwa T (2011) Influence of drying rate on clay fracturing by drying, European drying conference Palma. Balearic Island, Spain, pp 26–28
3. Arasan S, Yetimoglu T (2008) Effect of inorganic salt solutions on the consistency limits of two clays. Turkish J Eng Environ Sci 32(2):107–115
4. Mansour ZM, Taha MR, Chik Z (2008) Fresh-Brine water effect on the basic engineering properties of Lisan. J Appl Sci 8:3603–3611
5. Shariatmadari N, Salami M, Karimpour FM (2011) Effect of inorganic salt solutions on some geotechnical properties of soil-bentonite mixtures as barriers. Int J Civil Eng 9(2):103–110
6. Ajalloeian R, Mansouri H, Sadeghpour AH (2013) Effect of saline water on geotechnical properties of fine-grained soil. EJGE 18
7. Vanda Y (2014) Effect of pore water chemistry on hydro-mechanical behavior of compacted expansive clay, M.Sc. in Civil Engineering, Eastern Mediterranean University, Gazimağusa, North Cyprus
8. Khemissa M, Mahamedi A (2014) Cement and lime mixture stabilization of an expansive overconsolidated clay. Appl Clay Sci 95:104–110
9. Mitchell J, Soga K (2005) Fundamentals of soil behavior, 2nd edn. Wiley, New York
10. Al Najm H (2014) Effect of irrigation water salinity and magnetization and moisture depletion in some physical properties of soil, growth and yield of potatoes. Ph.D. Thesis, Anbar University
11. Karkush MO, Ahmed MD, Al-Ani SM (2019) Magnetic field influence on the properties of water treated by reverse osmosis. Eng Technol Appl Sci Res 9(4):4433–4439
12. IS 2720-40, Methods of test for soils, Part 40: Determination of free swell index of soils
13. ASTM Standards (2003) Soil mechanics and foundation engineering, building, stone; peats, Part 3
14. Bell FG, Culshaw MG (2001) Problem soils: a review from a British perspective. Nottingham, Problematic soils symposium, pp 1–35

15. Mostafazadeh-Fard B, Khoshravesh M, Mousavi SF, Kiani AR (2011) Effects of magnetized water and irrigation water salinity on soil moisture distribution in trickle irrigation, ASCE. J Irrigat Drainage Eng 137(6):398–402
16. Van Olphern H (1997) An introduction to clay colloid chemistry, 2nd edn. Wiley, New York

Effect of Rainfall and Dynamic Loading Conditions on Unsaturated Soil Slopes Stability

Tareq H. AL-Rahal, Suhail I. Khattab, and Bayer J. Al-Sulaifanie

Abstract This research was conducted to detect the effect of static load, dynamic condition, and rainfall on slope stability. The studied parameters are: rainfall intensity, static load, frequency, soil suction variation, and the form of slope failure. A laboratory model was manufactured in the form of a box with dimensions of (2000 × 1000 × 1450) mm that slides on a rail installed on the ground and moved by a dynamically generated system and provided with a controlled rainfall system. To clarify the rainfall effect with dynamic conditions on the slope stability, the effect of rainfall intensities of (20, 30, and 40) mm/h were studied on a series of three model tests. Two stages of a rainfall (duration of each one = 1 h) have been applied to the soil sample. After each stage, the model was left for 24 h to allow the water infiltration process into the soil layers. Dynamic conditions were enjoined as (20) mm of one direction horizontal vibration displacement at frequencies of (0.5, 1.0, and 2.0) Hz. The time duration for each frequency value is (3) minutes. The acceleration during vibrations was measured through the test by an accelerometer that was placed on a shaking table and inside the soil. The main result shows that the rainfall and dynamic load have negatively affected the slope bearing capacity. It decreased from (408) kPa at the static test to (204, 127, and 102) kPa at the three rainfall tests, respectively. On the other hand, results also revealed that the soil suctions were decreased after the application of rainfall on the slope model.

Keywords Slope stability · Dynamic condition · Rainfall · Numerical analyses

---

T. H. AL-Rahal · S. I. Khattab (✉) · B. J. Al-Sulaifanie
College of Engineering, University of Mosul, Mosul, Iraq
e-mail: suhailkhattab@yahoo.com

T. H. AL-Rahal
e-mail: tareqrahal73@gmail.com

B. J. Al-Sulaifanie
e-mail: civilengmosl@yahoo.com

M. O. Karkush and D. Choudhury (eds), *Modern Applications of Geotechnical Engineering and Construction*, Lecture Notes in Civil Engineering 112,
https://doi.org/10.1007/978-981-15-9399-4_6

1 Introduction

Slopes either existed naturally or by man-made. Man-made slopes involve roadway fill and cut slopes, dams, embankments, and other similar constructions. The problems of Slope stability and connected disasters happened in the past when the weak balance of slopes was disrupted by environmental conditions or human activity. Failure may occur due to inaccurate design of engineered slopes or unexpected natural hazards, which lead to disruption of engineered slopes because they were not expected through the design process. Moreover, failures occur in the slopes due to different forces types such as earthquakes, gravity forces, excavation work, strong leaks, and changes in surrounding conditions. So, weakening of the soil structure could be occurred by rainfall, added load on the slope crest, and cutting away at the toe for any purpose.

Stability of soil slope is a difficult problem due to different soil properties factors like permeability, initial moisture, and soil ability to retain water. At the unsaturated state of the soil, the behavior of soil is different and becomes further complex due to the nonlinearity behavior of soil hydraulic function [1]. So, the soil suction which contributes to increase the shear strength of unsaturated soil, and hence, improved slope stability [2]. Rainfall increases soil moisture, then changed soil suction value with time. In general, the magnitude of this change depends upon the duration of rainfall and intensity, the functions permeability coefficient for the soil (through an effect on suction value of unsaturated soil), and the physical unsaturated soil properties [2, 3].

The sources of a dynamic load may be natural, such as earthquakes, or they may be the result of human activities like vibration machines and railways. There are many instruments and types of equipment that can be used to monitor the response of slopes when exposed to earthquakes and to determine the degree of risk experienced by analyzing recorded data. In the absence of these instruments, laboratory models to simulate the behavior of slopes during earthquakes could be constructed as a centrifuge model and shaking table [4–6].

Instability of the slope caused by earthquakes is an important engineering problem that always produces risk. Earthquakes are the maximum danger to the long-term stability of the slope in the seismic activity zones, the earthquake landslides lead to damage and break homes and other buildings. Therefore, many researchers were carried out research works to spotlight on this problem. Yasin [7] executed a shaking table experiment utilizing a (900 × 400 × 500) mm mold on reinforced sandy soil slopes by two geotextile materials layers. Vyas et al. [10] calculated sandy soil slopes settlement that resulted by dynamic loads, at different percentages of gravity acceleration. The researchers [8, 9] studied the dynamic response for machine foundations. These foundations were put on piles inside sandy soils. Moreover, the effect of sand relative density and piles distribution patterns are taking into consideration.

In general, the researchers pointed an increase in acceleration value (within soils mass) with increasing soil relative density and time of vibration application on soil model. Moreover, the measured acceleration at the end piles was higher than that at piles tips. These results could be attributed to the damping phenomenon and wave dissipation within the soil mass. Moreover, they observed an increase in bending moments that exposed on piles with increasing frequency vibration. This study aims to show the effect of rainfall and frequency of horizontal vibration on the loading capacity of soil slopes. The parameters of the study are: the rainfall intensity, the displacement of horizontal vibration, suction values, frequency, and the failure form of the slope.

2 Experimental Tests and Working Program

Soil. The used soil samples in the model were brought from Mosul city (AL-Shmciatquarter). Table 1 explained some physical and chemical properties of this soil, it was classified as (CL) by the unified soil classification system.

Experimental Model. An experimental model was designed and executed by the authors and used to perform the laboratory tests. This model consists of soil tank (2100 × 1100 × 1450) mm dimensions, loading frame (100) tons capacity, shaking table with dimensions (2100 × 1500) mm, electric motor in (10) horse capacity,

Table 1 Engineering properties for used soil

Engineering properties of university soil			Value
Atterberg limit	Liquid Limit (%)		28
	Plastic Limit(%)		18
	Plasticity Index(%)		10
Soil classification			CL
Specific gravity (Gs)			2.73
Chemical tests	Total soluble salts (T.S.S %)		1.8
	Organic matter (%)		2.6
	Gypsum content (%)		1.82
Grain size analysis	Sand (%)		0
	Silt (%)		90
	Clay (%)		10
Activity=P.I/C			1.09
Compaction characteristic	Standard compaction	Max. dry unit weight (kN/m^3)	17
		Optimum moisture content (%)	14.63
Direct shear test		Cohesion, (c) kPa	24
		Angle of internal friction, (ϕ)	38°
Flow parameter		k_x (m/h)	8.5

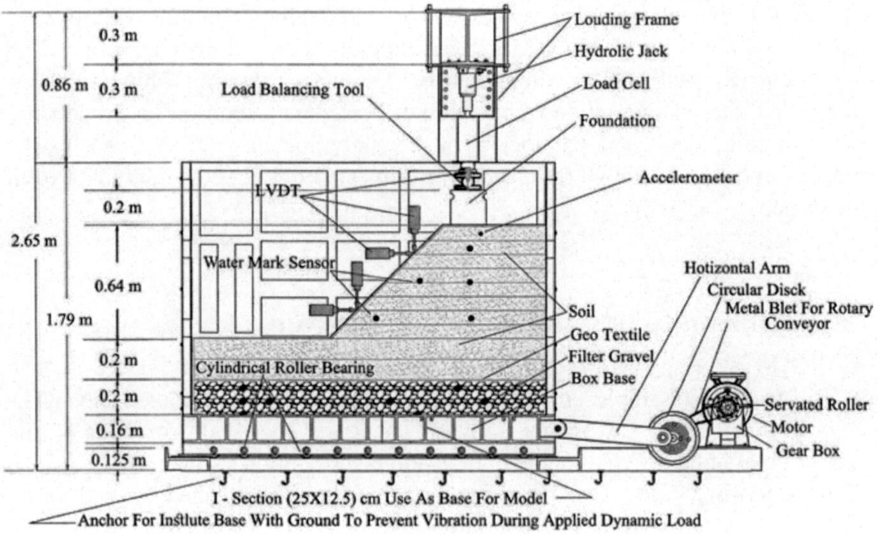

Fig. 1 Layout for the manufactured model plus all devices and tools (side view)

inverted drives control on the speed of the electric motor(vibration frequency of the shaking table), rotating disk and arm to translate the vibration to shaking table. A steel strip footing (1000 × 200 × 120) mm was placed on the slope crest which extends along the box width to simulate the condition of plane strain. The hydraulic jack is utilized to the application of vertical load on the center of this footing and the applied load magnitude was measured by utilizing the load cell. Aspherical seat (load balancing tool) was used to insure the applied stress remains vertical during shaking. The Linearly Variable Differential Transformers (LVDTs) were utilized to measure settlement of the slope and the horizontal displacement [10] which their locations are showing in Fig. 1.

Slope Preparation. Initially, by employing the Bishop method, the determining factor of safety (FS) for the suggested slope was (12.6). The soil was mixed with the optimum moisture content (14.65%), then the sample is left in nylon bags for 48 h for homogeneity. The soil sample was compacted using a vibrating compactor in 1-ton capacity to reach a maximum dry unit weight for standard compaction test (17 kN/m^3) and to simulate the real moisture and unite weight for the soil in the site. In the beginning, a well-graded gravel layer in (200) mm thickness is placed over the entire model area and then covered by a geo-grid mesh layer to produce a free drainage condition and permit a uniform water distribution below the soil. To decrease friction and simulate plane strain case, the inner face of the model walls were coated by polythene and a thin grease layer [11, 12]. The first two soil layers (each layer thickness is (100) mm after compaction) were placed on the total model area. The slope was consists of eight soil layers, thickness for each layer was (80) mm after compaction and with inclination angle equal to (45) degree with the horizontal.

Fig. 2 Equilibrium time for the water probes

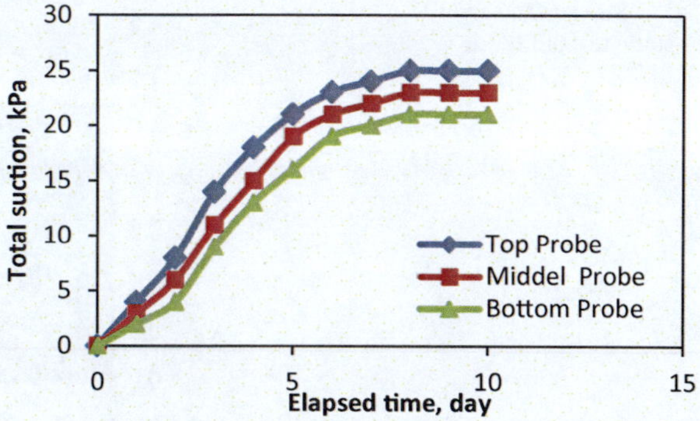

Five sensors of total suction are installed in the soil as shown in Fig. 1 and are joined to a watermark to record and save data. The soil slope model is left to reach the equilibrium total suction value of (21–25) kPa as seen in Fig. 2. The load cell and (LVDTs) were linked to the data logger to register and save the data.

3 Results and Discussion

Analysis of Slope Stability under Static stress. After soil slope was prepared at previous specifications, the static stress was applied on the slope crest in several steps (51, 102, 153, 204, 255, 306, 357, and 408) kPa at duration periods (1, 2, 3, 4, 5, 6, 7, and 8) days respectively or until settlement completion for each stage. It is worth pointing that the footing settlement was measured under each loading stage by the (LVDTs) that fixed on the strip footing. Then, the next stress stages are applied within the same above stated manner until the slope failure. It is clear from Fig. 3 that the slope settlement is increased with augmenting the applied vertical stresses. Also, the ultimate loading capacity for each slope model was calculated, at the ultimate load, the soil slope becomes unstable and then the failure happened. At failure, the

Fig. 3 Experimental model test and numerical

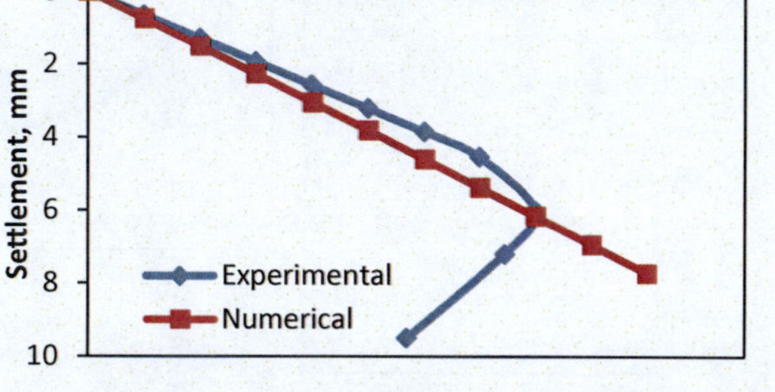

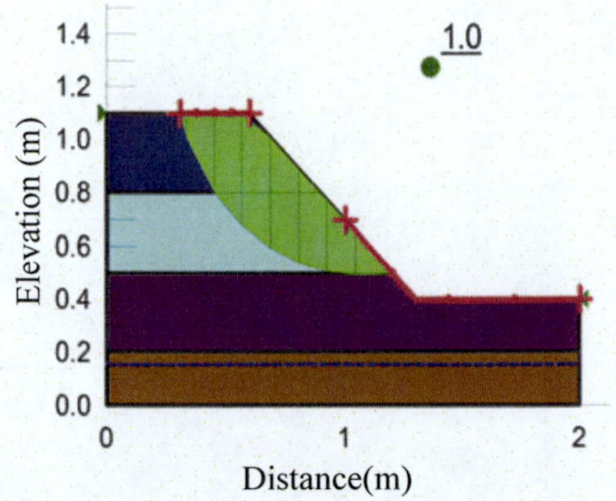

Fig. 4 Slip surface and FS for static test at failure

slip surface had a circular shape and extends from the footing edge and drops to (300) mm depth where the sliding soil mass was pushed downwards and outwards as presented in Fig. 4. The failure of slope has occurred at a vertical stress of (408) kPa with a (6.0) mm settlement.

The software (GEO-SLOPE) was utilized to execute theoretical analysis for the slopes at the same laboratory work conditions. The slope was simulated using the finite element method (FEM). The packages utilized in the analyses include: deformation modeling (SIGMA/W), dynamic modeling (QUAKE/W), and slope stability modeling (slope/w). Figure 3 reveals the experimental and numerical slope settlement with the applied vertical stresses. Figure 5 shows the FS values versus different static vertical stresses applied on the slope crest at the indicated periods. A good matching between the practical and theoretical results was found. Figure 3 clarifies that the maximum numerical settlement of slope is (7.9) mm, which occurred at a stress of (510) kPa. Otherwise, by using a watermark monitor, it is noted that the total suction for top layer decrease from (25 to 23) kPa when the vertical stress increases from (0 to 204) kPa, then total suction will become constant until failure at vertical stress (408) kPa. The total suction of the middle and bottom layers is decreased

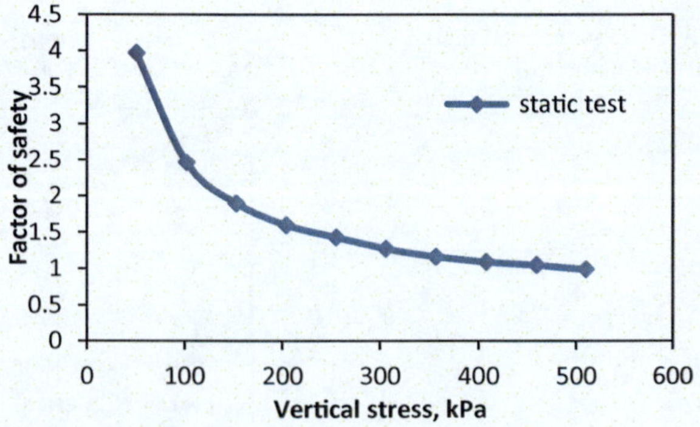

Fig. 5 Factor of safety value at different applied loading stages during a conducted static test

at a more quick rate comparing with the top layer. In these layers, the soil suction decrease from (23, 21) kPa to (14, 13) kPa respectively when the applied vertical stress on the slope crest increased from (0 to 306) kPa and then the suction remained constant until failure at vertical stress equal to (408) kPa. The soil suction variation is presented in Fig. 6. Figures 4 and 7 are displaying the critical slip line and the deformation mesh at the failure stage.

Analysis of Slope Stability under rainfall, dynamic condition, and vertical stress. Rainfall is an important factor that reduces slope stability around the world. Soil slopes are unsaturated in generally, when exposed to rainfall will cause a certain change in soil suction, water flow and reduce the shear strength of soil mass. These are the main important parameters related directly with slope stability analyses of unsaturated soil that its surface is influenced by infiltration and evaporation [13, 14]. Rainfall was applied to the slope model by using a controlled rainfall system that was installed at 500 mm above the slope model. To clarify the effect of rainfall with dynamic conditions on the soil slope stability, a series of three tests were performed. Heavy rainfall intensity of (20, 30, and 40) mm/h was used for (first, second, and

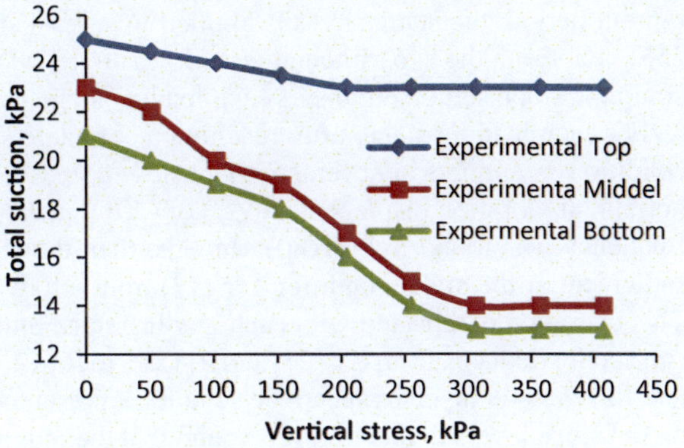

Fig. 6 Variation in total suction when increased vertical stress

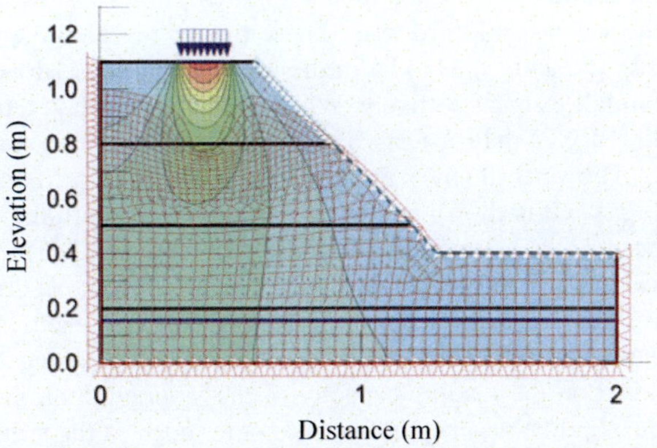

Fig. 7 Deformation mesh for a static test at failure

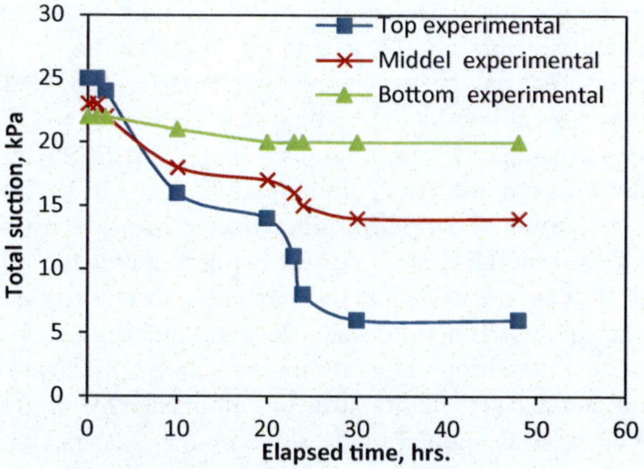

Fig. 8 Total suction-elapsed time relation after applied two rain stages at 20 mm/h intensity

third) rainfall tests, respectively. At each test, two rainfall stages (duration of each = 1 h) have been applied to the soil sample. After each stage, the model was left for 24 h to allow the water infiltration process into the soil layers. At the end of the rainfall period, the static stress is applied in several stages (51, 102, 153, 204, and 255) kPa. From the experimental tests, it was founded that the minimum duration to complete slope settlement under each loading stage is (2.75) hours.

For each loading stage, after reaching a complete slope settlement, the dynamic condition is applied in several frequencies (0.5, 1.0, and 2.0) Hz and a constant horizontal vibration displacement(Δ_x) of (20) mm. The vibration duration for each frequency was found to be (180) seconds, then the slope was left to complete the settlement at the static condition for (13) min before applying the next frequency. This operation is repeated with each loading stage until failure. At the first test, the soil suction reduces from (25, 23, and 21) kPa to (6, 12, and 19) kPa at upper, middle, and lower slope layers respectively. Figure 8 presented the suction variation in the model slope layers in this test. It is found that the reduction in suction was increased with increasing the rainfall intensity due to the augmentation in the amount of water infiltration into the soil layers. Figures 9 and 10 show the changes in soil suction for the second and third tests. These figures revealed the soil suction is reduced to (4, 12) kPa and (3, 11) kPa in the upper and middle slope layers for second and third rainfall tests respectively. While at the bottom layer, the total suction was dropped to (19) kPa in both tests. This is consistent with the results obtained by [15, 16].

The vertical stress and settlement relation for the first rainfall test are shown in Fig. 11. It is shown an increase of settlement with the vertical stress and the failure occurred at a stress of (204) kPa with about (3.543) mm settlement. While for the second and third rainfall testes, the failure was occurred at (127.55 and 102) kPa vertical stress (σ_v) and at (2.34 and 2.04) mm settlement respectively. Figure 12 explained the failure mechanism of these tests, a negative effect of the rainfall was noted on the loading capacity of the slope with an increase in the settlement. So, the slip line was begun from the outer edge of the strip footing to the lower part of

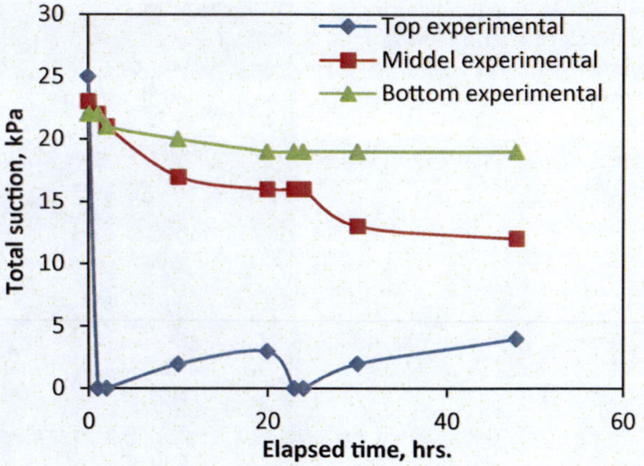

Fig. 9 Total suction-elapsed time relation after applied two rain stages at 30 mm/h intensity

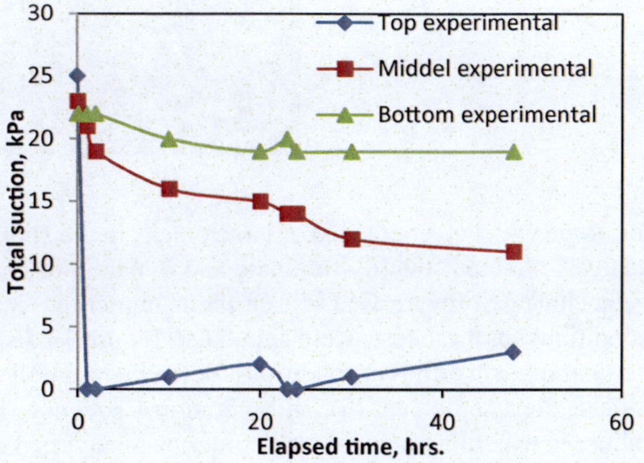

Fig. 10 Total suction-elapsed time relation after two rain stages at 40 mm/h intensity

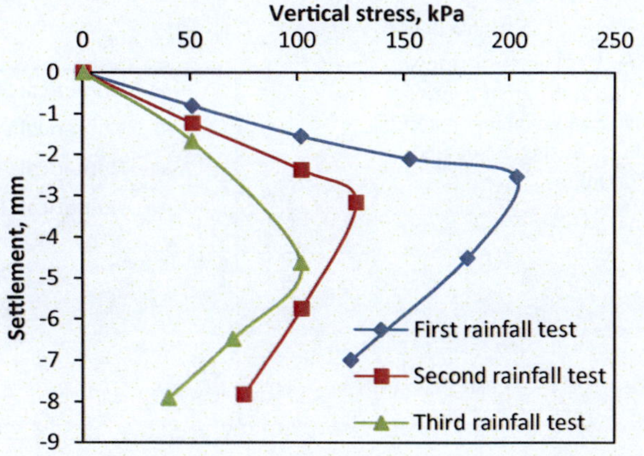

Fig. 11 Settlement of slope under static stress with dynamic condition following rain

Fig. 12 Failure mechanism of the slope after imposing rainfall, static stress, and dynamic condition

the slope with the emergence of large cracks in the sliding soil zone. The theoretical analysis of slope stability for static and dynamic tests was conducted by using the geotechnical software GEO-SLOPE. In numerical analyses, the same laboratory conditions for these tests were simulated. Figure 13 displays changes in FS with the vertical stress at different conditions for the first rainfall test. Table 2 presents the load failure, crest settlement (δ), and vibration frequency (Freq.) of the slope model at failure in experimental work and numerical analyses. From this table, a reduction in slope load-bearing capacity with increased rainfall intensity was noticed. Moreover,

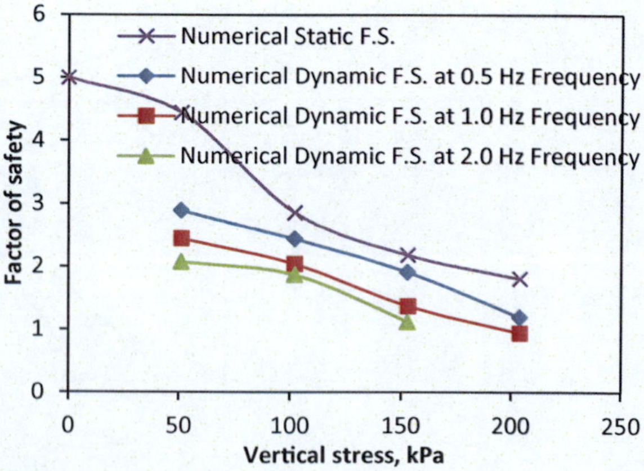

Fig. 13 Factor of safety value at different applied static and dynamic loading stages during conducted first rainfall test

Table 2 Comparison between the practical results and the numerical analyses of static, dynamic, and rainfall tests

Test name	Δ_x (mm)	Laboratory model			Numerical analysis		
		Condition at failure					
		σ_v (kPa)	Freq. (Hz)	δ (mm)	σ_v (kPa)	Freq. (Hz)	δ (mm)
Static test	0.0	408	0.0	6.0	510	0.0	7.9
1st rainfall test	20	204	0.5	3.56	204	1.0	3.0
2nd rainfall test	20	127.5	2.0	2.34	127.5	2.0	1.754
3rd rainfall test	20	102	2.0	2.04	102	2.0	1.15

the decrease of slope settlement at rainfall tests may be resulted due to a reduction period of loading stage from (24) to (2.75) h.

4 Conclusions

The main conclusions of experimental and numerical results are:

- A static test, the stress failure for the slope is (408) kPa and led to slope settlement (6.0) mm. The slip surface for this test had a rotating shape beginning from the interior edge of the footing and dropped to a depth equal to (300) mm.
- Applying rainfall and dynamic condition negatively affected on slope stability through.
- Reduction in the factor of safety value to about half especially at early stage loading.
- Reduction in bearing capacity of the slope decreased for the three rainfall-dynamic tests by (50, 68.75, and 75%,) respectively compared to the static test. The dynamic conditions are vibrations in (0.5, 1.0, and 2.0 Hz) of each loading step and horizontal displacement for these tests is (20) mm.
- decreases the slope settlement from (3.56) mm to (2.34 and 2.04) mm with the increase in the rainfall intensity from (20) mm/h to (30 and 40) mm/h, respectively.
- Reduction in the total suction of slope layers, it decreases by (76, 48, and 9.5%) for the upper, middle, and bottom slope layer, respectively, at the first rainfall test. So this reduction increased with increased rainfall intensity.
- Sliding curves of rainfall tests had a rotating form that began from the internal edge for the strip footing and continued to the lower slope end.
- There is a close convergence between the failure condition obtained by laboratory tests with the failure condition for numerical analysis by geo-studio software.

References

1. Fredlund DG, Rahardjo H (1993) Soil mechanics for unsaturated soils. Wiley, USA
2. Runqiu H, Lizhou W (2007) Stability analysis of unsaturated expansive soil slope. Earth Sci Front 14(6):129–133
3. Gavin K, Xue J (2008) A simple method to analyze infiltration into unsaturated soil slopes. Comput Geotech 35(2)
4. Srilatha N, Madhavi LG, Puttappa CG (2016) Seismic response of soil slopes in shaking table tests: effect of type and quantity of reinforcement. Int J Geosynth Ground Eng 33:2–13
5. Vyas M, Rukhaiyar S, Mittal S (2014) Behavior of slope under dynamic condition-an experimental study. J Civ Eng Res 5(2):122–128
6. Giri D, Sengupta A (2010) Dynamic behavior of small-scale model slopes in shaking table tests. Int J Geot Eng 4(1):1–11
7. Yasin ST (2017) Shake table tests to measure the dynamic performance of geotextile- reinforced embankment. Periodic Polytech Civ Eng 61(4):803–814
8. Fattah MY, Zbar BS, Mustafa FS (2019) Effect of soil saturation on load transfer in a pile excited by pure vertical vibration. In: Proceedings of the institution of civil engineers—structures and buildings
9. Fattah MY, Karim HH, Al-Recaby MKM (2016) Dynamic behavior of pile group model in two-layer sandy soil subjected to lateral earthquake excitation. Global J Eng Sci Res Manag 3(8):57–80
10. Sonnenberg R, Bransby MF, Bengough AG, Hallett PO, Davies MCR (2012) Centrifuge modeling of soil slopes containing model plant roots. Can Geotech J 49:1–17
11. Viswanadham BVS, Mahajan RR (2007) Centrifuge model tests on geotextile reinforced slopes. Geosynth Int 14(6):365–379
12. Mahajan RR (2007) Centrifuge model studies on geosynthetic reinforced soil slopes. Doctoral dissertation. Indian Institute of Technology Bombay, India
13. Cheng WJ, Nan GX, Guo SM (2014) Effects of pore-water pressure distribution on slope stability under rainfall infiltration. J Electr Geotech Eng 19(Bund H):1677–1685
14. Rahimi A, Rahardjo H, Leong EC (2010) Effect of hydraulic properties of soil on rainfall-induced slope failure. Eng Geol 114(3–4):135–143
15. Chuong SH (2019) Study of the rainfall infiltration effect and soil strength variation on unsaturated laterite terraces. In: Proceedings of the 4th world congress on civil, structural, and environmental engineering (CSEE'19), Rome, Italy, April, Paper No. ICGRE 184
16. Duong TT, Do DM, Ysuhara K (2019) Assessing the effects of rainfall intensity and hydraulic conductivity on riverbank stability. Water 11:741

Settlement Reduction of Reinforced Multi-Layered Sandy Soil Under Cyclic Loading of Machinery Foundation

Mohammed Ali Hussein, Hussein H. Karim, and Zeena W. Samueel

Abstract The present paper has been carried out to study experimentally the settlement of dynamic machine foundation on unreinforced and reinforced with multi-layered sandy soils (medium-dense sand MD). The relative density of the first and second layers is 50 and 85% corresponding to medium sand soil and dense sand soil respectively. The tests have been carried out on 30 models. The amplitude of the applied harmonic load is 0.25 and 2 ton with a frequency of 0.5 Hz. For each amplitude and frequency of load, the sand models were tested without and with reinforcement. Three types of reinforcements were used with three series of reinforcement configurations (0.5B, 1B using single reinforcement layer and 0.5B, 1B using double reinforcement layers) where B is the square footing width. It is found that the settlement of medium-dense soil will be in its maximum values in the first layer, then try to be more stable. While the settlement is decreased using geogrid fiberglass reinforcement which is in agreement to that with geogrid in most results. It was found that less values of surface settlement can be obtained with type II of reinforcement where load amplitude equal to 0.25 ton with percent of enhancement between 32 and 42% for different configuration of layers of reinforcement, while where load amplitude equal to 2 ton the best value of enhancement of settlement reached to 35–38% for type I.

Keywords Multi-layered soil · Dynamic load · Geogrid

M. A. Hussein (✉) · Z. W. Samueel
Civil Engineering Department, University of Technology, Baghdad, Iraq
e-mail: 40647@student.uotechnology.edu.iq

Z. W. Samueel
e-mail: 40104@uotechnology.edu.iq

H. H. Karim
University of Baghdad, Baghdad, Iraq
e-mail: husseinhkarim@uobaghdad.edu.iq

M. O. Karkush and D. Choudhury (eds), *Modern Applications of Geotechnical Engineering and Construction*, Lecture Notes in Civil Engineering 112,
https://doi.org/10.1007/978-981-15-9399-4_7

1 Introduction

The soil periodically undergoes shear stress on-site in many conditions such as earthquakes and wind forces in tall buildings, building piles, traffic loads, and vibrations caused by the machine. Typical tests have been reported to determine the permanent settlement of a shallow foundation subject to different types of dynamic loads had been a statement by [1]. The outcomes of the dynamic load against the displacement of strip lying on the dense sand are presented in reference [2]. The circular foundation on the granular soil undergoes vertical vibrations outcomes offered from the laboratory model on the constant displacement [3].

Assessment of the efficiency of geogrids in enhancing soil bearing capacity has been done in the latest years by many studies, due to engineers' desire to optimize geotechnical structures. Geosynthetics materials widely used, its economy, easy of installed, the using of reinforce soils in applications of geotechnical engineer have resulted in performance and dependability as in the construction of road, embankments of railway, slopes steadiness, weak ground amelioration, etc., for example [4] and [5]. This paper presents the results of laboratory model tests dealing with the settlement of rigid square machine foundation resting on surface and embedment of unreinforced and reinforced multilayered sand soil with different densities subjected to cyclic loads, with a comparison of the enhancement by using various types of geosynthetics materials.

2 Geometry of the Problem

In many practical engineering causes, it may be necessary to lay shallow foundations on stratified deposits. A layer of deposits below shallow footing affecting bearing capacity is named an underground. The thickness of the underground by a simplified analysis can be shown by Eq. 1 [6]:

$$H = \frac{B}{2}(\tan 45 + \frac{\phi}{2}) \tag{1}$$

where B is a width of a shallow foundation, and ϕ is the angle of soil internal friction. The subsoil displays a layered structure if the thickness of the deposit surface layer is less than H. In the most practical problems, the subsoil is two-layered as shown in Fig. 1.

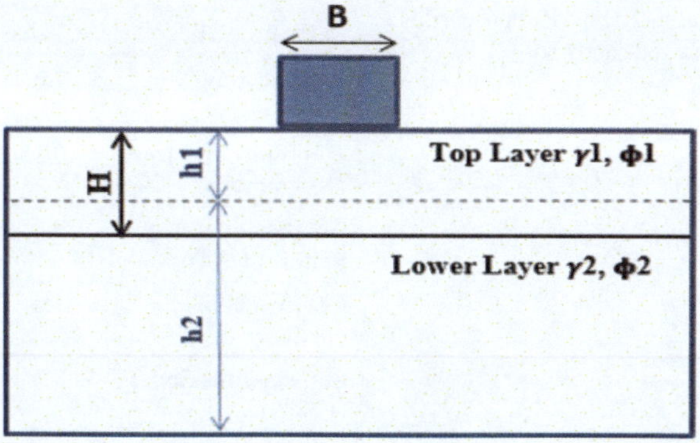

Fig. 1 Geometry of the problem, h1 depth for top layer, h2 depth for the lower layer

3 Testing Program

The tests were conducted in a total number of 30 models. All the model tests of dry sand under dynamic loads with relative density equal to 50% and 85%, respectively corresponding to medium-dense sand MD multi-layers soil. All the models were subjected to dynamic loads equal to 0.25 and 2-ton using a frequency of 0.5 Hz. The sand models testing without reinforcement and with three types of reinforcement of three series of reinforcement from the model surface (0.5B and 1B and 0.5B,1B where B is the width of foundation 200 × 200 mm), the foundation at surface, and embedment at 100 mm. In this study, Karbala sand is used and the physical properties of sand are accomplished through the standard tests. Table 1 shows the physical properties of sandy soil.

Reinforcement Material. Three types of reinforcement materials were used in this study as shown in Fig. 2, these are:

- Type I geogrid was manufactured by Al-Latifiya Factory for plastic mesh. The engineering properties provided by the manufacturing company are given by [7–9].
- Type II, geogrid fiberglass, consists of fiberglass filaments coated with an inorganic sizing agent. It has properties such as high tensile strength, low elongation, excellent temperature range, good anti-age, and alkali-resistance [10].
- Type III, geocells as reinforcement made from a form of non-woven geotextile, is used to stabilize the ground as an innovative approach. When the geocell is filled with soil or other mineral materials, the size of the geocell pocket remains constant at 50 mm [11].

Devices and Design Model. To research the influence of reinforcement material in reduce load settlement to the subsoil layers in sandy soil, it is needful to imitate the conditions closest is potential for those that occur in the field. To realize this goal, special testing devices and other parts were created by [12]. The device can apply various dynamic loads and various frequencies. Device overview at Fig. 3. The

Table 1 Physical properties of used sandy soil

Index properties	Value	Specification
Specific gravity	2.6	ASTM D 854 (2006)
D_{10} (mm), D_{30} (mm), D_{60} (mm)	0.15, 0.2, 0.5	ASTM D 422 and 2487 (2006)
Coefficient of uniformity (Cu)	3.3	
Coefficient of curvature (Cc)	1.2	
Soil classification (USCS)	SP	
Maximum void ratio	0.49	–
Minimum void ratio	0.33	–
Maximum dry unit weight (kN/m^3)	19.5	ASTM D 4253 (2006)
Minimum dry unit weight (kN/m^3)	17.4	ASTM D 4254 (2006)
Angle of internal friction (RD = 50%)	40°	ASTM D 3080 (2006)3080
Angle of internal friction (RD = 85%)	44°	ASTM D 3080 (2006)

Fig. 2 Types of used reinforcement material

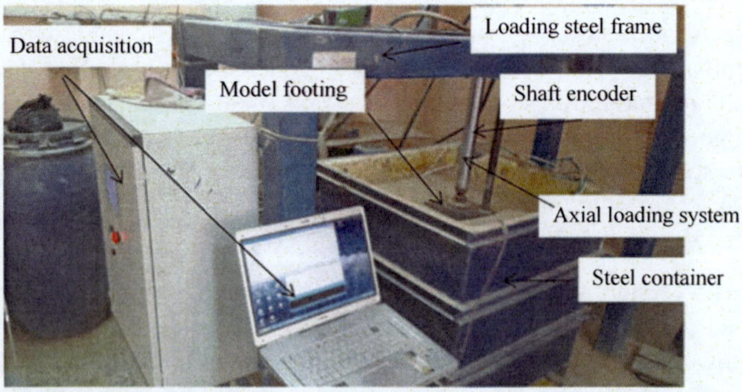

Fig. 3 The parts of the used device

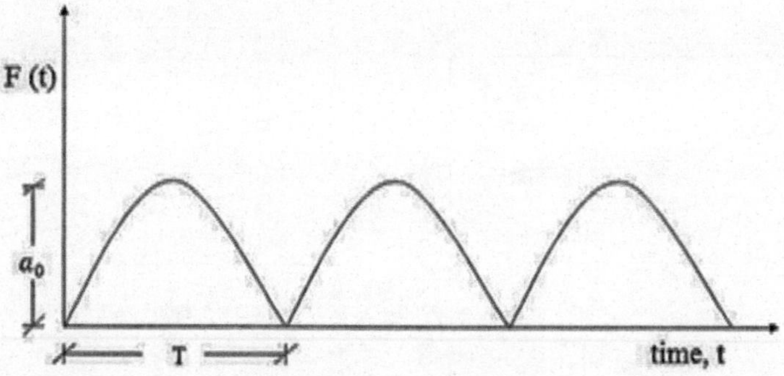

Fig. 4 Dynamic load wave

apparatus be composed of the next parts: Steel frame load, Shaft encoder, Model footing, Acquisition data, Axial loading system, and Steel container.

Preparation of Sand Deposited. Sand deposit is prepared by using for this purpose steel hummer manufactured, chosen relative density 50 and 85% (for medium and dense sand, respectively); the unit weight and the volume of the sand are predetermined, so the weight required to achieve the required relative density is determined.

Dynamic Loading Test. After placing the foundation on the sand layer surface, the dynamic load was applied during a preset sequence. The dynamic download application lasts for 20 min. The dynamic load function is represented by Eq. 2:

$$F(t) = a_0 * \sin \omega t \tag{2}$$

where a_0 = amplitude of load, ω = frequency of load, t = time, and T = period. The shape of the dynamic wave loading applied is of the form close to the sinusoidal compressive type as shown in Fig. 4.

4 Results and Discussion

Dynamic Settlement for Surface Foundation. Figure 5 shows an exemplary correlation of displacement against the number of cycles of different amplitudes of load with or without different levels and types of reinforcement. It can be noticed that the curves follow the same direction, and as the dynamic loading capacity increases, the surface settlement increases, besides, the rate of settlement increased in medium sand is greater than that of the dense sand. This may be due to an increase in particle pressure. This figure showed sudden settlement in the first layer then gradually increase until reach approximately stable of sample that means the second layer of soil (dense

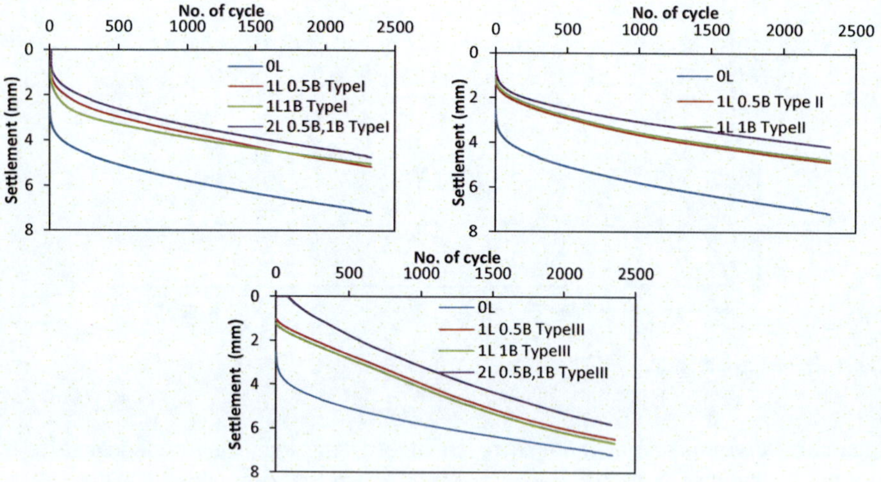

Fig. 5 Settlement of surface foundation for different levels of reinforcement, types I, II, and III

layer stay bearing load amplitude until it fails, that very clear with amplitude load 2 ton) depends on the type of reinforcement existence and the other studied parameters.

After that, fewer values of settlement can be obtained with type II of reinforcement where load amplitude equal to 0.25 ton with percent of enhancement between 32 and 42% for different configuration of layers of reinforcement, while where load amplitude equal to 2 ton the best value of enhancement of settlement reached to about (35–38)% for type I. From that, it can be concluded with high load amplitude the reinforcement type I increase the shear strength of soil compression with others types, but at load amplitude 0.25 ton, type II be the best because the small opening of reinforcement material increases the strength of soil and soil resistance to the low load.

Dynamic settlement for embedment foundation. In general, from Fig. 6, it can be observed for both unreinforced and reinforced models when the foundation depth raises to (100 mm), reducing the soil settlement. Also, it can be noticed during the time of the load, the sum displacement of a foundation continued to increase and at the end of time reached the ultimate value; so, the best results of less settlement were with type III.

Dynamic settlement of reinforced soil. Reinforcement placed in the soil in various configuration (0.5B, 1B, and both 0.5B and 1B together) as particular in the reinforced soil tests in a predestined position Fig. 7 providing the ideal relationship for the displacement against numbers of cycle for various reinforcement types locations and layers number for each type. Figures showed that very rapprochement on the magnitude of settlement in types I and II while type III always achieves the maximum values of settlement near to the reinforcement soils. Also, these figures showed the performance of reinforcement is a slight variation neither placing near the bottom of the foundation nor placing far away, two geogrid layers, or placing one

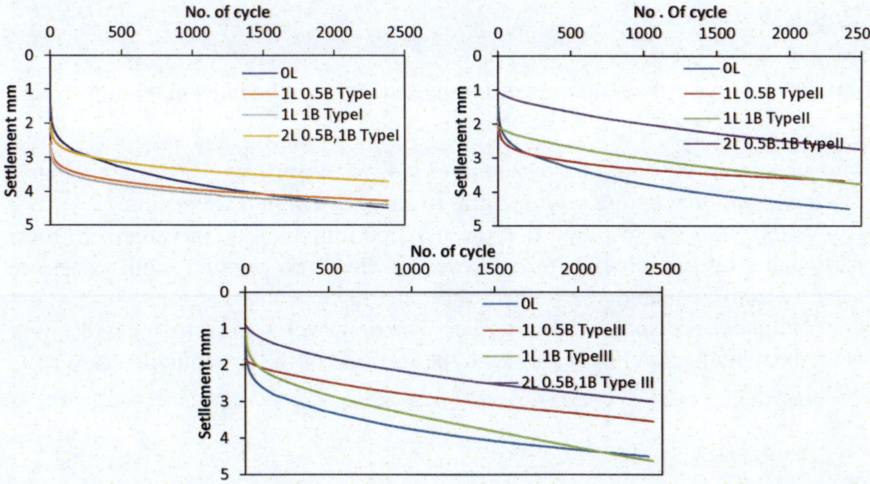

Fig. 6 Settlement of embedded foundation for different levels of reinforcement, types I, II, and III

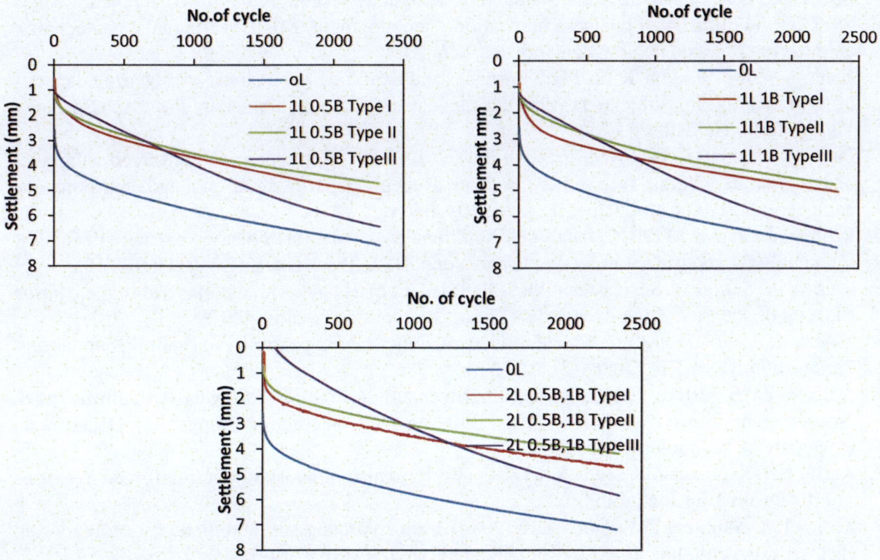

Fig. 7 Settlement of surface foundation for one layer of reinforcement at 0.5B,1B, and two layers at 0.5B and 1B with different types of reinforcement

geogrid layer. So, due that is achieved once effective interlocking; the whole layer will reinforce by the grid as a structure for a specific depth or area, as [13] detection corresponding results, who optimized that "a grid at mid-depth or a multi-layer system will not influence the performance of the composite in terms of settlement".

5 Conclusions

It is found that the settlement of medium-dense (MD) soil be in its maximum values in the first stage (layer) then try to be more stable.

- The settlement is decreased by about (32–42)% when using fiberglass geogrid reinforcement that agrees with geogrid to some extent in most results.
- It is noted from the test data that the maximum motivation of settlement for a particular sand-reinforcement system occurs when the optimum depth values are used for reinforcement.
- For continuous pressure and the number of loading cycles, the settlement resulting from the dynamic load increases with the increase of the dynamic pressure size.

References

1. McManus K, Cubrinovski M, Price C (2016) Earthquake geotechnical engineering practice -model 4: Earthquake resistant foundation design. In: News zealand geotechnical society (NZGS) and ministry of business innovation & employment (MBIE) earthquake geotechnical engineering practice in New Zealand
2. Liang L, Zong W, Arai K (2014) Numerical and experimental analyses for bearing capacity of rigid strip footing subjected to eccentric load. Central South University Press and Springer-Verlag Berlin Heidelberg
3. Puri VK, Kumar S, Das BM, Prakash S, Yeo B (2017) Settlement of reinforced subgrades under dynamic loading. In: Former Graduate Student, Department of civil and environmental engineering, SIUC Carbondale, IL, pp 62901–6603
4. Rostam V, Ghazav M (2015) Analytical solution for calculation of bearing capacity of shallow foundation on geogrid reinforcement sand slope. IJST, Trans Civ Eng 39(C1):167–182
5. Mudgal A, Sarkar R, Shrivastava AK (2019) Effect of 3D geogrid and glasgrid in the bearing capacity of square footing over soil slope. Int J Recent Technol Eng 8(2)
6. Moczar B, Szendefy J (2017) Calculation of presumed bearing capacity of shallow foundations. Periodica Polytech Civ Eng 61(3):505–515
7. Abbawi ZWS (2010) Evaluation of improvement techniques for ballasted railway track model resting on soft clay, Ph.D. Thesis, Building and Construction Engineering Department, University of Technology, Iraq
8. Karim HH, Samueel ZW, Huda KK (2017) Performance of geosynthetic-reinforced gypseous soil. Int J Curr Eng Technol
9. Karim HH, Samueel ZW, Huda KK (2016) Iraqi gypseous soil stabilized by ordinary and encased stone columns. Int J Civ Eng Technol (IJCIET) 7(6):179–192
10. Qingdao Chemetals Industries Co., Ltd (2001) https://www.bossgoo.com/product-detail/fiberglass-geogrids-15739735.html, No. 1, 41st Bldg., 1 Zhanghua Road, Qingdao, Shandong, China
11. Mohammed SA (2015) Cyclic loading on ring and circular footing resting on geocell reinforced dry sand. M.Sc. Thesis, Building and Construction Engineering Department, University of Technology, Iraq
12. Abd Al-Kaream KW (2013) The dynamic behavior of machine foundation on saturated sand. M.Sc. Thesis, Building and Construction Engineering Department, University of Technology, Iraq
13. Kwan CCJ (2006) Geogrid reinforcement of railway Ballast. Ph.D. thesis, University of Nottingham, UK

Experimental and Numerical Analysis of an Existing Single Pile Movement Due to Tunneling in Sandy Soil

Osamah Ibrahim Al-Zuhairi, Raid Ramzi Al-Omari, and Madhat Shakir Al-Soud

Abstract Tunneling may cause serious damage to the existing structures. This is because the tunneling is an unloading process which can change the stress along the existing pile. Therefore, studying the influence of tunneling on the existing pile foundation is important. This study uses an experimental model to evaluate the influence of the tunneling on the existing pile. A numerical analysis using Plaxis 3D software was used to back analyze the experimental results. The tunneling was simulated in the lab using the volume loss (V_L) technique. The study concluded that there is a zone in which most settlements of pile occurred. The study shows that when the lateral distance between pile and tunnel to tunnel diameter ratio (E/D) = 0.0, the pile settlement to tunnel diameter ratio $(\delta_{net}/D)_{max}$ is equal to 0.52%, and when E/D = 2.0 it was 0.17%, which means that $(\delta_{net}/D)_{max}$ is raised by 206% when (E/D) decreased from 2 to 0.0. These results indicate that the zone of influence in the lateral direction (E) is about 2.0D. The zone of the influence in terms of clearance distance (Z) is greater than 2.0D.

Keywords Piles · Soil-structure interaction · Tunneling · Experimental modeling · Numerical analysis

1 Introduction

Excavation of tunnels (tunneling) is one of the major problems in urban areas. The soil is moving toward the tunnel which can lead to serious movement in the existing pile.

O. I. Al-Zuhairi (✉)
Civil Engineering Department, College of Engineering, Al-Nahrain University, Baghdad, Iraq
e-mail: osamah.gatea@gmail.com

R. R. Al-Omari
Civil Engineering Department, Al-Farabi University College, Baghdad, Iraq
e-mail: tosharaid@yahoo.com

M. S. Al-Soud
Civil Engineering Department, Al-Mustansiriyah University, Baghdad, Iraq
e-mail: ms_madhat@uomustansiriyah.edu.iq

M. O. Karkush and D. Choudhury (eds), *Modern Applications of Geotechnical Engineering and Construction*, Lecture Notes in Civil Engineering 112,
https://doi.org/10.1007/978-981-15-9399-4_8

Tunneling underneath the pile will affect the stability of the pile and the soil movement around the pile will cause changes in axial forces of the pile. Several researchers studied the influence of tunneling on the pile foundation. Many researchers used field observations to analyze the effect of the tunneling on the existing pile [1–3], geotechnical centrifuge modeling [4–10], and numerical modeling [11–20]. In the current study, the behavior of a single pile in sandy soil due to tunneling was investigated. The behavior was studied using experimental and numerical analysis. The aims of the research are the evaluation of the longitudinal, lateral, and vertical distances of the tunnel face from the pile after which further tunneling progress would be risky.

2 Experimental Modelling

2.1 *Experimental Program and Setup*

To investigate the effect of the tunneling on a single pile, a total of ten model tests were conducted. In each test, the staged excavation of a tunnel was simulated through a series of rubber bags. The schematic diagram of the test setup is shown in Fig. 1. A single pile with a 250 mm pile length (L) and diameter (d) of 200 mm was embedded in the sand. A circular tunnel with an outer diameter (D) of 100 mm was used in this study. The tunnel model was made from an aluminum plate with 0.6 mm thickness. The effective soil width from tank walls to tunnel face is equal to the 2.5 tunnel diameter (D) which is recommended by Adachi et al. [21]. To eliminate the effect of grain size on soil-structure interaction of tunnel case Garnier et al., [22] suggested the following equation;

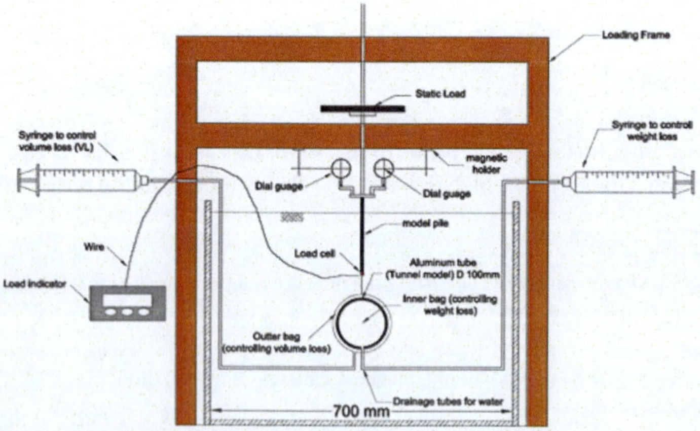

Fig. 1 Schematic diagram of the test set up for tunneling under the pile

Fig. 2 Cross-section of model test

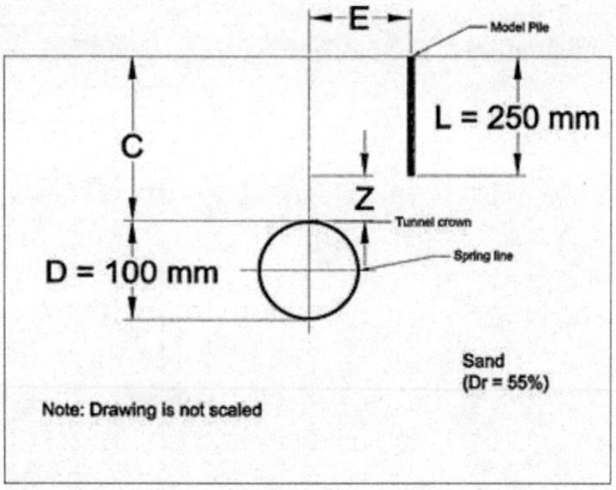

Table 1 Test program

Test ID	E/D	Z/D	Volume loss (%)	Test ID	E/D	Z/D	Volume loss (%)
P1	0	1	1	P7	0	2.5	1
P2	1	1	1	P8	0	1	1
P3	2	1	1	P9	0	1	3
P5	0	1.5	1	P10	0	1	5
P6	0	2.0	1	–	–	–	–

$$\left(\frac{D}{D_{50}}\right) > 175 \tag{1}$$

where D is the diameter of the tunnel, and D_{50} is the average particle size. The geometry of the model test is shown in Fig. 2. The test program is given in Table 1.

The plan view of the test setup is shown in Fig. 3. The tunnel model's length is 700 mm (7D). The tunnel excavation was simulated in ten stages as shown in Fig. 3 (numbered from V1 to V10). In each stage, the face of the tunnel was advanced by a distance of 0.5D. Figure 4 shows a photograph of the test setup.

3 Simulation of the Tunneling Advancement

To simulate the progress of the tunneling toward the pile, the tunnel model contains ten rubber bags equivalent to the ten-construction stage (Fig. 5). Each rubber bag was filled with water to simulate the volume loss (V_L) during the tunneling. The simulation of the three-dimensional tunneling progress was done by draining out a measured quantity of water gradually from each bag. An amount of water equivalent to 1.0% of the cylindrical rubber bag's total volume was drained away at each stage. The drainage is controlled and calibrated utilizing a syringe to achieve the desired

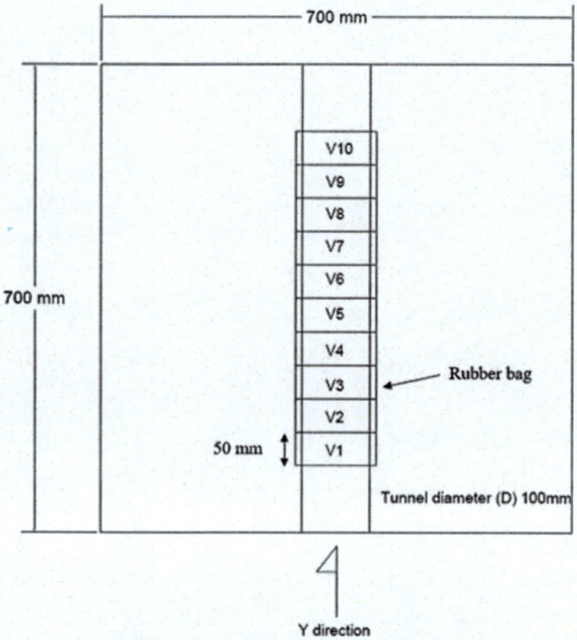

Fig. 3 Plan view of the basic configuration of the model

Fig. 4 Setup of the laboratory model

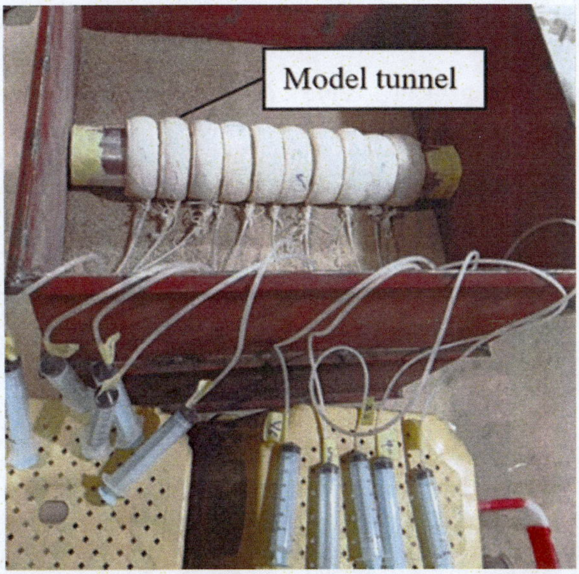

Fig. 5 Simulations of volume loss by a mean of the syringe

volume loss (V_L). This was designed to represent a volume loss (V_L) equal to 1.0% of the tunnel face excavated cross-sectional area during each tunnel construction phase. It is common to simulate the overall volume loss (V_L) Instead of simulating all tunnel excavations, such as tunnel lining construction [23, 24].

Model preparation. The sand with properties of (specific gravity, Gs = 2.655; maximum void ratio, $e_{max} = 0.67$ minimum void ratio, $e_{min} = 0.34$; friction angle φ = 34.55°, $D_r = 55\%$) was used in the present work. Standard tests were conducted to investigate the physical properties and classification of this soil. The relative density (D_r) of sand is 55% for each test. The dry tamping method was used to prepare and compact the dry sand. The model was divided into 8 layers, each layer has a height of 10 cm and compacted manually to the desired level. The process is repeated until the desired level (80 cm) is reached. Each layer is checked for the level and height.

Test procedure. After the model preparation, the pile was loaded with an allowable load which was calculated from the pile bearing capacity equations. Once the pile is loaded, the tunneling started as described above. The pile settlement for each tunnel advancement was observed using two dial gauges with a resolution of (0.01 mm).

4 Three-Dimensional Finite Element Back-Analysis

A three-dimensional finite element back-analysis to the experimental model was conducted to further investigate the effect of the tunnel construction on the existing pile foundation. The analysis was carried out using Plaxis 3D software [25]. The same properties of sand, tunnel geometry, and dimensions used in the experimental work were adopted in the numerical analysis.

Finite element mesh and boundary conditions. The 3D finite element mesh is shown in Fig. 6. The mesh consists of nearly 8118 elements (ten-node tetrahedrons) with over 12,779 nodes. The boundary conditions of the soil mesh in all cases were roller support on the vertical sides. Pin support was assigned at the mesh base. ten-node tetrahedral isoparametric elements were used to model the soil element. An embedded pile element was used for modeling the pile [25]. The embedded pile consists of beam elements that can be put in the sub-soil in an arbitrary direction and interact with the sub-soil through special interface elements [25]. The tunnel lining was modeled by deactivation of the soil inside it and activation of tunnel lining "wished-in-place". The plate element was used to model the tunnel lining. The plate element in Plaxis 3D consists of a 6-node elastic element [25].

Constitutive model. To simulate tunnel construction an elastoplastic analysis was performed using Plaxis 3D. Karbala sand was modeled using the elastoplastic model Hardening Soil (HS). The hardening soil model (HS) was described by Schanz et al. [26]. The soil properties were obtained from laboratory tests as summarized in Table 2. The pile was modeled using the embedded pile element in Plaxis 3D. The tunnel lining was modeled using the linear elastic model with properties of Young's modulus of 70×10^6 kPa, a Poisson's ratio of 0.33, and a thickness of 0.6 mm. The soil-tunnel interaction was simulated using interface elements. To eliminate the numerical difficulties, 1 kPa of cohesion c' was assigned.

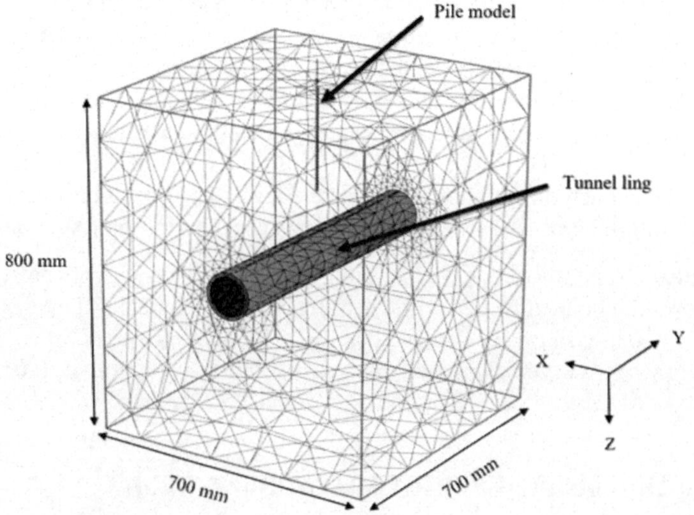

Fig. 6 Finite element mesh

Table 2 Soil parameters used in numerical analyzes (HS model)

Parameters	Value	Unit
Friction angle, φ	34.55	Degree
Cohesion, c	1	kN/m^2
Secant stiffness in the standard drained triaxial test, E_{50}^{ref}	35,000	kN/m^2
Tangent stiffness for primary oedometer loading, E_{oed}^{ref}	35,000	kN/m^2
Unloading and reloading stiffness, E_{ur}^{ref}	100,000	kN/m^2
Reference pressure, p^{ref}	100	kN/m^2
Power for stress-level dependency of stiffness, m	0.7	–
Failure ratio, Rf	0.9	–
Poisson ratio for unloading–reloading, V_{ur}	0.2	–
Soil weight above phreatic level, γ_{unsat}	17.5	kN/m^3
Strength reduction factor for interfaces in PLAXIS, R_{inter}	0.4	–

Numerical modeling procedures. The numerical analysis procedure follows the experimental work. A drained analysis was conducted for each experimental test. The numerical analysis steps are as follow:

(1) Create initial condition like the following:

- Initial geostatic equilibrium.
- Activating the pile element.

(2) Applying the static load to the pile.
(3) Begin the excavations of the tunnel using volume loss technique [27] as the following;

- Simulate the effect of volume loss by using the surface contraction technique (a feature in Plaxis 3D).
- Deactivate the soil element inside the tunnel to simulate the effects of weight loss.
- Activate the interface element around the excavated element.

(4) Repeat the above step by a distance of 0.5D at each excavation step for a total of ten steps.

5 Results and Discussion

The experimental and numerical results of the tunneling effect on the pile foundation are presented below.

Experimental and numerical results of pile settlement due to tunneling. The results of three-dimensional finite element analysis are compared to the experimental pile tests. Figures 7, 8, 9, 10, 11 and 12 shows the change in normalized pile head

Fig. 7 Distribution of the settlement of the pile due to tunneling for E/D = 0, Z/D = 1.0, and V_L = 1.0%

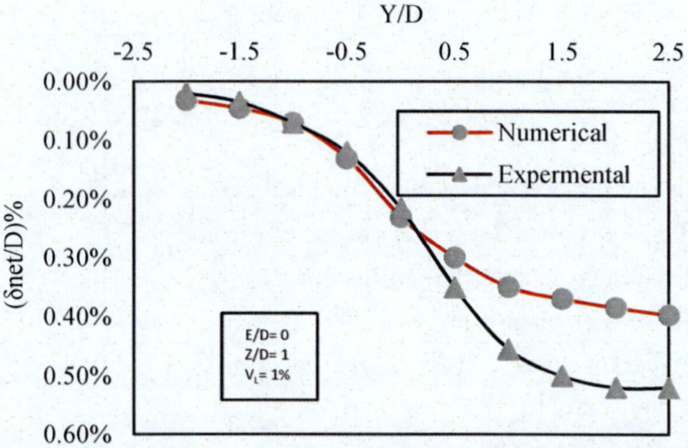

Fig. 8 Distribution of the settlement of the pile due to tunneling for test E/D = 1, Z/D = 1.0, and V_L = 1.0%

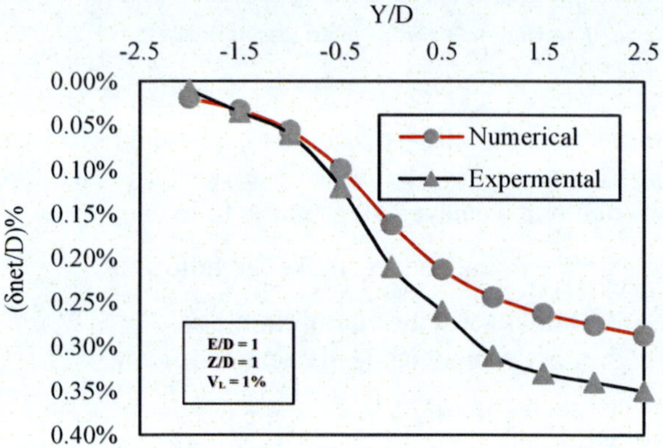

Fig. 9 Distribution of the settlement of the pile due to tunneling for E/D = 2, Z/D = 1.0 and V_L = 1.0%

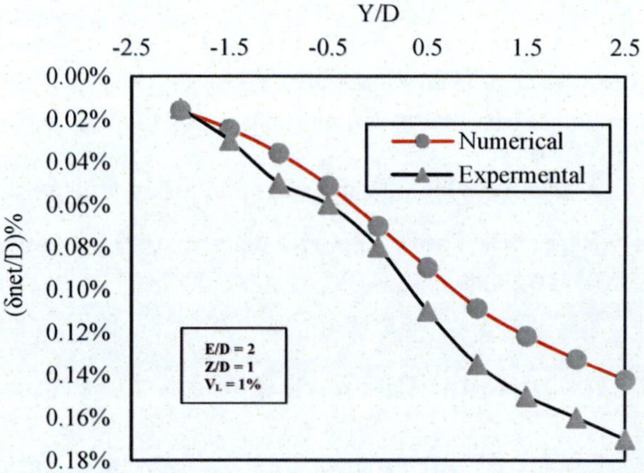

Fig. 10 Distribution of the settlement of the pile due to tunneling for E/D = 1, Z/D = 1.5 and $V_L = 1.0\%$

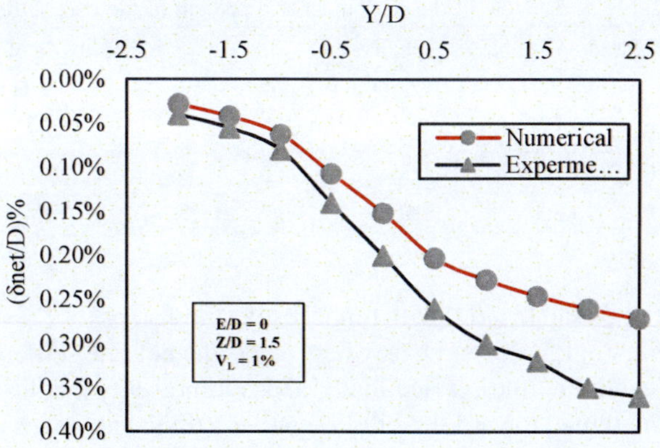

Fig. 11 Distribution of the settlement of the pile due to tunneling for E/D = 1, Z/D = 2 and $V_L = 1.0\%$

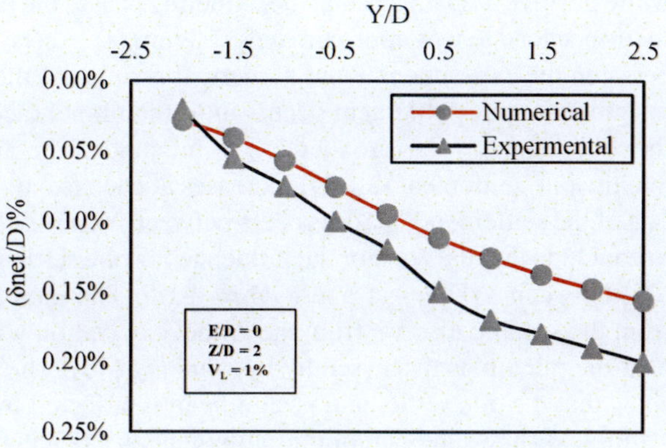

Fig. 12 Distribution of the settlement of the pile due to tunneling for test E/D = 1, Z/D = 2.5 and $V_L = 1.0\%$

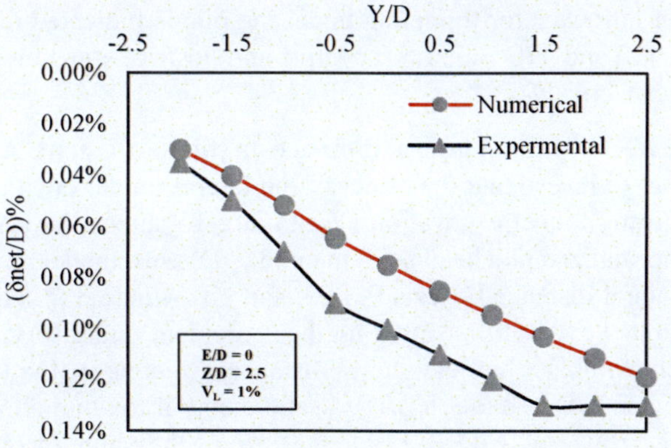

Table 3 The rate of the settlement as a percent of the max settlement at the end of the tunneling

Test id	% of δnet at specify Y/D			Test id	% of δnet at specify Y/D		
	Y/D = −1.0	Y/D = 0	Y/D = +1.5		Y/D = −1.0	Y/D = 0	Y/D = +1.5
P1	18	58	93	P5	20	60	90
P2	19	56	91	P6	35	59	86
P3	30	50	90	P7	43	63	86

settlement (δ_{net}/D) with tunnel excavation steps (Y/D = −2 to −2.5) for the tests (P1, P2, P3, P5, P6, and P7), where δ_{net} is the net pile settlement due to tunnel excavation (excluding the service load), and Y represents the direction of the advancement of the tunnel toward the pile. Negative Y/D value means that the tunneling progress is toward the pile, Zero Y/D value means that the tunnel face reaches the pile center, while positive Y/D means that the tunneling passes the pile center. The settlement due to service load δ_s was approximately 2.0 mm. The tests show a reasonable agreement between the experimental and numerical results. At the early stage of tunneling, an insignificant pile settlement occurs until the tunnel face reaches Y = −1.0D where the settlement δ_{net} starts to increase significantly. The 3D finite element results show that the pile settlement (δ_{net}/D) increases at each excavation step (Y). The maximum rate of the settlement (δ_{net}/D) occurs between Y/D = −1 to Y/D = +1.5, which leads to conclude that the zone of the influence for pile settlement in longitudinal direction Y is between Y/D = −1 and Y/D = +1.5. The increase in pile settlement results from the volume loss V_L from the tunnel excavation which leads to yield of the soil near the pile tip as discussed by Lee and Ng [13]. The 3D and experimental results show that the pile settlement is significantly affected by the tunneling process.

Table 3 shows the rate of the settlement as a percent of the max settlement at the end of the tunneling for Y/D = −1.0 and Y/D = +1.5. From the test results, the zone of influence for tunneling under the pile is indicated to be roughly between Y/D = −1.0 and Y/D = +1.5 (negative and positive signs means before and after the pile, respectively).

Effect of tunnel lateral distance on pile settlement. A numerical study was carried out to investigate the effect of the lateral tunnel position (E) on the pile settlement during tunnel excavation (Y). Figure 13 shows the results for the development of normalized pile head settlement (δ_{net}/D) with tunnel advancement (Y/D) for various lateral distance E (tests P1, P2, and P3), where δ_{net} is the net pile settlement due to tunnel excavation excluding the settlement under service load (δ_s is 2.0 mm). It is shown that when the lateral distance E increases, the (δ_{net}/D) is decreased. In case when E/D = 0 the $(\delta_{net}/D)_{max}$ at the end of the tunneling was 1.4 times the value of $(\delta_{net}/D)_{max}$ at E/D = 1, and 2.86 times the value of $(\delta_{net}/D)_{max}$ at E/D = 2.0. The maximum pile head settlement $(\delta_{net}/D)_{max}$ at the end of tunneling is plotted in Fig. 14. It can be seen that $(\delta_{net}/D)_{max}$ decreases with the increase of lateral distance E which gives an idea that the zone of the influence for pile settlement in the lateral direction is approximately more than E = 2D. The total soil settlement (U_Z) around the pile

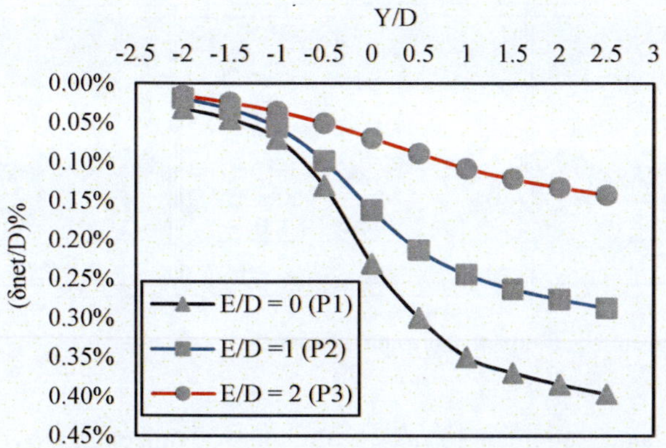

Fig. 13 Variations of pile head settlement during the tunneling process for various E/D (tests P1, P2, and P3)

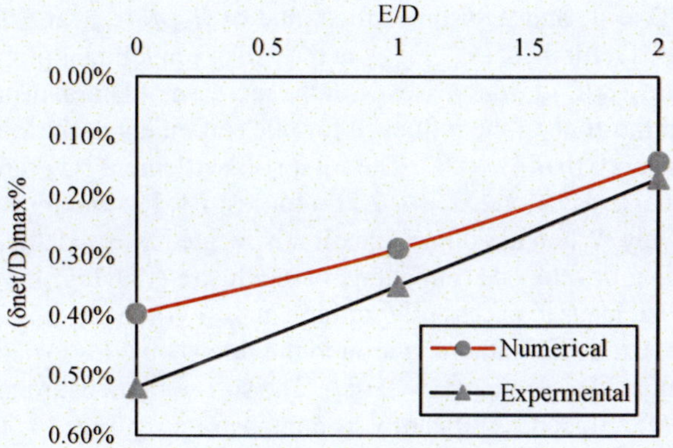

Fig. 14 Variation of normalized net pile head settlement with E/D at the end of the excavation of the tunnel (Y/D = +2)

at the end of tunnel excavation (Y = +2.5D) for test P1, P2, and P3 is illustrated in Fig. 15, respectively. When the tunnel location is located at E = 0.0D, a significant soil settlement U_Z is developed around the pile as illustrated in Fig. 15. This is due to volume loss V_L that leads to excess pile settlement δ_{net}. When the tunnel is located at E = 1.0D (test P1), the soil settlement around the pile U_Z begins to decrease but not significantly compared to test P1 (E = 0.0D). The soil settlement around the pile for test P3 (E = 2.0D) reduced significantly as demonstrated in Fig. 15. This indicates that the zone of influence is located in E > 2.0D.

Effect of tunnel lateral distance on pile settlement. A numerical study was carried out to investigate the effect of the lateral tunnel position (E) on the pile settlement during tunnel excavation (Y). Figure 13 shows the results for the development of

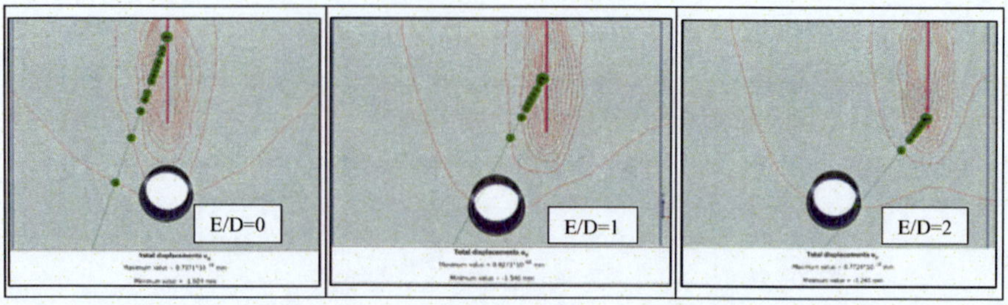

Fig. 15 Soil settlement U_Z around the pile and the tunnel

normalized pile head settlement (δ_{net}/D) with tunnel advancement (Y/D) for various lateral distance E (tests P1, P2, and P3), where δ_{net} is the net pile settlement due to tunnel excavation excluding the settlement under service load (δ_s is 2.0 mm). It is shown that when the lateral distance E increases, the (δ_{net}/D) is decreased. In case when E/D = 0 the $(\delta_{net}/D)_{max}$ at the end of the tunneling was 1.4 times the value of $(\delta_{net}/D)_{max}$ at E/D = 1, and 2.86 times the value of $(\delta_{net}/D)_{max}$ at E/D = 2.0. The maximum pile head settlement $(\delta_{net}/D)_{max}$ at the end of tunneling is plotted in Fig. 14. It can be seen that $(\delta_{net}/D)_{max}$ decreases with the increase of lateral distance E which gives an idea that the zone of the influence for pile settlement in the lateral direction is approximately more than E = 2D. The total soil settlement (U_Z) around the pile at the end of tunnel excavation (Y = + 2.5D) for test P1, P2, and P3 is illustrated in Fig. 15, respectively. When the tunnel location is located at E = 0.0D, a significant soil settlement U_Z is developed around the pile as illustrated in Fig. 15. This is due to volume loss V_L that leads to excess pile settlement δ_{net}. When the tunnel is located at E = 1.0D (test P1), the soil settlement around the pile U_Z begins to decrease but not significantly compared to test P1 (E = 0.0D). The soil settlement around the pile for test P3 (E = 2.0D) reduced significantly as demonstrated in Fig. 15. This indicates that the zone of influence is located in E > 2.0D.

Effect of clearance distance (Z) between the crown of the tunnel and the tip of the pile. The comparison between the numerical results for tests P4, P5, P6, and P7 is plotted in Fig. 16 which demonstrates the development of the normalized net pile settlement (δ_{net}/D). The numerical results show that as the clearance distance Z between the pile tip and tunnel crown becomes closer, the (δ_{net}/D) is increasing. Figure 17 shows the maximum $(\delta_{net}/D)_{max}$ at the end of the tunneling for various clearance distance to tunnel diameter ratio Z/D for both the experimental and numerical analysis. In numerical results, the percentage of the increase in $(\delta_{net}/D)_{max}$ of test P4 (Z/D = 1) is 48%, 150%, and 233% larger than test P5 (Z/D = 1.5), P6 (Z/D = 2.0), and P7 (Z/D = 2.5), respectively. The increase in $(\delta_{net}/D)_{max}$ for test P5 (Z/D = 1.5) is 69% larger than test P6 (Z/D = 2.0). While the increase in $(\delta_{net}/D)_{max}$ for test P6 (Z/D = 2.0) is 33% larger than test P7 (Z/D = 2.5). It may be concluded that the pile settlement after clearance distance Z more than twice the tunnel diameter is not significant.

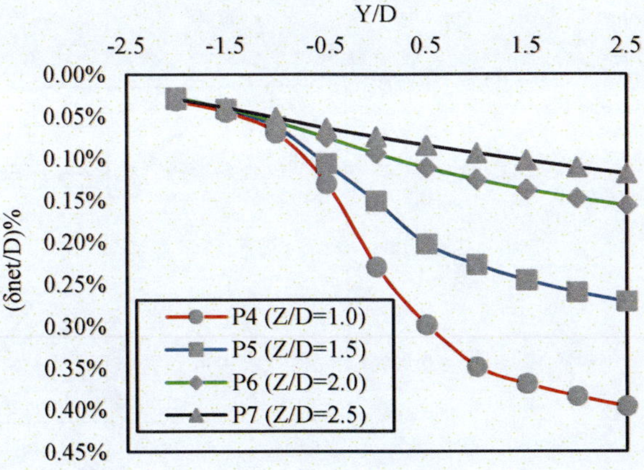

Fig. 16 Distributions of normalized net pile head settlement with tunnel advancement for various Z/D

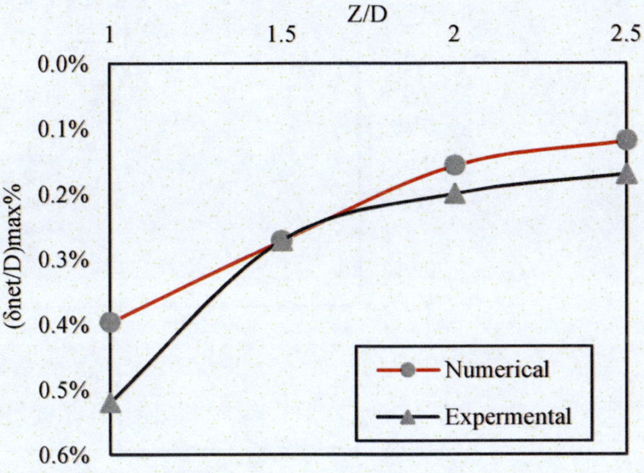

Fig. 17 Maximum pile head settlement at the end of tunneling for various clearance distance to tunnel diameter ratio Z/D

Effect of volume loss V_L on the settlement of the pile. Figures 18, 19 and 20 showed the development of normalized pile head settlement δ_{net} with tunnel advancement Y for different values of $V_L = 1\%, 3\%$, and 5%, respectively. The direct relation between the V_L and the pile settlement is clearly shown in these Figures. Figure 21 demonstrates the comparison of the results using different values of V_L. The maximum pile settlement $(\delta_{net}/D)_{max}\%$ against the volume loss V_L is plotted in Fig. 22.

The numerical results show good agreement with the experimental results. It is observed that (δnet/D) is increased with the increase of VL. For the numerical results (δnet/D) for VL= 3% was increased by 85% compared to (δnet/D) for VL= 1.0%.

Fig. 18 Distributions of normalized net pile settlement with tunnel advancement (Y) for test P8 (E/D = 0, Z/D = 1, VL = 1.0%)

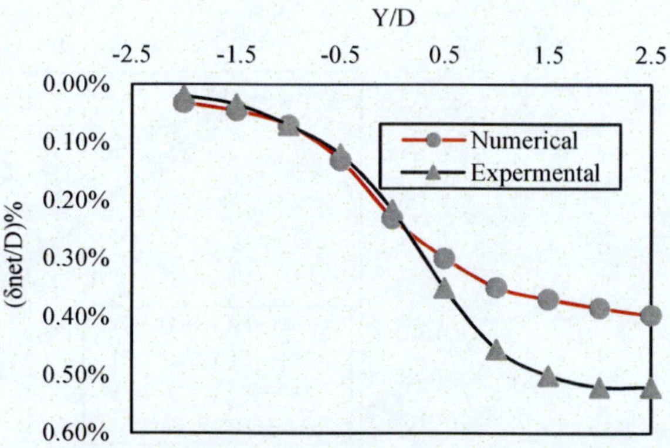

Fig. 19 Distributions of normalized net foundation settlement with tunnel advancement (Y) for test P9 (E/D = 0, Z/D = 1, VL = 3.0%)

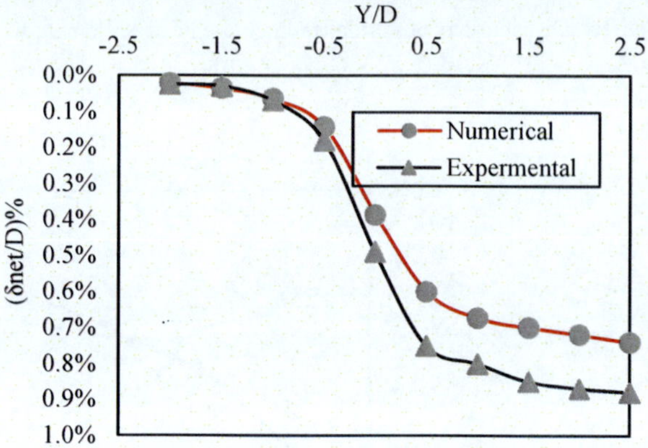

Fig. 20 Distributions of normalized net foundation settlement with tunnel advancement (Y) for test P10 (E/D =0, Z/D =1, VL = 5.0%)

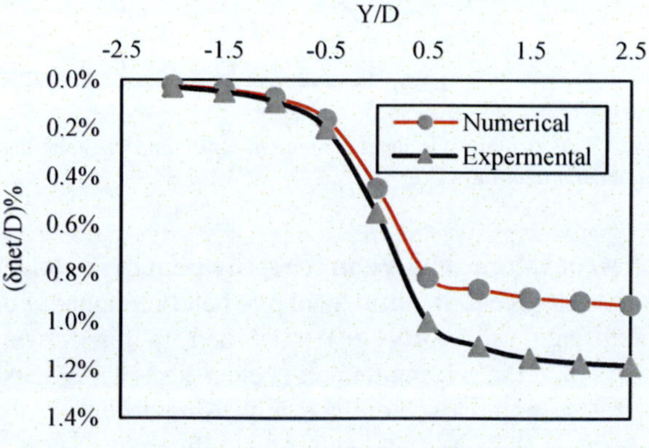

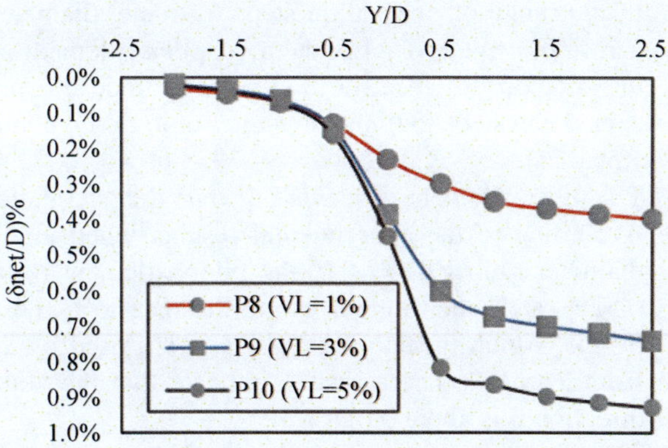

Fig. 21 Variations of normalized net pile head settlement during the tunneling process for various VL (tests P8, P9, and P10)

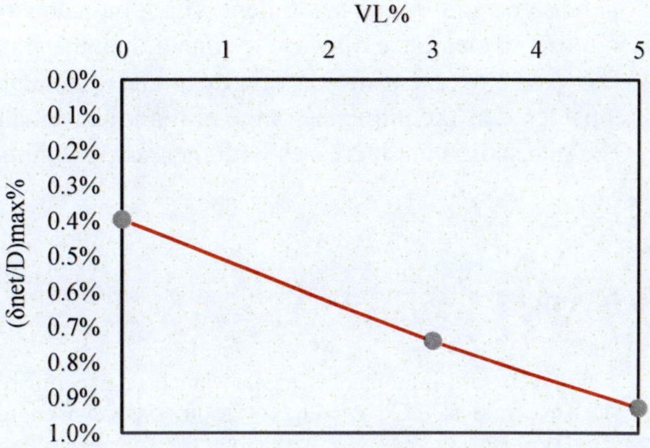

Fig. 22 Maximum pile settlement at the end of the tunneling for various volume loss

While (δnet/D) for VL= 5% was increased by 132.5% compared to (δnet/D) for VL= 1% (Figs. 21 and 22).

6 Conclusions

The following conclusions can be drawn based on the obtained results;

- As the tunnel advances toward the pile, the pile settlement to tunnel diameter ratio δ_{net}/D increases.
- In the range of excavation steps to tunnel diameter ratio (Y/D) equal to −1.0 and (Y/D) equal to +1.5, the most pile settlement is observed which means that the zone of influence for the tunneling under the pile is approximately between excavation steps to tunnel diameter ratio (Y/D) equal to −1.0 and (Y/D) equal to +1.5.
- The pile settlement is decreased when the lateral distance (E) is increased.

- In the range of excavation steps to tunnel diameter ratio (Y/D) equal to −1.0 and (Y/D) equal to +1.5, the most pile settlement is observed which means that the zone of influence for the tunneling under the pile is approximately between excavation steps to tunnel diameter ratio (Y/D) equal to −1.0 and (Y/D) equal to 1.5.
- The pile settlement is decreased when the lateral distance (E) is increased.
- According to the experimental results, when tunnel lateral distance to tunnel diameter ratio E/D = 0.0, the pile settlement to tunnel diameter $\delta_{net}/D)_{max}$ is 0.52%, and when tunnel lateral distance to tunnel diameter E/D = 2.0 it was 0.17%, which means that $\delta_{net}/D)_{max}$ is raised by 206% when (E/D) decreased from 2 to 0.0. These results indicate that the zone of influence in the lateral direction E is about 2.0D.
- The pile settlement is decreased by increasing the clearance distance (Z).
- After the clearance distance to tunnel diameter ratio Z/D = 2.0 no significant increase occurs in pile settlement which indicates that the zone of the influence in terms of clearance distance to tunnel diameter ratio is Z/D = 2.0.
- The range of excavation steps to tunnel diameter ratio (Y/D) = ±2 can be considered as the influence range of tunnel advancement.
- The pile settlement increases with increasing volume loss.

References

1. Mair RJ (1993) Developments in geotechnical engineering research: application to tunnels and deep excavations. In: Proceedings of institution of civil engineers: civil engineering, vol 93, pp 27–41
2. Forth RA, Thorley CBB (1996) Hong Kong Island line-predictions and performance. In: Proceedings of international symposium on geotechnical aspects of underground construction in soft ground, pp 677–682
3. Selemetas D (2006) Response of full-scale piles and piled structures to tunnelling. University of Cambridge
4. Hergarden H, Van der Poel JT, Van der Schrier JS (1996) Ground movements due to tunneling: influence on pile foundations. In: Proceedings of the international symposium on geotechnical aspects of underground construction in soft ground. AA Balkema, London, pp 519–524
5. Loganathan N, Poulos HG, Stewart DP (2000) Centrifuge model testing of tunnelling-induced ground and pile deformations. Geotechnique 50(3):283–294
6. Jacobsz SW, Standing JR, Mair RJ, Soga K, Hagiwara T, Sugiyama T (2001) The effects of tunneling near single driven piles in dry sand. In: Proceedings of the regional conference on geotechnical aspects of underground construction in soft ground. Tongji University Press, pp 29–35
7. Feng SH, Leung CF, Chow YK, Dasari GR (2002) Centrifuge modelling of pile responses due to tunnelling. In: The 15th KKCNN symposium on civil engineering, Singapore
8. Xia RAN (2005) Tunnel pile interaction in clay
9. Ong CW, Leung CF, Yong KY, Chow YK (2007) Performance of pile due to tunneling-induced soil movements
10. Ng CWW, Lu H (2013) Effects of the construction sequence of twin tunnels at different depths on an existing pile. Can Geotech J 51(2):173–183

11. Vermeer PA (1991) Pile settlements due to tunneling. In: Proceedings of 10th European conference on soil mechanics and foundation engineering, Florence, vol 2, Balkema, pp 869–872
12. Mroueh H, Shahrour I (2002) Three-dimensional finite element analysis of the interaction between tunneling and pile foundations. Int J Numer Anal Methods Geomechs 26(3):217–230
13. Lee GTK, Ng CWW (2005) Three-dimensional numerical simulation of tunnelling effects on an existing pile. In: Geotechnical aspects of underground construction in soft ground: proceedings of the 5th international symposium TC28, 15–17 June 2005. CRC Press, Amsterdam, The Netherlands, p 139
14. Pang CH, Yong KY, Chow YK (2005) Three-dimensional numerical simulation of tunnel advancement on adjacent pile foundation
15. Lee CJ, Jacobsz SW (2006) The influence of tunnelling on adjacent piled foundations. Tunn Undergr Sp Technol 21(3–4)
16. Lee C-J, Jun S-H, Yoo N-J (2007) The effect of tunnelling on the adjacent single pile. J Korean Soc Civ Eng 27(2C):121–127
17. Lee CJ (2012) Three-dimensional numerical analyses of the response of a single pile and pile groups to tunnelling in weak weathered rock. Tunn Undergr Sp Technol 32:132–142. https://doi.org/10.1016/j.tust.2012.06.005
18. Lee CJ (2013) Numerical analysis of pile response to open face tunnelling in stiff clay. Comput Geotech 51:116–127. https://doi.org/10.1016/j.compgeo.2013.02.007
19. Yoo C (2013) Interaction between tunneling and bridge foundation–A 3D numerical investigation. Comput Geotech 49:70–78
20. Al-Omari RR, Al-Soud MS, Al-Zuhairi OI (2019) Effect of tunnel progress on the settlement of existing piled foundation. Stud Geotech Mech 41(2):102–113. https://doi.org/10.2478/sgem-2019-0008
21. Adachi T, Tamura T, Yashima A, Ueno H (1985) Behavior and simulation of sandy ground tunnel. Doboku Gakkai Ronbunshu 1985(358):129–136
22. Garnier J, Gaudin C, Springman SM, Culligan PJ, Goodings D, Konig D, Kutter B, Phillips R, Randolph MF, Thorel L (2007) Catalogue of scaling laws and similitude questions in geotechnical centrifuge modelling. Int J Phys Model Geotech 7(3):1
23. Mair RJ (2008) Tunnelling and geotechnics: new horizons. Geotechnique 58(9):695–736
24. Taylor RN (1995) Centrifuges in modelling: principles and scale effects. Geotech Centrifuge Technol: 19–33
25. Brinkgreve RBJ, Engin E, Swolfs WM (2013) PLAXIS 3D 2013 user manual. Plaxis Bv, Delft
26. Schanz T, Vermeer PA, Bonnier PG (1999) The hardening soil model: formulation and verification. Beyond 2000 Comput Geotech: 281–96
27. Potts DM, Zdravkovic L, Zdravković L (2001) Finite element analysis in geotechnical engineering: application, vol 2, Thomas Telford

Geotechnical Piling Construction and Testing on Problematical Soil Ground of Kazakhstan and Russia

Askar Zhussupbekov, Rashid Mangushev, and Abdulla Omarov

Abstract The paper presents the results of the monitoring of construction for the unique residential building of 310 m high with 4-storey underground parking. One of the unique projects is the housing estate called "Abu-Dhabi Plaza" which started in 2011 in Nur-Sultan city. High-rise buildings pose new challenges for engineers, especially in the design of above-ground structures, bases, and foundations. The basement walls are formed with reinforced concrete diaphragm wall panels, which were supported with ground anchors. The shoring system analyses had been designed using PLAXIS 2D, which can assess in detail of the deformations that occur in soil. In the course of excavating a thick layer of soft, weak soil in St. Petersburg, Russia, instabilities were encountered in a 21 m long sheet pile excavation fence made of AU Arcelor 18. In light of the rather large depth (12 m) and areal extent (150 m by 80 m) of the excavation, such instabilities presented a significant challenge for the geotechnical engineers. This paper reviews the results of soil property tests performed in conjunction with the construction of the aforementioned large excavation, as well as the neighboring buildings. The need for continuous monitoring of geotechnical conditions using the methods of static sounding, to assess the zone of construction influence and to forecast the mutual influence of the structure being constructed on the surrounding buildings, is discussed. Finally, the raft foundation has recommended for future high-rise buildings to be constructed on competent residual soils.

Keywords Plaxis 2D · Piles · Diaphragm wall · Monitoring · FLAC3D

A. Zhussupbekov (✉) · A. Omarov
Nur-Sultan, Kazakhstan
e-mail: astana-geostroi@mail.ru

A. Omarov
e-mail: omarov_01@bk.ru

R. Mangushev
St. Petersburg, Russia
e-mail: ramangushev@yandex.ru

M. O. Karkush and D. Choudhury (eds), *Modern Applications of Geotechnical Engineering and Construction*, Lecture Notes in Civil Engineering 112,
https://doi.org/10.1007/978-981-15-9399-4_9

1 Introduction

High-rise buildings (buildings with a height of more than 75 m) pose new challenges for engineers, especially in the designing of above-ground structures, bases, and foundations. Therefore, designers of both above-ground and underground parts of the building are requiring to resort to more complex methods of analysis. This applies to geotechnical engineers, who are involved in the design of foundations for high-rise buildings. Due to complexity, problematic design, erection, operation, impact on the environment and people, high-rises can be attributed to the structures of increasingly dangerous and complexity. Kazakhstan has its own modern experience in designing and erecting unique high-rise buildings, including the "Emerald Quarter" (210 m), "Northern Lights 1" (180 m) in Nur-Sultan, Railways Building (174/156 m), Transport Tower (the building of the Ministry of Transport and Communications) (155 m).

After completion, Abu Dhabi Plaza (ADP) will become the tallest building in Kazakhstan and Central Asia, and the highest block of the complex will be 320 m. It represents the construction site—in the center of which the skyscraper should rise. This grandiose skyscraper will be the fourteenth tallest building in the world. ADP is a high-rise building with a retail and leisure podium and a hotel cluster at the base that rises to form a series of office and residential towers to the north—creating a new landmark on Nur-Sultan's skyline. Recommended soil parameters relevant to the analysis used for the design of the pile foundation system are summarized in Table 1.

Table 1 Geotechnical soil parameters for analysis

Soil	Clay/Loam	Sand and gravel	Loam and clay	Rock debris	Sandstone	Hard sandstone
Thickness (m)	4.0	4.5	5.5	1.5	2.0	Below
Natural unit weight Υ (kN/m^3)	18.5	19	18.5	20.5	23.5	24.0
Internal friction angle φ' (°)	25	35	26	35	38	38
Cohesion c' (kN/m^2)	1	1	30	1	40	50
Modulus of elasticity E (MPa)	15	25	30	50	65	100
Permeability k, (m/day)	1 × 10E-5	1 × 10E-2	1 × 10E-5	1 × 10E-2	1 × 10E-5	1 × 10E-5

2 The Analysis of Base Deformation Using FLAC3D

The analysis of the deformations of the base is dependent on the load of its weight for a single design in the horizontal plane of the vertical deformation, as shown in the Fig. 1 [1–5]. The horizontal grid plan is a form of deformation reference and used to estimate column settlement in various places. The settlement for each stage of construction is calculated using the ratio of the vertical reaction at each stage to the vertical reaction at the full design load. Block R and the basement tile of the general basement are modeled by taking into account the changes in thickness. Model physical boundaries are set within grid 9–21 and from A to M. Figure 1 and Table 2 show the settlement contour of the site using FLAC 3D programmer. The settlements from FLAC 3D (Program 1) are based on characteristics of soil properties and so on advice from BLP can be taken as long-term settlements.

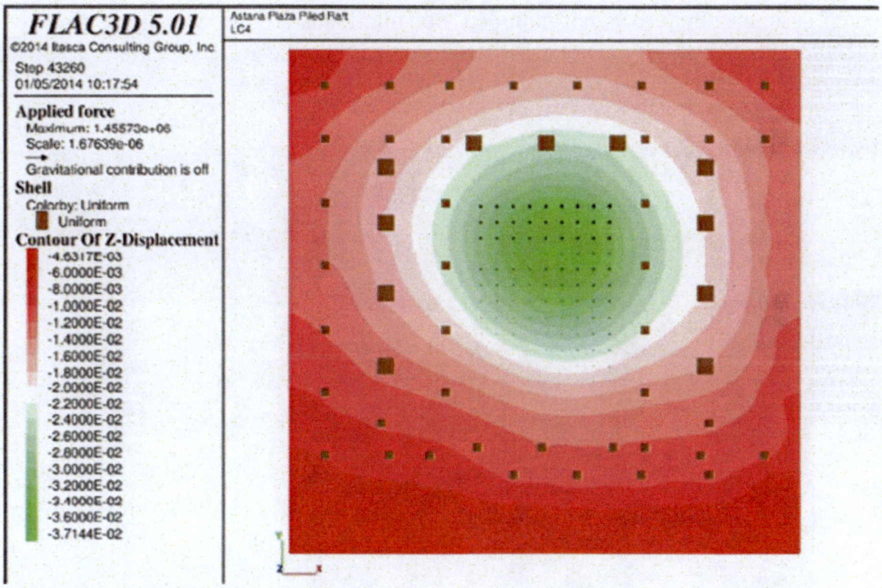

Fig. 1 Basement 4 (Raft Foundation) deformation contours (Model 1)

Table 2 Raft Settlements under full dead and live loads

Location	FLAC 3D (Model 1) (mm)	Program (Model 2) (mm)	% Variance
Core	37.4	39.6	5.6
Column RC1	14	13.4	4.3
Differential settlement	23.4	26.2	10.7

Adjustments to the vertical displacement of the tower columns and core walls will be made to allow for the raft deformation. The raft deformation analysis based on full design gravity load produces a vertical deformation contour plan as shown in Fig. 1 below. Displacement units are in meters. The contour plan represents the reference deformation shape and is used for estimating the settlement at the various core and column locations. The settlement at each construction stage is calculated using the ratio of the vertical reaction at each stage to the vertical reaction at full design load. The analysis uses in-house software Geotechnical Analysis of Raft and Piles developed and used by Program 2 to model the soil-structure interaction of the piled raft. The stiffness parameters are established using software. The Block R and general basement raft are modeled taking into account the variation in thickness. Refer to Fig. 2 for the settlement contour plot using the Program (Model 2).

Table 2 lists the maximum and minimum settlement results for FLAC 3D and Program models for the full design gravity load.

Hydrostatic pressure under the raft foundation is not considered relevant to the studied case, and therefore, not included. Seismic loads are not considered for this project as Nur-Sultan city is not classified as an earthquake zone.

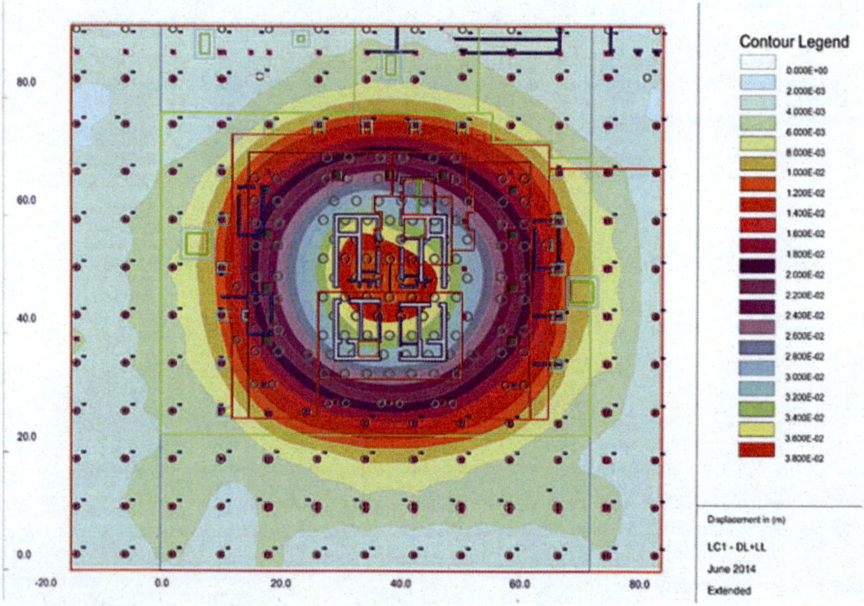

Fig. 2 Basement 4 (Raft Foundation) deformation contours (Model 2)

3 Anchor Design Calculations

Pre-stressing loads (tension: T forces) of ground anchors are shown on the shoring section drawing as determined 'Pre-tensioning force in Plaxis inputs dividing to the tension forces by spacing (distance) between adjacent anchors on the same level for the anchored section (Fig. 3). After these pre-stressing stages, the tension forces on the ground anchors generally increase until the capacity of the anchor tendons with a desired safety factor [1]. All manuscripts must be in English. Please keep a second copy of your manuscript in your office (just in case anything gets lost in the mail). When receiving the manuscript, we assume that the corresponding authors grant us the copyright to use the manuscript for the book or journal in question. Should authors use tables or figures from other Publications, they must ask the corresponding publishers to grant them the right to publish this material in their paper.

The Finite Element Method is based on a model in which the behavior of soil and structure is integrated (see Table 3). The properties of soil are introduced utilizing

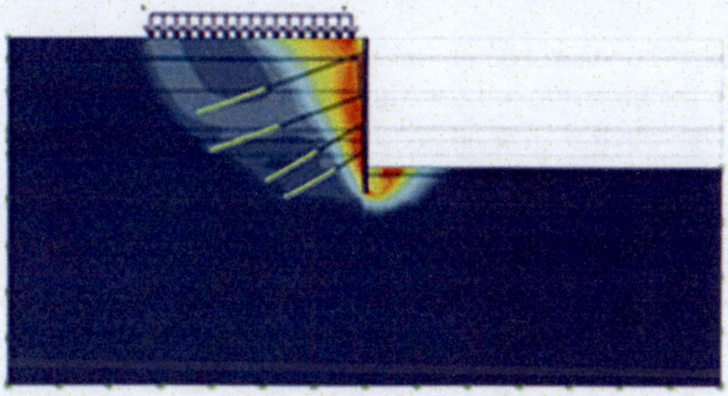

Fig. 3 Calculation scheme of underground construction [1]

Table 3 Results of calculation by Plaxis 2D [1]

Anchorage summary	Number anchors			
	1	2	3	4
Level (–)	3.0	9.0	13.0	17.0
Spacing, m	2.20	1.65	1.65	1.65
Free length, m	15	13	8	5
Grout length, m	10	10	8	8
Angle, degree	20	20	30	30
Total length, m	25	23	16	13
Prestressing load, kN	460	530	570	640
Calculated anchorage load, kN (:Cal = Tw)	607	672	670	675
Test load, kN (=Cal*1.25) (=Tw*1.25)	759	840	838	844

stress-deformation relations. With this method, fundamental calculations of stresses and deformations of soil and structural members can be made.

The method can be used to calculate the sectional forces of the structural members, verify the global stability of the structure, and to calculate and verify deformations. Due to loading, the increase in pressure within the pore water will cause drainage (flow out of the soil), and transfer to the solid grains. The rate of drainage depends on the permeability of the soil. The strength and compressibility of the soil depend on the stresses within the intergranular parts called effective stresses [1]. The effective stresses are a measure for the forces in the contact points of the grains. Thus, the total stress is a summation of the effective stress and porewater pressure. After these pre-stressed stages, the tension forces on the ground anchors generally increase up to the capacity of the anchor tendons with a safety factor (see Table 3 and Fig. 3).

4 Monitoring Diaphragm Wall in Construction Site ADP, Nur-Sultan

Excavations can be produced without problems by using of diaphragm wall stabilize by the underground parking lots. Depending on the depth of the pit, they are extensively anchors. The advantage of the construction method of the walls in the ground—a large depth of possible pits. Excavation—Circuit of hard-integrated process for the construction of the partition anti-filtrating wall is made of special mechanisms: Stein-Greifer for diaphragm wall and chisel. Depending on the soil we use grapples weighing up to 25 tons and chisel weighing up to 16 tons. During the development of the bare flat smooth vertical soil wall, the trench is formed by the method of rectification (a kind of deck from the ground for concreting). During the development of the soil by Greifer Grabber, the trench is systematically filled by bentonite suspension that ensures the stability of the surface of the surface of the wall.

Concreting—Bentonite tube, equipped with a tremie pipe, is collecting from an individual, each is 3 m length. In the process of concreting, as the level of concrete rising, pipes sequentially removing links. Concrete mixture, fed through tremie pipes enters the bentonite tube and slowly displaces bentonite slurry that is pumped. One of the most important challenges at this stage is to review the work, which has been done by the previous contractor. The inspection of the existing construction helps the engineers to decide whether to carry on or to modify it. The objective was to inspect the structure of the diaphragm wall in comparison to is the theoretical position. In this case, the level of tolerance of the deviation from the theoretical position must be minimal since the design involves the construction of the inner wall (or parking wall) along the diaphragm wall, within a distance of 15 cm. More specifically, the distance between the inner sides of the diaphragm wall to the outer side of the parking wall must be 15 cm, according to the design [5].

The high level of detail, which is required, made it necessary to utilize a method, which allows designers to get the as-built situation as it is and to analyze into an

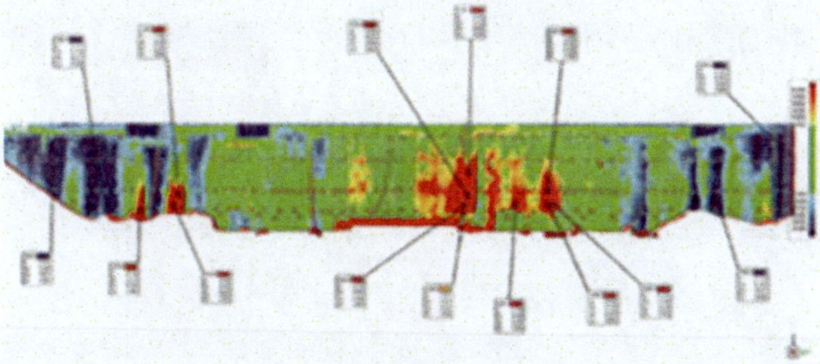

Fig. 4 Model of the west wall [1]

advanced CAD system [1–5]. Recent advances in generating 3D environments using laser-scanning technologies, and acquiring quality information about built environments using embedded and other advanced sensors, provide the capability to the frequently gather an integrated and accurate three-dimensional and material quality-related as-built data. This information is combined with the design model to create an integrated model, which is dynamically updated during the construction period [1–5]. The resulting model constitutes a color-based deviation model, as is indicated in Fig. 4. The site scanning, with the application of laser scanner technology, was completed in four days with 400 scans in total. Advanced computer software gives us the ability to generate vertical and horizontal cross-sections, along with the diaphragm wall, allowing the user to define the interval between the consecutive cross-sections. The decision was based on the length of the panels, which constitute the diaphragm wall. Therefore, an interval of five meters between two consecutive cross-sections allows us to check the as-built condition of every panel.

The internal dimensions of the sub-structure itself are approximately 220 m × 195 m on plan and are typically 17.4 m deep. The basement walls were formed with reinforced concrete diaphragm wall panels which are supported with ground anchors. The basement slab is a pile-supported raft which is made up of discrete sections to accommodate thermal and lateral movements. The piles are 1200 mm and 1500 mm diameter reinforced concrete bored piles that carry compression, tension, and lateral loading. The engineers were also requested to inspect the existing rafts (Fig. 5) which were constructed by the previous contractor. The raft inspection involves a review of the slopes and a comparison of the elevation with the current design. Figure 6 indicates the profile of the slopes of the foundation. The dark red shows the areas where the distance from the theoretical raft position exceeds 10 cm while the green indicates the areas where the deviation is lower than 5 mm [5].

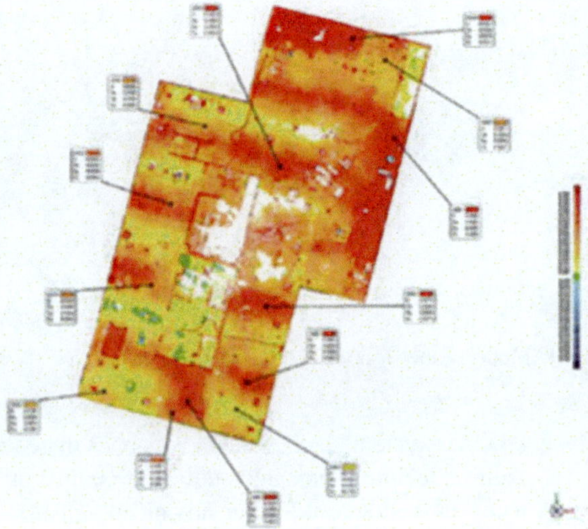

Fig. 5 Results of raft foundations ADP [5]

Fig. 6 Static sounding of soil at site № 1

5 Second Project in Construction Site in St. Petersburg City, Russia Federation

The modern large-volume underground construction in the conditions of the compact urban development requires considering the effect of technogenic factors on the change of soil properties during the construction. These changes occur in the soil under the foundation of the construction building and outside this building in the zone affected by this construction. It is most important to consider these changes during the excavation of large volume pit in difficult geotechnical conditions. Such information allows to make geotechnical calculations and to specify the stress and deformation properties of the soil to forecast the settlement of the erecting and adjacent buildings during the crucial stages of the pit excavation [6–10]. The technogenic effect on the soil under the foundation may be of local and general (territory) character. Local changes in the soil under the foundation during the piling work made according to different technologies are discussed in previous articles. In the present article, the change of soil properties are discussed while the construction of the excavation pit of the size 150 × 80 m and the depth 12,5 m in the soft soil conditions of St. Petersburg and adjacent buildings.

The Main Constructive Elements of Structures in the Underground Part and the Adjacent Buildings. The protective and strutting devices of the excavation pit include [11–13]:

- metal sheet piling made from AU Arcelor 18 penetrated to the depth of 21 m by vibration;
- protective “wall in soil” made from piles and reinforced by steel shape structures installed on distance 1.0 m from one to another. The Jet grouting “wall in the soil” with installed steal shape structures and the sheet piling on top are connected by reinforced concrete beam 2.3 m wide and 1.15 m high;
- longitudinal Jet grouting diaphragm on the depth of 11.5–14 m made by jet-grouting technology. The project strength of the diaphragm material measured by axis compression is 1.0 MPa min, deformation modulus—400 MPa min;
- temporal bored piles 29 m long, diameter 600 mm made inside the excavation pit in the form of 6 m x 6 m net;
- constantly bored piles having the diameter from 800 mm up to 1200 mm which present the constant constructive elements of the plate grillage made lower the bottom of the excavation pit;
- strutting reinforced concrete plates made during excavating the pit on the marks 4,41 and 11.1 m (top of the plate). The plate at the mark 4.41 is 400 mm thick and rests on the temporal piles. The plate thickness at the mark 11.1 is 1200 mm and the constant bored piles serve as its base;
- Concrete beams at the mark 8.11 m and fragments of cross walls for additional rigidity of the pit protection at this level.

The adjacent 4–7 floor buildings were constructed earlier in 1917 and have cross and longitudinal bearing walls and strip foundations. The nearest buildings to the pit have 4–5 floors and are located 14 m from the pit. The effect of Vibro penetration of the sheet pile, temporal and constant piles mounting, Jet grouting vertical and longitudinal diaphragms construction, might cause the serious changes of physical and mechanical properties and strength of soil in the pit and outside the pit.

Based on investigations made before the projecting in 2005 the construction sites are characterized by the following soil layers:

- fill soil layer (L1) 1.5–3.4 m deep composed from small debris, sand of different grain size, and plastic sand clays;
- under the fill soil layers, there are sea-lake deposits, composed from silty sand (L2) 2.4–4.4 m thick and clay-sand (L3)1.2–2.0 m thick. The sea-lake sand and clay-sand are characterized by thixotropic properties. The underlying sea

glaciers deposits are represented by belt, fluid clay sand (L4). The thickness of lake–glaciers deposits is 3.4–8.3 m. These soils and sea–lake deposits refer to thixotropic soil;

- glaciers deposits are represented by sand-clays with sand inclusions (L 5) 0,7–4,9 m thick, light sand-clays with sand intermediate layers and inclusions with different size grain (L6) 1,7–9,3 m thick, light grey-greenish sand clays with intermediate layers of sand and gravel of semi-hard consistency (L7) 0.9–6.1 m thick.

The glaciers deposits are laid on stiff proterozoy blue-grey, green-grey and blue-green clays, whose roof is opened to the depth of 22.4–28 m [14]. The roof of the stiff clay is touched by glaciers till the depth of 27.2–32 m and is characterized by rear gravel inclusions (L8). Lower there are fine stiff clays. (L9).

6 Soil Test Methods

The estimation of the effect of the underground construction on soil properties was done by a comparative analysis of the results of static sounding received in 2005 and 2009. In 2005 the static sounding test was done by the "Iziskatel" Company during the projecting work. The sounding was done by the sound (probe) "CPT', of Swedish company "Geotech AB", which corresponds to type II in Russian classification [15]. In 2009 the static sounding was done by research and development center of Geotechnology in St. Petersburg State University of Architecture and Civil Engineering. as a part of the technical supervision program. The sounding was done after the installation of the piles, sheet piling and excavation of the pit up to 2/3 its depth, i.e., before the most important part of the work—opening of the lower part of the pit. Taking into account the intensity of the pit excavation according to "top-down" method and confined construction site area the tests were done using modern multi-channel sound CPT-U, of Swedish "Geotech AB" PROBE No. 3531. The sound penetration was done by small-size geotechnical RIG 204 D of Swedish "Geotech AB" Company (Figs. 6 and 7).

For testing the soil by static sounding method there were chosen 4 sites—one inside the excavation pit and three outside. At each site, there were done 6–14 points of static sounding at the distance of 1 m minimum from one another (see Fig. 8).

The locations of sounding points were linear and done along the sheet piling to measure average soil characteristics at the cross points on different distances from sheet piling. The main parameters used for testing physical and mechanical properties were the resistance of the soil to cone penetration q_c and friction along the side surface f_s. The additional parameters of soil characteristics were porous pressure u, temperature t, inclination angle from the vertical position of the sound T. These parameters were used for estimating the sounding conditions, inhomogeneity, and variability of soil properties. Taking into account the possibility of changing the composition and soil type under technogenic effect, the test of the stability of the

Fig. 7 Penetrations of the sound at the site № 3

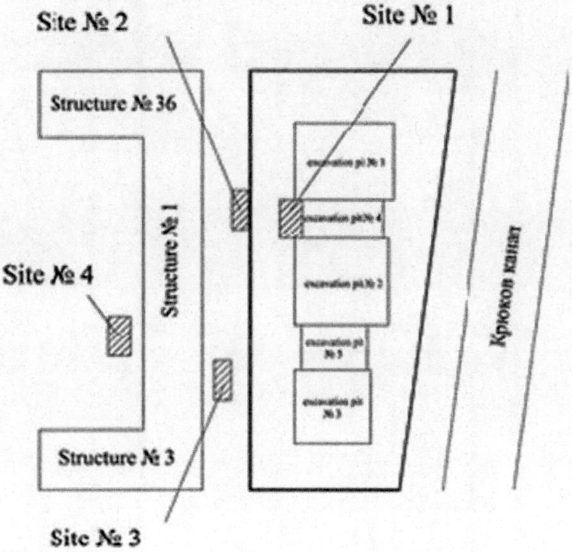

Fig. 8 Scheme of location of static sounding sites

soil type was done for each engineering and geological element according to P.K. Robertson et al. using q_c and $R_f = (f_c/q_c) \times 100\%$ [9]. The physical and mechanical characteristics of the soil were estimated using the tables and monogrammed [9, 10].

7 Test Results of the Change of Soil Characteristics

The registered numeric parameters at the sounding points, their comparison with previous results of static sounding made in 2005, and statistical treatment of data in total allowed to make diagrams of the change of the average value of the investigated parameters for each site of investigation. At the site № 1 located inside the pit at the depth of 4 m, there were done 14 sounding points. Here the soil was affected more significantly by vibrato penetration of the sheet pile during the construction of constant and temporal drilled piles made under casing pipe by screw according to SOB and DDS methods. Besides, the longitudinal and vertical soil–cement diaphragm in soil was done here by "jet grouting" technology. In Fig. 9 the changes of the parameters q_c and fs depending on the depth is given for this site. The roof of the longitudinal soil–cement diaphragm was fixed at the depth of 7.7 m.

At site № 2 located 2.6 m from the pit, there were done 10 sounding points. The sounding was done at 0.3 m distance from the side surface of the reinforced

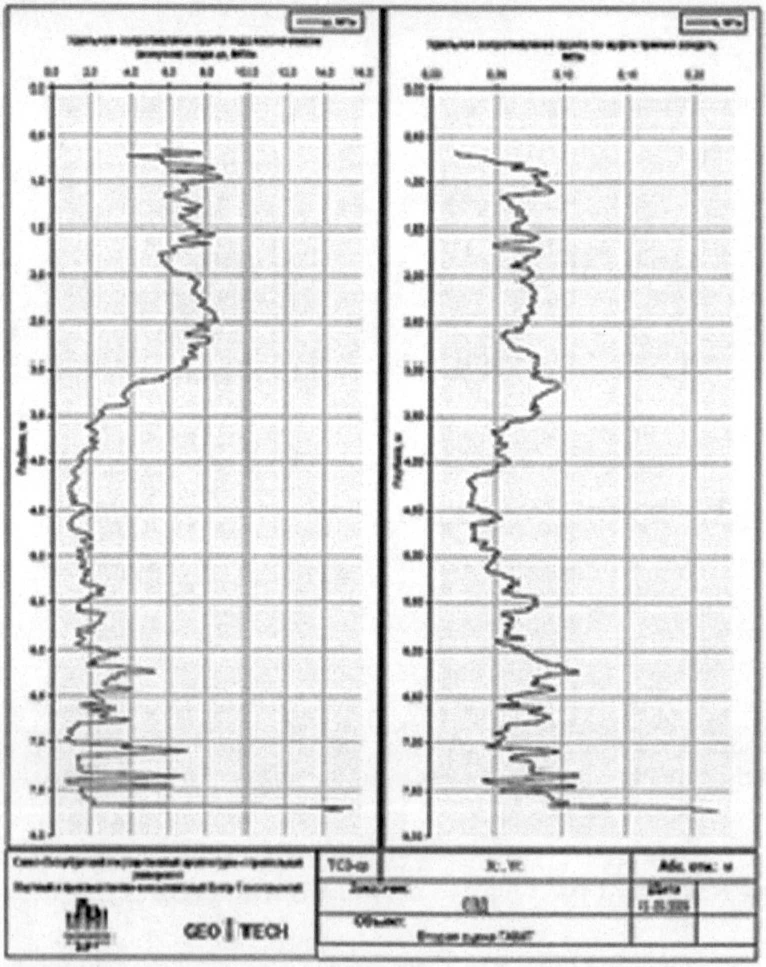

Fig. 9 The diagram of the average value q_c and f_s

concrete binding beam of the outside pit protection. The soil here was affected by vibro penetration of the sheet pile and by the construction of the vertical wall in the ground made by jet grouting technology. Besides the loading of the adjacent building affects the soil at this site.

At the site № 3 located 5.6 m from the pit, there were done 8 sounding points located 3.3 m from the outside face of the reinforced concrete binding beam of the outside pit protection.

At the farthest site № 4 located inside the yard of the block of flats 25 m from the pit there were done 6 sounding points [16–20].

The average value of q_c and f_s at the depth of 2.3–21.4 m at this site are given in Fig. 10.

For the comparative analysis of the changes in soil due to the construction, the average values of q_c, f_s, and R_f for each type of soils were received based on sounding test results.

The diagrams of the average value of these parameters for each soil type are given in Fig. 11 in comparison with engineering and geological data.

The average values of q_c and f_s were received in 2005 from the sounding in 8 points located between sites № 1, 2, and 3.

The analysis of the mentioned data showed that the upper part of the soil inside the pit at soil layers L2, 3, and 4 received the biggest technogenic effect. The comparison of the diagrams shows that soil resistance to the static sounding at site 1 increased by more than 2 times due to construction works. Outside the pit on the distance of 0.3 m from reinforced concrete binding beam the most significant increase of the soil resistance q_c, f_s occurred in clay sand at soil lays L5 and L6. At the distance of 3.3 m from the pit protection the resistance of soil to sounding increased by not more than 50% against soil resistance measured before the start of the construction.

At the distance of 3.3 m from the pit protection the resistance of soil to sounding increased by not more than 50% against soil resistance measured before the start of the construction. The registered change of the soil properties in terms of increasing occurred only up to the depth 17.5 m at soil lays L2…0.6. The increase of strength and deformation soil properties received in the second sounding can be explained by soil consolidation at soil lays L2…0.6 due to technogenic effect caused by installing a great number of piles, restoring of structural links and also partial reinforcement of soil due to construction of cement soil diaphragm by Jet grouting method. In the lower glaciers sand clays at L7 the change of the soil resistance is insignificant (Fig. 12).

For L 2…0.7 the possible change of the soil type was checked on average q_c and R_f values. The results showed the absence of any change in soil type that could be caused by construction. This conclusion allowed to use in further calculations the data received in former tests and new data with different values of physical and mechanical properties. After statistical treatment of the results of soil static sounding using tables and specifications nomograms [10, 11] the specified values of deformation modulus E, specific cohesion c and angle of inner friction φ of soil in the construction site were received (Tables 4, 5, and 6).

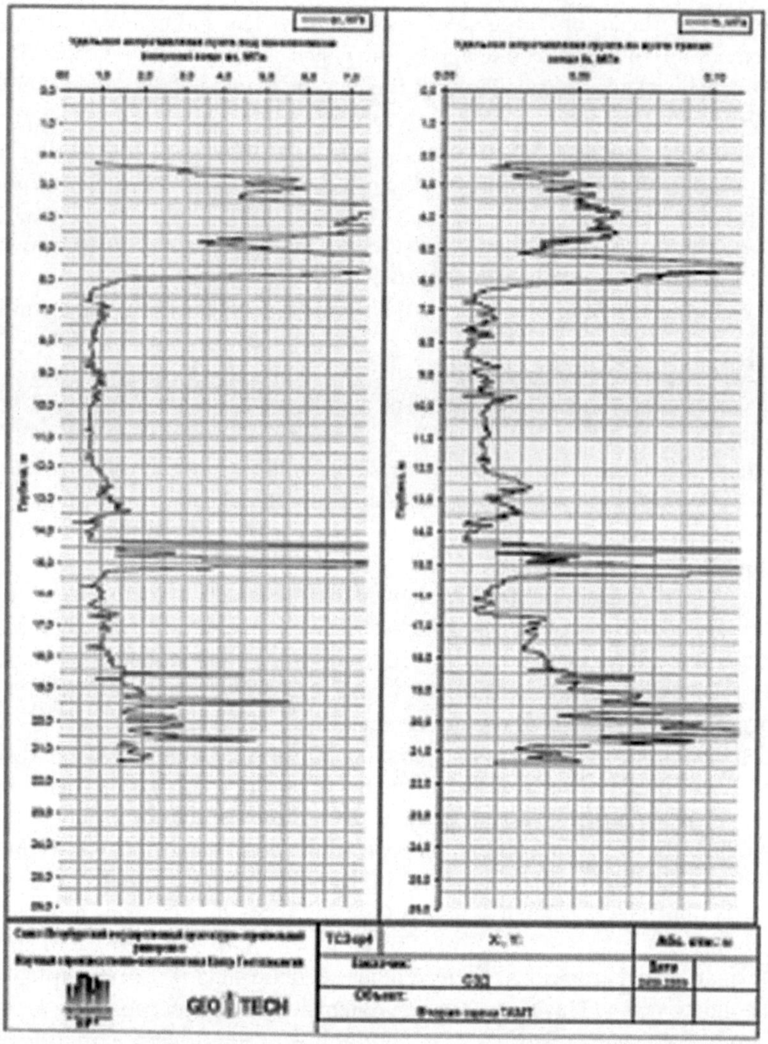

Fig. 10 The diagram of the average value q_c and f_s

The values of soil properties changed under the technogenic effect received for each soil layer and investigated zones were used for detailed geotechnical forecast of the stress deformation condition of the soil in the pit basis and adjacent soil during the construction and estimation of the effect on the adjacent buildings. Calculations were done by FEM programmer "PLAXIS" using 24 engineering and geological elements (Fig. 13).

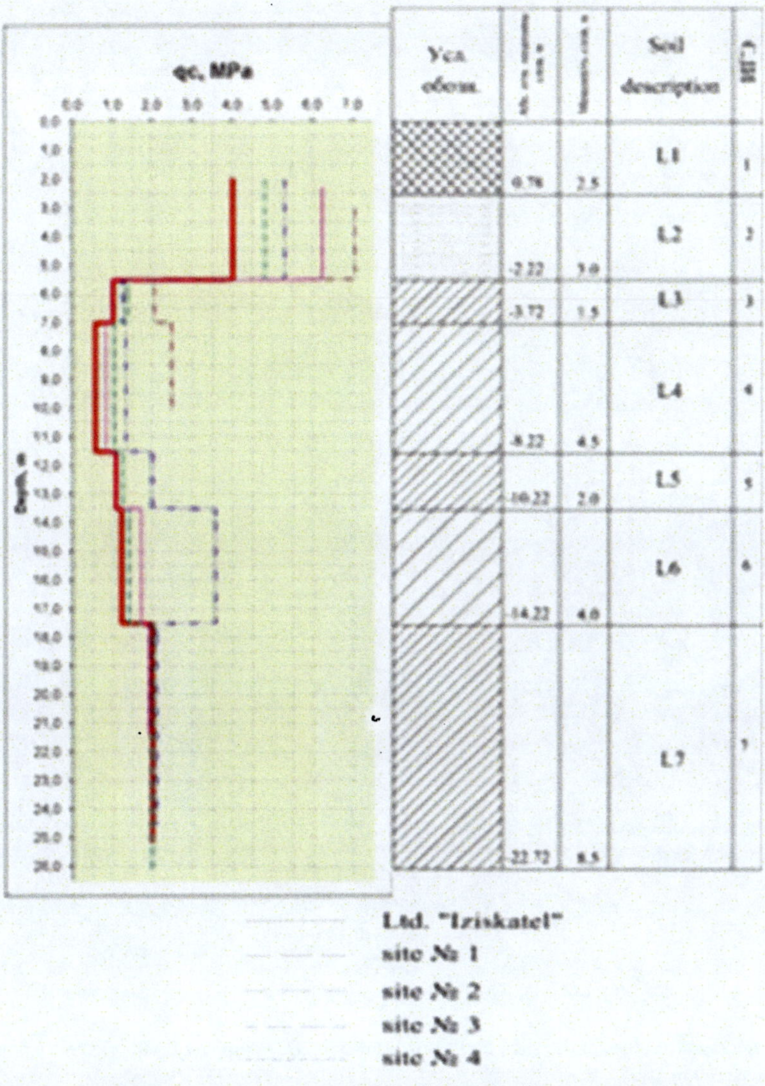

Fig. 11 Average q_c values at the sites № 1–4 measured in 2009 compared with the corresponding values measured in 2005

8 Conclusions

- There is a need in considering the anticipated responses of the adjacent buildings during construction. This will require the design professional to survey the condition of the adjacent properties to understand their present condition and fragility, establish acceptable limits, conduct soil-structure analyses of various

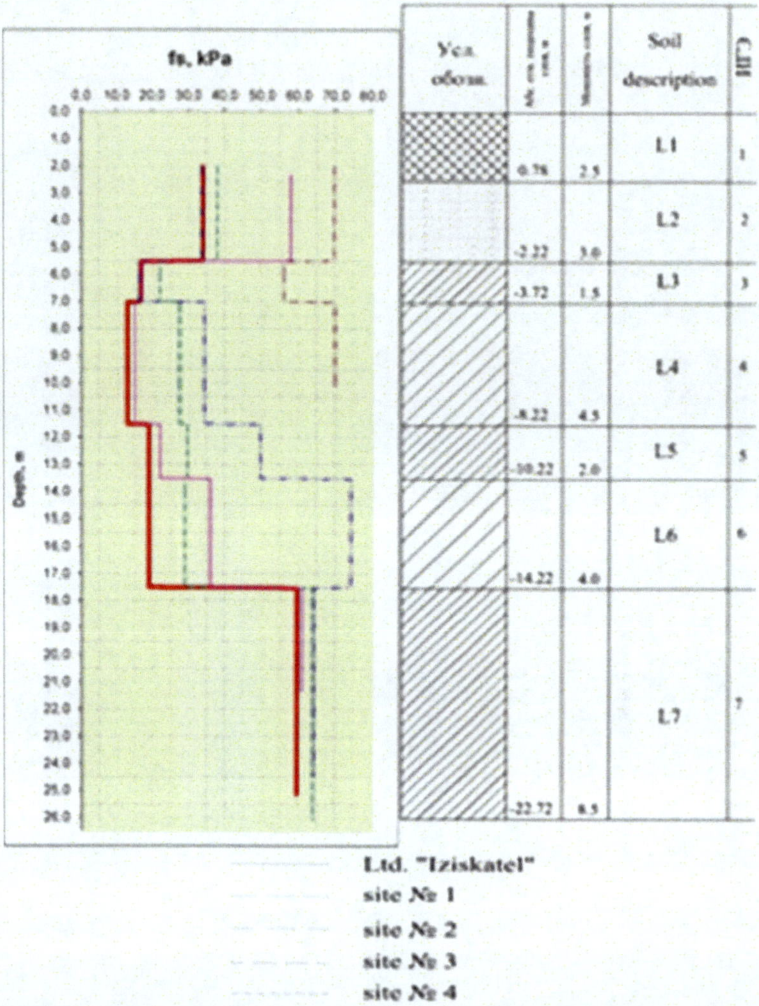

Fig. 12 Average f_s values at the sites № 1–4 measured in 2009 compared with the corresponding values measured in 2005

support systems and develop limits on their respective movements, and develop a monitoring strategy.

- The code is assigned according to the value of the deformation between the as-built measured data and the design model. The areas, which are colored red, represent the highest deviation with values, which exceed 150 mm, in contrast to the green-colored areas, which present the lowest deviation with the values, which fluctuate between 0 and 150 mm. The positive pattern of deviation is illustrated in red while the negative deviation is in blue. The red-colored areas indicate the areas where the wall penetrates the parking wall and therefore must be trimmed.

Table 4 Specified values of deformation modulus E, MPa based on results of static sounding

No. L	*E*, MPa			
	2005 г	Number of sites		
		No. 1	No. 2	No. 3
2	12.0	21.0	16.0	14.4
3	7.0	14.0	9.0	10.0
4	3.5	17.5	10.0	7.0
5	9.3	–	11.5	9.8
6	9.3	–	15.0	9.6
7	12.0	–	12.0	12.0

Table 5 Specified values of specific cohesion *c*, kPa based on results of static sounding

No. LL	*c*, kPa			
	2005 г	Number of sites		
		No. 1	No. 2	No. 3
2	–	–	–	–
3	20.0	30.0	23.0	23.0
4	17.0	29.0	23.2	20.5
5	25.0	–	34.0	27.0
6	25.0	–	43.0	30.0
7	35.0	–	35.0	36.0

Table 6 Specified values of angle of inner friction φ based on the results of static sounding

No. L	*φ*, degrees			
	2005 г	Number of sites		
		No. 1	No. 2	No. 3
2	29.0	31.4	30.2	29.8
3	15.4	15.7	15.5	15.5
4	9.8	10.0	10.0	9.9
5	25.0	–	25.0	25.0
6	25.0	–	25.5	25.0
7	26.0	–	25.0	25.5

The green areas indicate the parts of the wall where the deviation fluctuates within acceptable limits while the blue color indicates the areas where the wall deviates in the opposite direction.

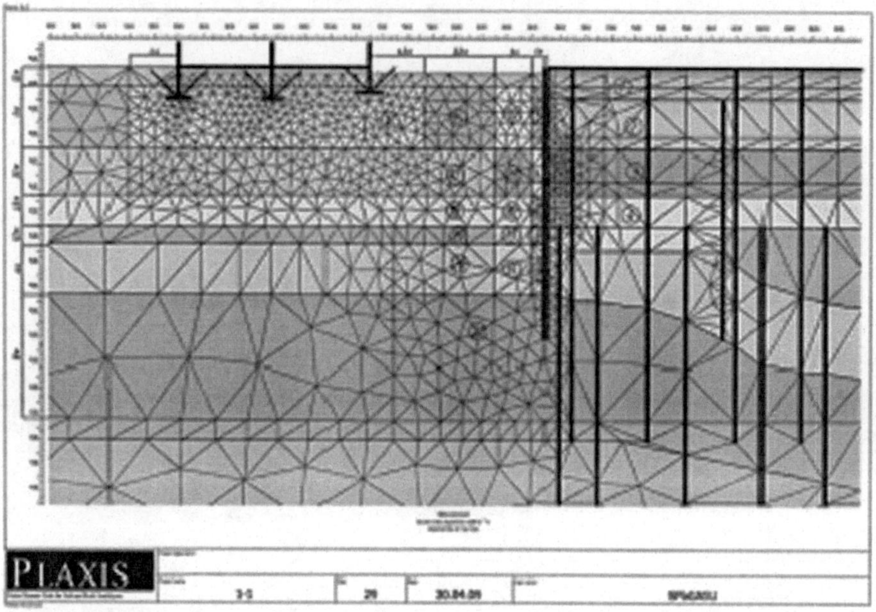

Fig. 13 The settlement diagram of the changed earth basis

- The method of static sounding and the analysis of the results allows in proper time to estimate the changes in soil properties in the pit of the building under construction and around caused by the technogenic effect.
- The increase of the strength and deformation values of soil registered at the second sounding can be explained by soil compaction caused by technogenic effect during the erection of a great number of drilled piles, restoration of the structural binds, and partial compaction of soil while constructing cement-soil diaphragm by jet grouting method.
- The changed values of soil properties received during calculations by PLAXIS program allowed to make reliable estimations of the stress deformation condition of the soil and to forecast stability and deformation of the building under construction and on the adjacent territory.

References

1. Zhussupbekov A, Omarov A (2018) Piling construction and testing of megastructures on problematical soil ground of Kazakhstan. In: Proceeding 20th SEAGC-3rd AGSSEA conference in conjunction with 22nd annual Indonesian national conference on geotechnical engineering. Jakarta, Indonesia, pp 15–18
2. Zhussupbekov AZh, Lukpanov RE, Omarov AR (2016) Bi-directional static load testing. In: Selected paper from the proceedings of the fourth geo-China international conference, pp 35–42

3. Zhussupbekov AZh, Lukpanov RE, Omarov AR (2016) Experience in applying pile static testing methods at the Expo 2017 construction site. Sci J Soil Mech Found Eng, 251–256
4. A Zhussupbekov. A Omarov G. Tanyrbergenova. Design of anchored diaphragm wall for deep excavation. International Journal of GEOMATE, Vol.16, Issue 58, (2019), pp.139 – 144.
5. Uakhitov AB, Igilmanov ZA (2014) Analysing deformations using laser scanning in Astana. In: Proceedings of the IX international scientific conference for students and young scholars science and education-2014, 4635–4639
6. Mangushev RA, Oshurkov NV, Lgoshin AV (2005) Use of mobile installation for the purposes of reconstruction and construction in the constrained conditions of Saint Petersburg. In: The collection of reports of the international scientific-practical conference "Reconstruction of Saint Petersburg—2005", the Part 1, SPb, pp 214–218
7. Mangushev RA, Oshurkov NV, Arutjunov IS, Ershov AV (2007) Research of mechanical characteristics грунтов at the device of chisel piles checkpoint screw. The interuniversity collection of works "Scientific—practical and theoretical problems of geotechnics". Saint Petersburg, SPbGASU, pp 91–104
8. Lunne T, Robertson PK, Powell JM (2001) Cone penetration testing in geotechnical practice. Sponpress, 312
9. Xu SY, Kannangara KP, Taciroglu E (2018) Analysis of the stress distribution across a retaining wall backfill Kannangara, Ertugrul Taciroglu. Comput Geotech 103, 13–25
10. Hosseinzadeh S, Joosse JF (2015) Design optimization of retaining walls in narrow trenches using both analytical and numerical methods. Comput and Geotech 69, 338–351
11. Tung S-H, Weng M-C, Shih M-H (2013) Measuring the in-situ deformation of retaining walls by the digital image correlation method. Eng Geol 166:116–126
12. Ramezani MS, Ghanbari A, Hosseini SAA (2017) New mathematical model for computing natural frequencies of retaining walls considering soil-structure interaction. Appl Math Model 45:179–191
13. Butterfield R, Marchi M (2017) On the displacement of a traditional retaining wall when first loaded. Soils Found 57(6):1083–1087
14. Mohammad Hossein Khosravi, Thirapong Pipatpongsa, Jiro Takemura. Theoretical analysis of earth pressure against rigid retaining walls under translation mode. Soils and Found, Volume 56, Issue 4, (2016). P. 664–675.
15. Mangushev RA, Konyushkov VV, Sapin DA (2016) Geotechnical engineering surveys in construction and reconstruction in dense urban areas. Ind Civ Const (5):47–54
16. Mangushev RA, Nikiforova NS, Konyushkov VV, Osokin AI (2013) Design and construction of underground structures in open pits. Textbook. Publishing house DIA, M., St. Petersburg, p 256
17. Ilyichev VA, Mangushev RAM (2016) Handbook of geotechnical engineering. Bases, foundations and underground structures, p 1040
18. SP 11-105-97 (2001) Engineering—geological researches for construction. A part 1. The general (common) rules of construction works. GOSSTROY of Russia
19. TCH 50-302-2004 (2004) Designing of the bases of buildings and constructions in Saint Petersburg. GOSSTROY of Russia
20. GOST 19912-2001 (2001) A method of field tests by static sounding. Soils, Moscow

Bearing Capacity of Embedded Strip Footing Placed Adjacent to Sandy Soil Slopes

Mahmoud Qarmout, Firas Ghrairi, Arash Lavasan, Diethard König, and Torsten Wichtmann

Abstract In some infrastructure projects, the strip footing adjacent to the slope is usually constructed at some depth below the ground surface. For practical engineers, estimating the bearing capacity of such footing is one of the challenging tasks. In this paper, the Finite Element Limit Analysis (FELA) is employed to determine rigorous lower and upper bounds for bearing capacity of embedded strip footing resting adjacent to a sand soil slope. The results obtained using FELA are compared with analytical approaches available in the literature. The parametric studies have been performed to examine the effect of slope angle, friction angle of the soil, footing location, and embedded depth of the footing on the bearing capacity. The results indicate that, when the strip footing is built adjacent to the sandy soil slope, the bearing capacity is reduced depending on the location of the footing relative to the crest of the slope. Also, the bearing capacity of the slope increases with the increase in the embedded depth of footing and the soil strength. Whereas, it decreases with an increase in the slope angle. The numerical outcomes are presented in design charts which can be used by the practicing engineers.

Keywords Bearing capacity · Strip footing · Slope · Finite element limit analysis

---

M. Qarmout (✉) · F. Ghrairi · A. Lavasan · D. König · T. Wichtmann
Ruhr-Universität Bochum, Universität str. 150, 44780 Bochum, Germany
e-mail: Mahmoud.Qarmout@rub.de

F. Ghrairi
e-mail: Firas.Ghairi@rub.de

A. Lavasan
e-mail: arash.alimardanilavasan@rub.de

D. König
e-mail: Diethard.Koenig@rub.de

T. Wichtmann
e-mail: Torsten.Wichtmann@rub.de

M. O. Karkush and D. Choudhury (eds), *Modern Applications of Geotechnical Engineering and Construction*, Lecture Notes in Civil Engineering 112,
https://doi.org/10.1007/978-981-15-9399-4_10

1 Introduction

In practice, strip footing constructed on the slopes is often encountered in different engineering projects. For example, mobile phone towers, bridge abutments, and buildings include construction of the strip footing on the slopes. In such a geotechnical structure, both the slope stability and the footing bearing capacity should be taken into account in evaluating the load-carrying by the footing. Therefore, the estimation of the bearing capacity for the strip footing resting on the slope is an intricate stability problem for geotechnical engineers. During the last two decades, several methods are adopted for calculating the bearing capacity of the footings resting on the slopes, including limit equilibrium method (e.g., [1–4]), upper bound method (e.g., [5–7]), lower bound method (e.g., [8–10]), finite element method (e.g., [11–13]). The previous researches are useful to determine the bearing capacity of footing on slopes. However, the bearing capacity of footing on slope, much remains to be investigated using new methods, such as the Finite Element Limit Analysis methods (FELA). This method does not require assuming a prospective collapse mechanism or a settlement to calculate the bearing capacity of the footing. FELA can bracket the true collapse load from above (upper bound) and below (lower bound). The upper bound solution of FELA leads to an upper bound estimate of collapse load by considering a statically kinematically admissible velocity field. Whereas, the lower bound solution of FELA leads to a lower bound estimate of collapse by considering a statically admissible stress field.

This paper uses the finite element limit analysis method to determine the bearing capacity of footing resting on sand soil slope. An extensive finite element limit analysis has been performed to study the effect of the setback distance on the bearing capacity of the footing for various possible slopes angle, friction angle, and embedded depth of the footing. The upper bound and lower bound results obtained by the present solution are compared to the results from the previous studies. Also, the failure mechanism of the slope is discussed providing a more realistic understanding of the slope failures. For practical purposes, design charts are provided.

2 Problem Definition

The geometry of the slope and footing details are presented in Fig. 1a. The geometric parameters include the footing width (B), which is located at a distance (D_e) from the crest of the slope, the embedded depth (D_f) below the ground surface and the angle of inclination to the horizontal surface (β). The Finite Element Limit Analysis (FELA) using OptumG2 commercial software is employed to calculate the collapse load from above (upper bound) and below (lower bound) [14]. 2D analysis was performed to simulate the embedded strip footing placing adjacent to the slope. In numerical analysis, the soil obeys the associated flow rule and Mohr–Coulomb yield

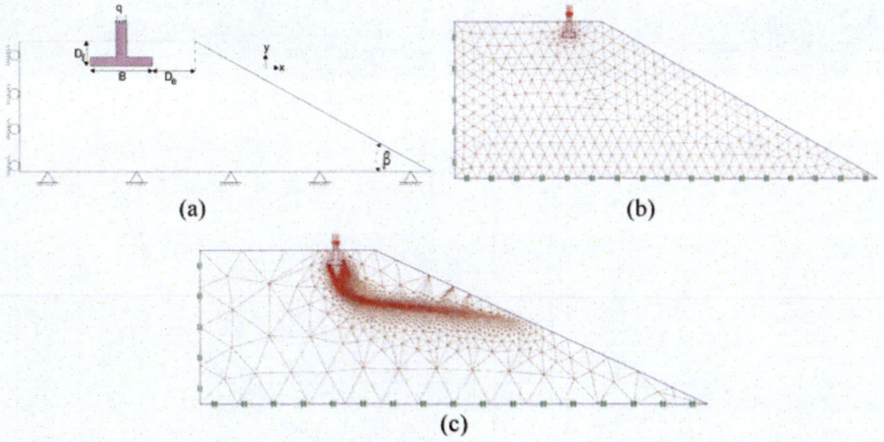

Fig. 1 **a** Schematic diagram of the model and the boundary condition; **b** initial non-adaptive mesh; **c** final adaptive mesh

criterion, in which the dilatancy angle (ψ) is equal to the friction angle of the soil (φ).

The boundary conditions were defined such that the left boundary of the model is allowed to move only in a vertical direction, while the bottom boundary of the model is fixed in vertical and horizontal directions. In both upper and lower bound calculations, the soil mass was discretized as triangular elements. Fig. 1b shows a typical 2D finite element model used in this study. Five iterations of mesh adaptively refinement were adopted in both upper and lower bound analysis, starting with 1,000 elements and increased to a final mesh with the number of 10,000 elements, see Fig. 1c. The distributed multiplier load is applied on the footing which is automatically increased during the calculation until the failure has occurred. The soil is modeled with Young modulus equal to 30 MPa, a Poisson's ratio of 0.2. Moreover, the lateral earth pressure coefficient at rest is adopted $Ko = 1 - \sin\varphi$, according to Jaky [15].

The present study considered different rang of friction angles varying from 15° to 45°. To ensure the stability of the soil slope, the slope angle (β) is always assumed to be less than the friction angle of the soil (φ). The dry unit weight of the soil is (γ) 18 kN/m^3 and the footing width is 1.5 m. The soil is assumed to be homogeneous cohesionless soil. Also, the hydrostatic pore water pressure effects are not considered in this study. The strip footing is considered as a rigid body.

3 Comparisons with Previous Studies

The results obtained from the analytic solution of Meyerhof [1] and Graham et al. [16] were used to validate the FELA model (upper and lower bound solutions). The case to be used in the validation has embedded depth to footing width ratio of 1, the

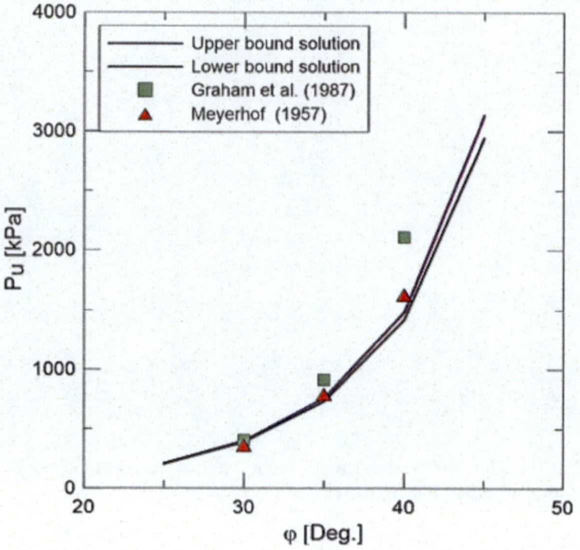

Fig. 2 Comparison of the ultimate bearing capacity between the present study and existing solutions

setback distance to the footing width ratio is 0, the dry unit weight of the soil is 18 kN/m^3, the slope angle is 20°, the footing width is 1.5 m and the friction angle of the soil is 30°, 35°, and 40°.

The comparison of the bearing capacity calculated from Meyerhof [1], Graham [16], and upper and lower bound solutions is shown in Fig 2. It can be seen that the results obtained from the upper and lower bound solution follow a similar trend to the analytic solutions. The results obtained from upper and lower bound solutions are mostly close to the results of Meyerhof [1]. The maximum difference between the lower bound solution and Meyerhof [1] results was found to be 12.1%.

4 Bearing Capacity of the Strip Footing Resting on the Slope

The ultimate bearing capacity of the strip footing resting on the slope can be calculated by the following equation [1, 16]:

$$q_u = c \cdot N_{cq} + 0.5 \cdot \gamma \cdot B \cdot N_{\gamma q} \tag{1}$$

where N_{cq} and N_q are the bearing capacity factors, B is the width of footing, c is the cohesion of soil, and is the dry unit weight of soil. For cohesionless soil (c = 0 kPa), Eq. 1 could be reformulated as follows:

$$q_u = 0.5 \cdot \gamma \cdot B \cdot N_{\gamma q} \tag{2}$$

Within this study, the upper and lower bound solutions for the bearing capacity factor $N_{\gamma q}$ have been computed using FELA. The difference between the upper and lower bounds solutions provides an approximated measure of the error in the analysis. This relative error can be used to refine the meshes until a suitably accurate estimation of the true collapse load is obtained. The maximum relative error is found to be 4%. Therefore, the average value of the upper and lower bound for $N_{\gamma q}$ will be used in the following discussions.

5 Parametric Studies

For cohesionless soil, the major parameters that affect the normalized bearing capacity of footing located on the slope can be seen in the following form:

$$\frac{q_u}{\gamma \cdot B} = f\left(\varphi, \beta, \frac{D_f}{B}, \frac{D_e}{D_f}\right) \quad (3)$$

Understanding the effect of these parameters is very important in the design of the strip footing built near the slope. Hence, to highlight the effect of various parameters in Eq. 3, the parametric studies are conducted into the effect of setback distance to the footing width ratio D_e/B, embedded depth to the footing width ratio D_f/B, the friction angle of the soil φ and the slope angle β. The results of this study are discussed in the following sections.

Effect of Embedded Ratio (D_f/B) and Setback Ratio (De/B) on Bearing Capacity Factor (Nγq). To investigate the influence of D_f/B and D_e/B ratios on the bearing capacity, three different embedded depth ratios were considered for analysis, namely, 0.5, 1, and 1.5. Therefore, the footing can be considered to behave as a shallow footing. The numerical analysis was performed for eleven ratios D_f/B ($D_e/B = 0.0$, …, 14.0). In this case, $\varphi = 30°$ and $\beta = 25°$ were kept constant.

The average values of the upper and lower bound solutions of the bearing capacity factor Nγq with different combinations of D_f/B are presented in Fig. 3a. The results in Fig. 3a indicate that for $\varphi = 30°$ and $\beta = 25°$, Nγq increases with the increase in De/D_f ratio. It is obvious from Fig. 3a that once a footing is built at a sufficient setback distance from the crest of the slope, its bearing capacity is similar to the bearing capacity for the footing resting on the horizontal ground. The critical setback distance required to reach this bearing capacity is dependent on De/B. It can be seen from Fig. 3a that for D_f/B ratio equal to 1.5 and De/B ratio over 7, the stability problem is no longer a footing on slope.

Figure 3a shows the percent of decrease in bearing capacity for three values of the D_f/B ratio. The major reductions in the bearing capacity occurred for D_f/B ratio between 0 and 3, which represents almost from 70 to 40 % reduction in the bearing capacity. Moreover, the rate of decrease in the bearing capacity is same for the all D_f/B ratios.

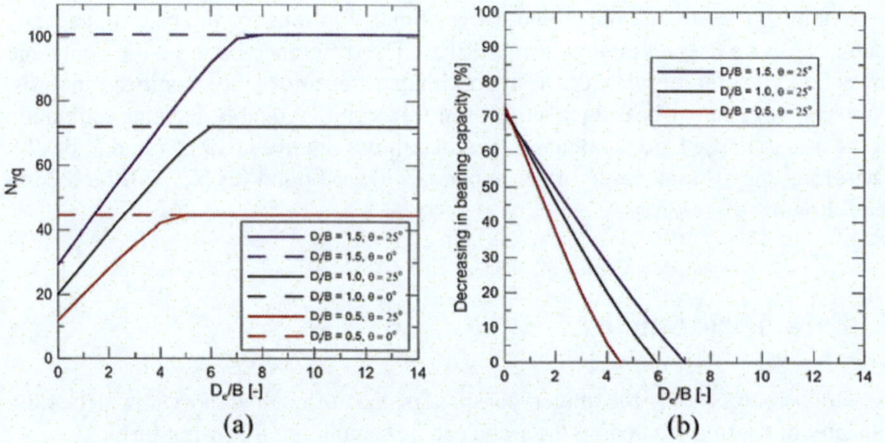

Fig. 3 Effect of D_f/B on bearing capacity factor $N_{\gamma q}$, $\varphi = 30°$ and $\beta = 25°$: **a** Variation of $N_{\gamma q}$ with D_f/B; **b** Percent of decrease in ultimate bearing capacity

Effect of Friction Angle (φ) and Slope Angle (φ) on Bearing Capacity Factor Nγq. To study the effect of combined variation of friction angle of the soil and the slope angle on the ultimate bearing capacity, the numerical analysis was performed for 15°, 30°, and 40°. Also, for β of 15° and 25° have been considered. In this case, D_f/B= 1 was kept constant. Figure 4 shows that the variation in the friction angle of the soil can significantly change the bearing capacity. The increase in β resulted in the decrease in the magnitude of Nγq, the effect is more prominent at higher friction

Fig. 4 Effect of friction angle of the soil and slope angle on bearing capacity factor Nγq

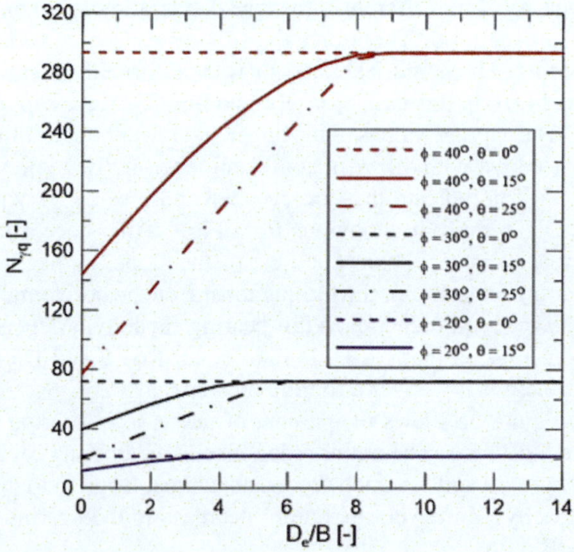

angle of the soil ($\varphi = 30°$ and $40°$). For the higher friction angle of the soil, the reduction in $N\gamma q$ with an increase in the angle of the slope is considerable when the footing is resting on the slope crest.

From Fig. 4, it can be concluded that the bearing capacity of the footing resting on the slope is governed by the slope angle about the friction angle of the soil. The effect of the setback distance is becoming more pronounced with an increase in the friction angle of the soil. For example, considering the case of $\beta = 15°$, $N\gamma q$ reaches the constant value for $De/D_f = 4$ at $\varphi = 15°$, $De/D_f = 5$ at $\varphi = 30°$, and $De/D_f = 9$ at $\varphi = 40°$

6 Failure Mechanism of Footing on Slope

Figure 5 shows the strain rate of the slope for $\varphi = 30°$, $\beta = 25°$, and different De/B ratios, which indicates the soil movement and the failure mechanism. From Fig. 5, it can be observed that when the footing is located at the crest of the slope De/B = 0, the soil below the footing tends to move toward the side of the slope, which is attributed to the dominant free deformation of the soil at the slope face. This behavior results in a considerable decrease of the ultimate bearing capacity and an unsymmetrical failure mechanism is formed. As De/B ratio increases, the dominant influence of free deformation for the soil at the slope face gradually reduces causing symmetrical failure mechanism at specific setback distance.

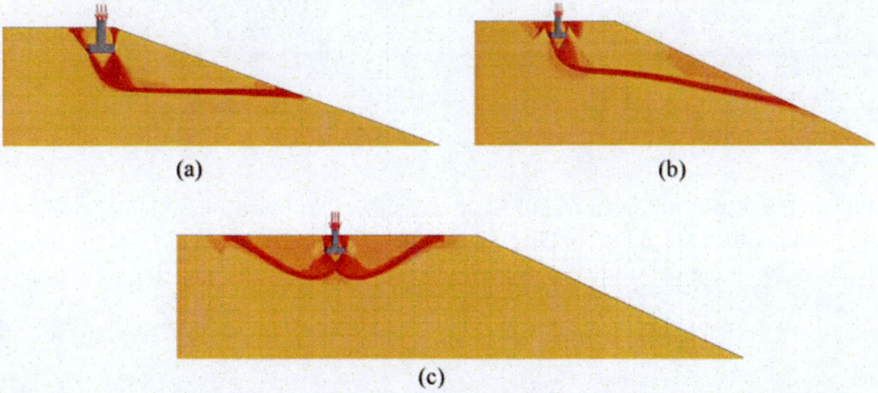

Fig. 5 Incremental shear strains (failure mechanism) De/B ratio (**a**) De/B = 0; (**b**) De/B = 3; (**c**) De/B = 7

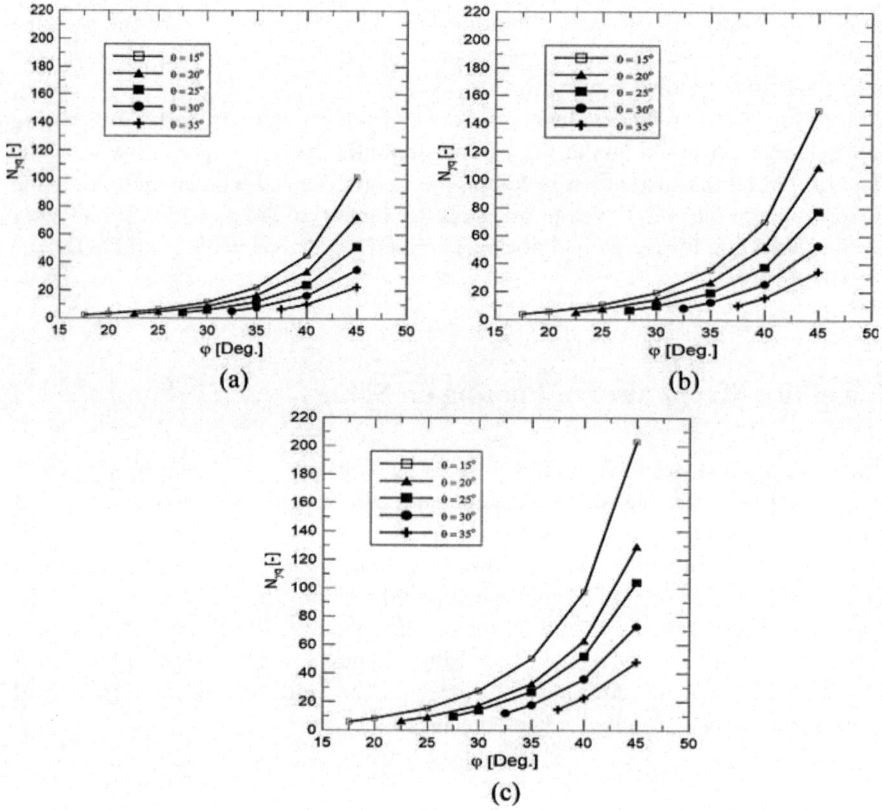

Fig. 6 Design charts, De/B = 0 (**a**) $D_f/B = 0.5$; (**b**) $D_f/B = 1.0$; (**c**) $D_f/B = 1.5$

7 Design Charts

Based on the results obtained from FELA, a design chart for different D_f/B, φ and β have been produced, see Fig. 6. Given the value of friction angle and D_f/B, the user can easily calculate the ultimate bearing capacity for the footing resting on a slope with $D_f/B = 0$.

8 Conclusions

This study aims to investigate the effect of the friction of soil angle, setback distance, angle of slope, and the embedded depth of the footing, on the ultimate bearing capacity for the strip footing resting on the slope. To achieve this goal, the upper and the lower solutions using two-dimensional FELA are obtained. The main conclusions are as follows:

- The bearing capacity decreases with the increase in the steepness of the slope, the reduction of the bearing capacity is relatively pronounced for higher slope angle.
- The bearing capacity increases with the increase of setback distance. Beyond a critical setback distance, the footing on the slope behaves similarly to that footing built on horizontal ground.
- The embedded depth of the footing has a considerable effect on the bearing capacity. A significant gain in ultimate bearing capacity can be achieved by increasing the embedded depth of the footing.
- The friction angle of the soil plays an important part in determining the bearing capacity of a slope. In general, the bearing capacity increases with the increase in the friction angle of the soil.

References

1. Meyerhof G (1957) The ultimate bearing capacity of foundations on slopes. In: Proc., 4th Int. Conf. on Soil Mechanics and Foundation Engineering, vol 1
2. Saran S, Sud VK, Handa SC (1989) Bearing capacity of footings adjacent to slopes. J Geotech Eng. ASCE 115(GT4), 553–573
3. Castelli F, Motta E (2008) Bearing capacity of shallow foundations near slopes Static analysis. In: Proceedings of the second international British geotechnical association conference on foundations, ICOF
4. Shukla RP, Jakk RS (2014) A critical review on bearing capacity of a footing on sloping ground. In: Proceedings of Indian geotechnical conference, Kakinada, India, 1993–1996
5. Michalowski RL (1989) Three-dimensional analysis of locally loaded slopes. Geotechnique 39:27–38
6. Askari1 F, Totonchi A, Farzaneh O (2008) Three-dimensional stability analysis of convex slopes in plan view. J Geotech Geoenviron Eng 134, 1192–1200
7. Shiau JS, Merifield RS, Lyamin AV, Sloan SW (2011) Undrained stability of footings on slopes. ASCE, Int J Geomech 11:381–390
8. Lysmer J (1970) Limit analysis of plane problems in soil mechanics. J Soil Mech ASCE
9. Davis EH, Booker JR (1973) Some adaptations of classical plasticity theory for soil stability problems. In: Published in the proceedings of the symposium on the role of plasticity in soil mechanics, Cambridge, England, No. Conf Paper, September 1973 13–15
10. Chakraborty D, Kumar J (2013) Bearing capacity of foundations on slopes. Geomechanics and Geoengineering 8:274–285
11. Ip. Kai Wing (2005) Bearing capacity for foundation near slope. Diss. Concordia University
12. Alamshahi S, Hataf N (2009) Bearing capacity of strip footings on sand slopes reinforced with geogrid and grid-anchor. Geotextiles and Geomembranes 27:217–226
13. Georgiadis K (2010) Undrained bearing capacity of strip footings on slopes. J Geotech Geoenviron Eng 136:677–685
14. Krabbenhoft K, Lyamin A, Krabbenhoft J (2015) Optum computational engineering (OptumG2). Comput Softw. Retrieved from https://www.optumce.com
15. Jaky J (1944) The coefficient of earth pressure at rest. J Soc Hung Archit Eng, 355–358
16. Graham J, Andrews M, Shields DH (1988) Stress characteristics for shallow footings in cohesionless slope. Can Geotech J 25:238–249

Comparing the Axial Performance of Screw Pile with Ordinary Piles in Soft Clay Layer Overlaying Sandy Soil

Hassan Obaid Abbas and Omar Kareem Ali

Abstract Soft soil covers large parts of central and southern Iraq. As many locations in these regions can be used for important projects; many types of research had been carried out to find a suitable technique to solve the problem of this soil which is characteristic with a low bearing capacity, problems of settlements, and stability. An appropriate, fast, efficient, and economical solution has been proposed to solve the problems of this soil using a screw pile. The present study is concerned with screw piles utilized in the foundations of structures exposed to compressive loads. The behavior of 10 mm^2 circular shaft screw pile models, embedded 300 mm in soft clay layer overlaying 200 mm thickness of well compacted sandy soil were studied and compared with pile without helix (conventional steel pile). Model tests are carried out with screw lengths of 300, 350, and 400 mm and helical plate diameters of 15 and 30 mm. Single and double helical screw pile have been considered (30 and 50) mm spacing between helix is used in this study. This study revealed that the screw pile has a higher compressive capacity on soft clay soil than a conventional pile (steel pile without helix). The augment in compressive strength can reach 2–8 times larger than the value obtained using the ordinary steel pile. The results showed that the screw piles with single and double helix can resist pile settlement 45–700% more than ordinary piles when screw piles are embedded to dense sandy soil through soft clay soil.

Keywords Screw pile · Compressive capacity · Ordinary pile · Soft clay · Deep foundation

H. O. Abbas (✉) · O. K. Ali
Department of Civil Engineering, College of Engineering, Diyala University, Baqubah, Iraq
e-mail: temimi71@yahoo.com

O. K. Ali
e-mail: omar.alzuhairy.oa@gmail.com

M. O. Karkush and D. Choudhury (eds), *Modern Applications of Geotechnical Engineering and Construction*, Lecture Notes in Civil Engineering 112,
https://doi.org/10.1007/978-981-15-9399-4_11

1 Introduction

Vast areas in central and south governorates of Iraq covers with soft clay soil. Since the technical properties of soft clay are lower than that of other soils, this is often considered problematic [1]. These types of soils are typically characterized by extremely high compressibility, high creep rates, poor strength characteristics, poor workability, and poor bearing capacity [2]. Over the last few decades, there have been unrelenting efforts to understand and solve engineering problems in soft soils. Various methods can be utilized to lessen the effects of soft soil harm. These involve soil replacements, chemical and physical treatment, and the usage of special systems. The use of these methods remains for a long time. However, many of them have certain limitations and can be very expensive. To address these shortcomings, try to develop an effective, simple, fast, and easy install and low-cost alternative foundation system, this study submits a simple foundation method under the name of screw piles as a reliable solution for suppressing problems caused by soft soils [3]. Screw piles are structural parts used in civil engineering applications to provide compressive, tensile, and uplift resistance [4]. The screw pile is ground anchors built-in spiral (helix) form circular blades (plates) soldered to a circular or square steel shaft at a particular spacing [5]. Screw piles have more than one significant advantage compared to ordinary piles. Screw piles in high clay depths are more effective than other piles due to difficulties in drilling [6]. Some of the advantages of screw piles are, alternative high capacity foundation, predictable capacity, well-established torque-capacity relationship, lead sections, and extensions can be configured to achieve design depth and capacity, quick installation without generating spoils, installed in restricted or narrow access areas, installation in all weather conditions, clean installation, and low mobilization costs [7].

This study aims to examine the performance of screw piles installed in soft clay overlaying sandy soil with variable depth of anchorage (depth of embedment) L, helix number, diameter of the helix D_h, and spacing ratio s/D_h under compressive load with comparison to an ordinary pile.

2 Materials Used

Soft clay soil. Soil samples utilized in this research were fetched from the village of AL-Bawia located in Diyala/Iraq from 2 to 5 m depth from the ground surface. The soil was subjected to routine laboratory tests to determine its properties. Table 1, shows the results of the laboratory test. The soil is classified (CL) according to the Unified Soil Classification System USCS.

Sandy soil. 200 mm sandy soil layer is used as a stable zone under the layer of 300 mm soft clay soil. Fine, clean sand was brought from the location in Karbala governorate about 110 km south of Baghdad, Iraq. Before the testing step, the sandy sample is desiccated in a desiccating oven for 24 h at 105 °C in the laboratory. Afterward

Table 1 Geotechnical properties of soft clay

Soil property	Value	Standard	Soil property	Value	Standard
LL (%)	39	ASTM D 4318-00	Sand (S) (%)	3.3	ASTM D 422
PL (%)	22		Silt (M) (%)	44	
P (%)P (%)	17	–	Clay (C) (%)	51.7	
Gs	2.81	ASTM D 854-2	USCS classification	CL	

Table 2 Sandy soil properties

Soil property	Value	Standard
Uniformity Coefficient (Cu)	2.61	(ASTM D-422) and ASTM D 2487 (2006)
Coefficient of Curvature (Cc)	0.85	
(USCS) Classification	SP	
Specific Gravity (Gs)	2.65	/
Drained Cohesion (kN/m^2)	0	ASTM D3040-04(2006)
Angle of Internal Friction (φ°)	35.75	ASTM D3040-04(2006)
Maximum void ratio, e_{max}	0.75	–
Minimum void ratio, e_{min}	0.464	–

sieved on Number 40 sieve to exclude the rough particles. The Lab experiments were accomplished to determine the geotechnical properties of the sandy soil. The laboratory results are demonstrated in Table 2. The drained direct shear test was performed at a relative density of 70% that corresponds to a dry unit weight of 15.9 kN/m^3.

Pile models. In this study, twenty-one models of steel piles measuring 30, 35, and 40 cm long were used. A circular high-resistance steel section with a diameter of 1cm^2 was used to fabricate the pile models as presented in Fig. 1. The helical plate diameters D_h are used 15, 30 mm with a thickness of 0.2 cm. The pitch-to-plate diameter p/Dh is used 1/3 [8]. The helix plates were made from stiff steel and joined (welded) tightly and precisely to the pile rod (shaft). 3 and 5 cm plate spacing is utilized. The tip of the shaft was 45% to facilitate the installation of the pile in the soil. The laboratory pile models were scaled to the field prototype with 1:10 percentage, respectively. The experimental program was done on a screw pile with various lengths, number of helices, the diameter of the helix, and spacing between helix also ordinary pile (steel pile without helix).

Test tank. Two test tanks (soil containers) were manufactured using a thick steel plate (4 mm in thickness) possessing an interior diameter of 30 and 55 cm in height. The containers base of the square form contains two angle sections which help to fix the tank during the process of installation and pile test. The container base was

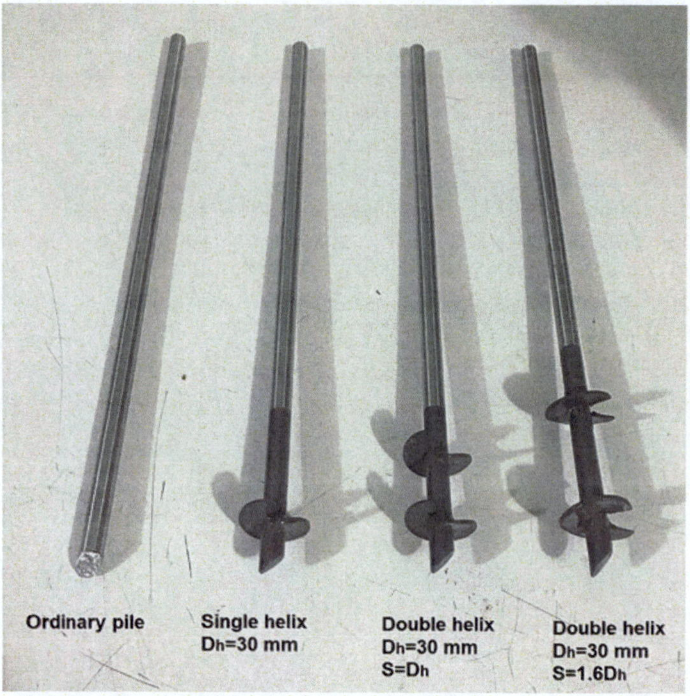

Fig. 1 Different types of pile models used in this study

supplied with 4 wheels for free movement; the tank (container) was firm enough to prevent side deformity during the process of soil preparation and pile test. Figure 2, shows the test tank details. The containers plated with three layers of anti-corrosion paint and double layers of conductive base to withstand rust, through the period of

Fig. 2 Soil test tank utilized in this investigation

the test. In general, the dimensions of the test tank are determined to depend on the effectively stressed zone of the mass of soil from the side foundation [9]. It is considered that the failure stress extends about (2–4 pile diameter) under the pile tip for a single pile (D is the). The spread-out load distribution on the single pile side is 2:1 from the depth of L/3 from the pile tip [10].

3 Testing Procedure

The experimental program, proposed for this investigation include twenty-one pile models (18 models of screw pile and 3 models of ordinary pile). The tests were performed in the laboratories of Civil Engineering Department/Collage of Engineering/University of Diyala.

Sandy soil bed preparation. The following preparation procedure of soil bed is followed for all models test:

- The required quantity of natural oven-dried sand was prepared at a 15.9 kN/m3 drying unit weight, which corresponds to the dense state.
- The sandy layer was prepared by pouring the sand into four (5 cm thick) sub-layers in the test tank and compacted statically so that the final thickness of the sandy soil layers reaches 20 cm.
- To reduce the effect of friction, all insides of the containers are coated with petroleum jelly.

Soft clay control tests. Before the preparation stage of soft clay, trial tests were performed to check the effectiveness of the method of preparation. These tests were conducted to establish the variation in shear strength at diverse liquidity indices (different water contents), numerous trials were performed and exemplary results are presented in Fig. 3.

Soft clay soil bed preparation. The following procedures were followed to prepare the soft clay soil in the test tank so that it is ready to perform model pile tests:

(1) After drying and pulverizing the untreated soft soil, soft clay soil was produced with a moisture (water) content of 30%, corresponding to undrained shear strength = 31 kPa.
(2) Thirty kg of natural soft clay soil was mixed with sufficient water to obtain the desired consistency.
(3) The mixing was carried out using an electric mixer; each 10 kg of dehydrated soft clay was mixed (blended) separately until the entire amount was completed.
(4) After being well mixed, the wet (moist) soil was stored in tight polythene bags for 48 h to obtain a uniform moisture content.
(5) The inside of the container is coated with lubricating oil to reduce the frictional effect that can occur between the inside surface of the container and the soil.

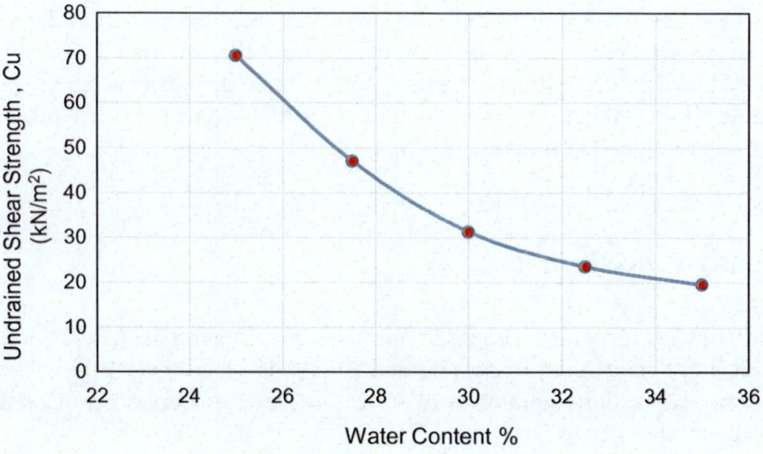

Fig. 3 Variation of c_u versus water content for remodeled soft clay after 2 days

(6) The soil was placed in a circular steel tank above the sandy layer in 6 layers; every layer was gently leveled with a wooden rammer, thereafter the leveling layer was carefully leveled with a metal hammer (having a diameter of 150 mm and weight of 10 kg) to eliminate trapped air. This procedure is continued for the six layers until the thickness of the bottom in the test tank is reached 30 cm.

Installation of pile model. The model of piles installed slowly into the soil by directing a torque (T) with a suitable force (N). The torque was applied by using a hydraulic torque motor that supplied the gyratory and perpendicular forces required to screw (installed) screw pile in the center of soil bed surface to the depth wanted. The amount of perpendicular speed and the number of revolutions per minute (rpm) are used which depends on the helical pitch (P). The installation of a screw pile has been carried out so that the screw pile penetrates the soil at each complete revolution with a value equaling to the helical pitch (P) to minimize the disturbance of the soil. Therefore, for all tests, a controlled displacement installation with a rate of penetration depends on the 1/3 pitch to helix diameter ratio [8], and rotational speed was used as shown in Table 3. Figure 4 shows the hydraulic torque motor that had been used in the installation of screw pile.

Pile model test. In the case of loading, the vertical load (compression) is applied to a single screw pile post by a three-ton load cell (S-shape). Use of the timer with

Table 3 Penetration rate used in the pile installation process

Diameter of Helix, D_h (mm)	Pitch, P (mm)	Penetration rate (mm/min)	Rotational speed (rpm)
15	5	5	1.80
3	10	10	3.60

Fig. 4 Hydraulic torque motor (screw pile installation device)

a constant load, the penetration rate of 1 mm/min, in the full test, procedure-based, on Prakash (1990) and ASTM (2018) standard-D1143, for compression testing [11, 12], that are concluded, for cohesive soils, the range of penetration rate is (0.25–1.25 mm/min). The test is carried out until a continuous settlement of the screw pile up to 12 mm (depth) is recorded. A digital weighing indicator connected to the load cell gauges the load. Two digital dial gauges with a sensitivity of 0.001 mm used to read the central settlement of the screw piles cap. Figure 5 shows the pile testing device.

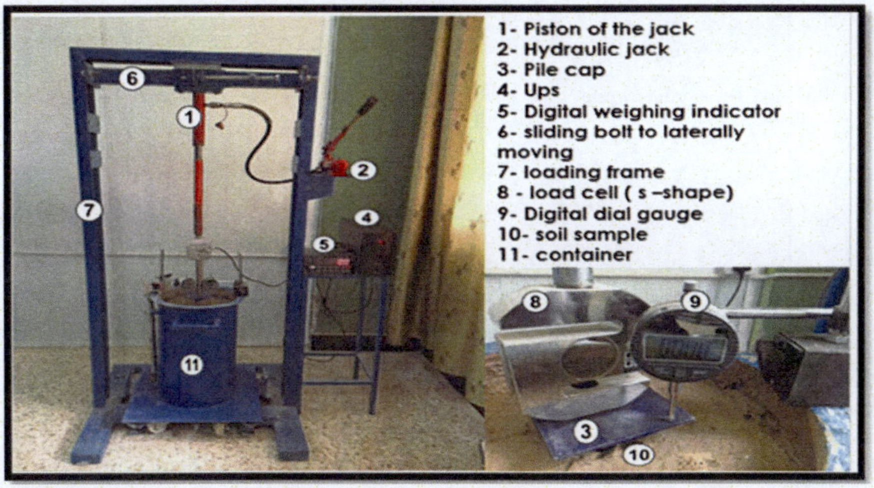

Fig. 5 Pile test device

4 Results and Discussion

The results of characteristic tests are carried out on the 30 cm clay layer overlaying 20 cm sandy layer, on three embedment depth 30, 35 cm (5 cm in dense sand), and 40 cm (10 cm in dense sand) are discussed in the current section. The effect of parameters, namely; the diameter of the helix plate (3D and 1.5D), number of helix plate, the spacing between the helices (3 and 5 cm), and other factors such as slenderness ratio (L/D) of 30, 35, and 40 are studied in details. Also, ordinary steel piles with a diameter similar to the diameter of the shaft of the screw pile are studied for comparison purposes. The results of the experimental work using the setup device are reviews and discussions of (21) models test. The tests contain static pile load tests for a single pile. As indicated in the literature, various methods of failure criteria used to predict the ultimate bearing capacity of screw piles from the load test. The method, ISSMFE, is adopted in this study, which determines the ultimate load at a settlement (displacement) level of 10% of the helix plate diameter [13, 14]. Figures 6, 7, and 8 show that the increase of L/D ratio for conventional and screw piles decreases the settlement that results from the axial compression load. This is due to the anchoring resistance of long piles in the deep soil layer even if this layer is in the active zone of the soil. The percentage of increase obtained from the compression capacity of the conventional pile of 14.3% when increasing L/D from 30 to 35 and the compressive capacity of the ordinary pile with L/D 40 up to as much as 40.5 times the compressive capacity of the pile with L/D 30 while for screw piles with single helix plate were 37.4.8% for $D_h = 3D$ and 26.7% for $D_h = 1.5D$ when growing L/D ratio from 30 to 40 as shown in Fig. 6.

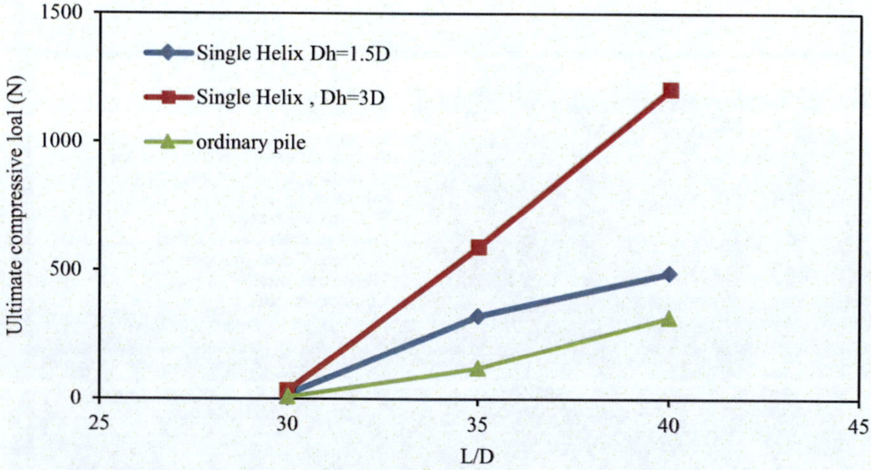

Fig. 6 The relation between the ultimate compressive capacity and l/d ratio for ordinary and single helix screw pile

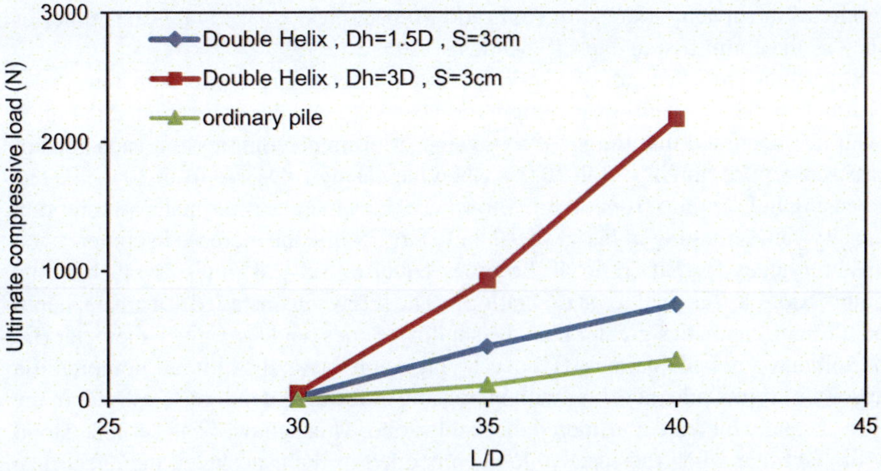

Fig. 7 The relation between the ultimate compressive capacity and l/d ratio for ordinary and double helix screw pile (s = 3 cm)

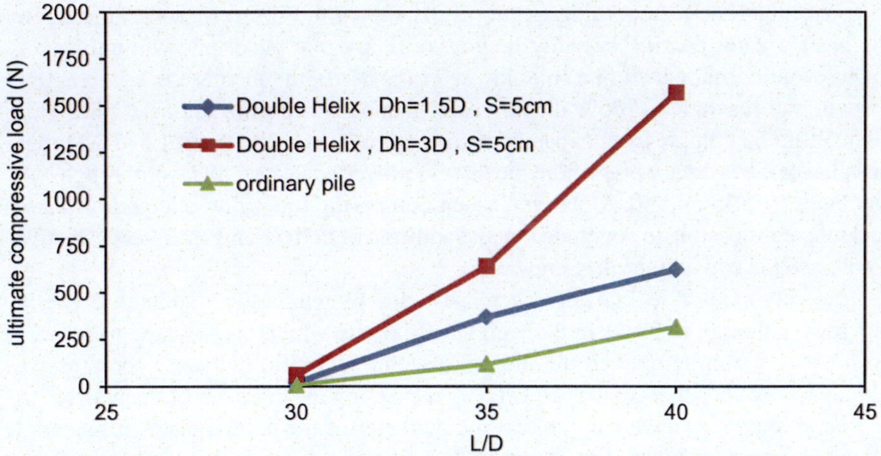

Fig. 8 The relation between the ultimate compressive capacity and l/d ratio for ordinary and double helix screw pile (s = 5 cm)

Figure 7, shows the effect of L/D ratio for double helix plates screw pile with inter helix spacing (s = 3 cm) and for different helix diameter on ultimate compression capacity. It can be noticed the amelioration in value of ultimate compression capacities of these screw pile have reached (38 times for 3D helix diameter) and (29 times when the diameter of helix plate equal to 1.5D) when increasing L/D ratio from 30 to 40. The increasing in L/D ratio from 30 to 35 for these piles increases ultimate compression capacity values ranging of between 15 and 31% while the increase in compression capacity values for L/D ratio 40 becomes equal to (1.7–2.3) times as much as the compression capacity values at L/D ratio 35.

Regarding the effect of L/D ratio for double helix screw pile with inter helix spacing (s = 5 cm) on ultimate compression capacity which can be noticed in Fig. 8. It can be concluded that the growth in value of ultimate compression capacities of these screw piles ranging 23.9–30.9% when increasing L/D ratio from 30 to 40. The increasing in L/D ratio from 30 to 35 for these piles increases ultimate compression capacity values ranging of between 9.2 and 18.6% while the increase in compression capacity values for L/D ratio 40 becomes equal to 1.5–2.4 times as much as the compression capacity values at L/D ratio 35. The screw pile installed at a greater depth had ultimate capacities greater than the shallower ones, and for a given diameter (D) and soft clay soil bed thickness (H) overlaying sandy layer, the ultimate compression capacity of screw pile increases with increasing embedded depth of screw pile in the layer of sand which are considered the stable zone. This behavior can be understood by the fact that when the screw pile is embedded in the sand layer, the anchorage becomes more mobilized, thereby increasing the compression capacity of the pile.

In general, increasing the length of screw piles leads to an increase its ultimate compressive capacity, that agrees with [15–17], from laboratory tests, the researchers found that the compressive capacity of screw piles increased with increasing depth of penetration (L/D). According to the results shown in Figs. 6, 7, and 8, the screw pile has a greater bearing capacity than an ordinary pile (steel pile without helix). For screw piles embedded in a soft clay layer (L/D 30), the increase in compressive strength can reach 125–800% of the value obtained using the ordinary pile. This augment in the compression capacity of the screw pile can be up to 1.5–7.75 times the value of the ordinary pile where the pile is embedded in a sandy layer through soft clay layer L/D 35 and 40. Moreover, when comparing the screw pile and ordinary steel pile, it is possible to conclude that the addition of helix diameter is very effective to enhance the capacity of the screw pile.

However, axial pile capacity increases clearly while the diameter of helix increases, although it grows in the weight of the pile which is estimated to be 3.2–9.6%. Also, it can be noticed that the increase in L/D ratio increases compressive pile capacity values that can be achieved by increasing the diameter of the helical. To sum up all the results, we can see that the diameter of the helical plates is essential influencing screw pile capacity, for any L/D ratio and pile geometry, we can see how the compressing capacity of the screw pile changes while changing the diameter of the helical plate. So the larger helical plate diameter provides greater compression capacity and can be attributed to the increased surface area that provides greater bearing capacity and greater anchorage for the screw pile when the diameter of helix increases, and that agrees with [18–21]. This behavior can be attributed to augment in surface area of the part of the screw pile embedded in sandy soil (augment in end bearing in addition to skin friction) that causes an increase in anchorage resistance.

Moreover, it is vital to conduct the compression tests on the model pile without a helical plate (conventional or ordinary steel pile) and with varying numbers of helix plates (single and double helix). Figure 9, clearly shows the influence of the number of helix plates on the ultimate compressive capacity of the ordinary and screw pile

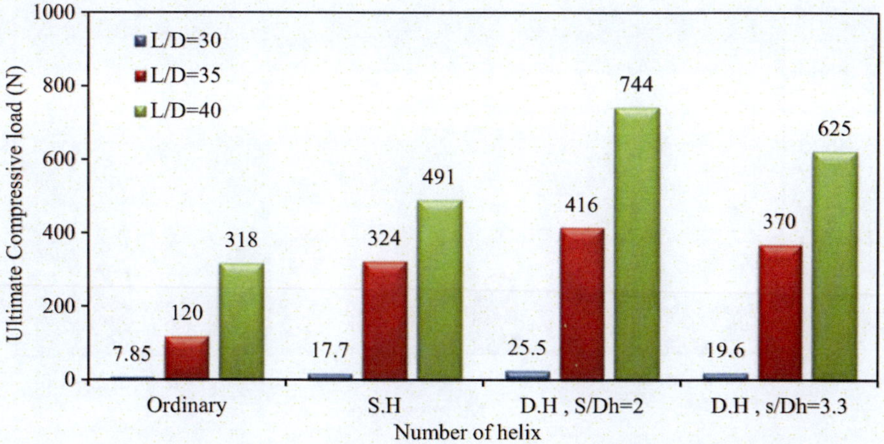

Fig. 9 The effect of helix number on the ultimate compressive capacity of ordinary and screw pile model with ($D_h = 1.5D$)

model with helix diameter $D_h = 1.5D$. It was found that the ultimate compressive capacity for L/D ratio 30 or for pile model embedded in soft clay soil rise to 2.25 times the value of ultimate compressive capacity of ordinary pile (steel pile without helix) when using single helix and 3.25 and 2.5 times when using double helix with s/D_h ratio 2 and 3.3. For L/D ratio 35 or pile model embedded 5 cm in the sandy layer through 30 cm soft clay later, the ultimate compressive capacity of the ordinary pile is 120 N and the growing in ultimate compressive capacity can reach to 2.7 times capacity of ordinary pile when using single helix pile and 3.5, 3.1 times when using double helix with s/D_h ratio 2 and 3.3. On the other hand, increasing the helix number of pile model embedded 10 cm in sandy layer through 30 cm soft clay layer L/D ratio 40 from ordinary to single and then to double s/D_h 3.3 and 2 leads to an increase the pile capacity to 1.5, 2.3 and 1.96 times the ultimate capacity of ordinary pile (without helix), respectively. Also, it can be deduced that the increase in number of the helix from single to double for all screw pile model with $D_h = 1.5D$ and $s/D_h = 3.3$ produces an increase in ultimate compressive capacity reached 44, 28, and 51% for L/D ratio 30, 35, and 40, respectively. While for the same diameter and s/D_h ratio 2 10.7, 14, and 27% for L/D ratio 30, 35, and 40, respectively.

The influence of the number of the helix on the ultimate compressive capacity of the ordinary and screw pile model with $D_h = 3D$ for different L/D ratio 30, 35, and 40 is shown in Fig. 10. It can be noticed that the double helix plate screw pile s/D_h ratio = 1.6, double helix plate s/D_h ratio = 1, and single helix exhibits increased ultimate compression capacity estimated at (8, 7.2, and 4) times compared to the ordinary pile capacity when L/D ratio 30. For the same condition and L/D ratio 35, the value of the ultimate compression capacity of the ordinary pile is 120 N and can be seen as this value increases to 5 times when using a single helix screw pile. When increasing the number of helix to double with s/D_h ratio 1 and 1.3 the values of

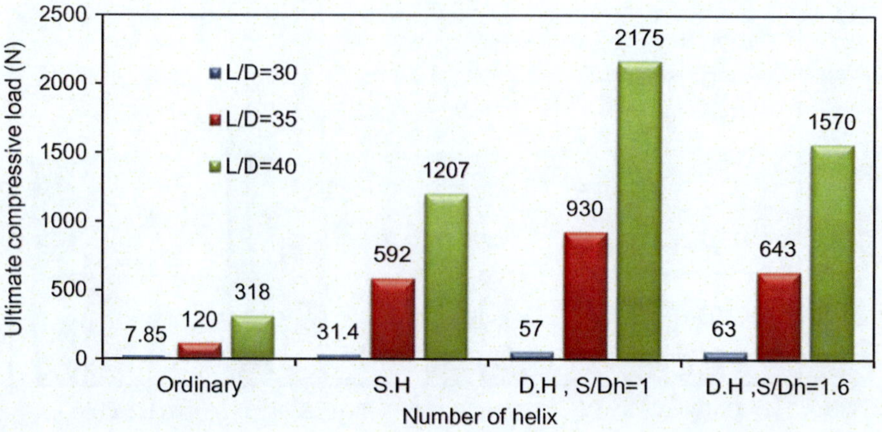

Fig. 10 The effect of helix number on the ultimate compressive capacity of ordinary and screw pile model with ($D_h = 3D$)

ultimate compressive capacity increased to 7.75 and 5.35 times the value of ultimate compression capacity of ordinary pile. For model pile embedded 10 cm in sandy layer through 30 cm soft clay layer L/D ratio = 40 the increase in number of helix from ordinary pile to single helix screw pile the ultimate capacity will raise 1207 N and this value equal to 3.79 times the value of ordinary pile capacity.

When using a double helix screw pile the ultimate compressive capacity will raise to 1570 N for s/D_h ratio 1.3 and 2175 N for s/D_h ratio 1 and that's mean 4.93 and 6.84 times the value of ultimate compressive capacity of the ordinary pile. Also, it can be deduced that the increase in number of the helix from single to double for all screw pile model with ($D_h = 3D$ and $s/D_h = 1$) produces an increase in ultimate compressive capacity, 81.5% for L/D ratio 30, 57% for L/D ratio 35 and 81.2% for L/D ratio 40 while for the same diameter and (s/D_h ratio 1.6) 100%, 8.6%, 30% for L/D ratio 30, 35 and 40 respectively. These values are higher than the values obtained when the diameter of the helix is 1.5 cm is approximately double in the case of ($s/D_h = 1$), and approximately similar when $s/D_h = 1.6$, except for the pile model that is embedded in soft clay soil, the larger spacing ratio s/D_h gives greater resistance to compressive loads and this can be attributed to the soil was confined in the space between the helical plates and gave a larger anchor resistance. The increase in the area of the cylindrical surface of double helix plates gives large resistance in compressive loading.

5 Conclusions

- The screw pile has a compressive capacity on soft clay soil greater than an ordinary pile.

- Single and double helix plates screw piles with can resist pile settlement 45–700% more than ordinary piles when screw piles embedded in dense sandy soil through soft clay soil.
- Adding a spiral plate to the pile shaft is very effective to improve the compression capacity of the screw pile compared to ordinary steel piles.
- Despite the increase in the weight of the screw pile when changing its geometry, the compressive capacity of the screw pile increases as compared with the ordinary pile.

References

1. Saride S, Puppala AJ, Chikyala SR (2013) Swell-shrink and strength behaviors of lime and cement stabilized expansive organic clays. Appl Clay Sci 85(1):39–45
2. Jegandan S, Al-Tabbaa A, Liska M, Osman AA-M (2010) Sustainable binders for soil stabilization. Proc ICE Ground Improv 163(1):53–61
3. Ali OK, Abbas HO (2019) Performance assessment of screw piles embedded in soft clay. Civil Eng J 8(5):1788–1798
4. Abbase HO (2017) Pullout Capacity of screw piles in sandy soil. J Geotech Eng 4(1):8–12
5. CHANCE® Civil Construction Helical Systems CC Catalogue, CD/Centralis, Missouri, USA, 2003.
6. Chapel TA (1998) Field investigation of helical and concrete piers in expansive soil. MSc thesis, Colorado State University
7. Pack JS (2009) Design and inspection guide for helical piles and helical tension anchors. Inc., Denver, Colorado, U.S.A.
8. Byrne BW, Houlsby GT (2015) Helical piles: an innovative foundation design option for offshore wind turbines. Phil Trans R Soc A373:20140081. https://doi.org/10.1098/rsta.2014.0081
9. Hamed M, Canakci H, Khaleel O (2019) Performance of multi-helix pile embedded in organic soil under pull-out load. Trans Infrastruct Geotechnol 6(1): 56–66. https://doi.org/10.1007/s40515-018-00069-0
10. Bowles JE (1996) Foundation design and analysis, 5th edn. McGraw-Hill, USA
11. Prakash S, Sharma HD (1990) Pile foundations in engineering practice. Wiley
12. ASTM D1143M-07 (2018) Standard test methods for deep foundations under axial compressive load. ASTM International, West Conshohocken, PA
13. Sakr M (2011) Installation and performance characteristics of high capacity helical piles in cohesion less soils. DFI J J Deep Found Inst 5(1):39–57. https://doi.org/10.1179/dfi.2011.004
14. Zhang DJY, Chalaturnyk R, Robertson PK, Sego DC, Cyre G (1998) Screw anchor test program (Part I & II): instrumentation, site characterization and installation. In: International proceedings of the 51st Canadian geotechnical conference, Edmonton
15. Mohajerani A, Bosnjak D, Bromwich D (2016) Analysis and design methods of screw piles: a review. Soils Found 56(1):115–128. https://doi.org/10.1016/j.sandf.2016.01.009
16. Javad K, Eslami A (2016) Geotechnical behavior of helical piles via physical modeling by Frustum Confining Vessel (FCV). Int J Geogr Geol 5(9):167–181. https://doi.org/10.18488/journal.10/2016.5.9/10.9.167.181
17. Abdel-Rahim HHA (2013) The compression and uplift bearing capacities of helical piles in cohesionless soil. J Eng Sci 41(6):2055–2064
18. Mahmood MR, Humaish AH, Khalaf MK (2018) Design and manufacturing loading rig machine for testing screw pile models. Int J Eng Technol 7(4.20):420–425s

19. Likhitha H, RaghavendraH, Rakesh K, Shrihari U (2017) The compression bearing capacity of helical piles in black cotton soil. Int J Innov Res Sci Eng Technol 6(7)
20. Hamza M, Ali Jawaid SM (2017) Helical pile subjected to vertical loading: theoretical study. Int J Sci Res Dev 5(2)
21. Sprince A, Pakrastins L (2009) Helical pile behavior analysis in different soils. Sci J Riga Tech Univ Constr Sci 10:121–130. https://doi.org/10.2478/v10137-009-0012-2

Experimental and Numerical Study on the Pullout Resistance of a Single and Group of Granular Pile Anchor (GPA) in Soft Soils

Karrar A. Alsaadi and Ala N. Aljorany

Abstract Granular pile anchor (GPA) is an innovative technique to resist pullout and uplift loads. In a (GPA) system, the footing is tied by a steel rod passes through the granular pile and connected to an anchor plate located at the bottom of the granular pile. In this study, a series of experimental tests have been performed on an individual and groups of GPAs that are installed in very soft clay with undrained shear strength $c_u < 25$ kPa. The pullout behavior of individuals and groups of GPAs are studied. The effects of length to diameter L/D ratio, length to the thickness of soft clay soil ratio L/H, the spacing between GPAs, and the number and the configuration of the GPAs in the group on its efficiency, are investigated, as well. The results showed that for individual GPA, the pullout resistance increases significantly as the L/D ratio increases from 2 to 10. The rate of increase becomes slower for L/D ratios greater than 6. For a group of GPAs, the results showed that the group pullout resistance increases significantly with increasing the GPA spacing and numbers of GPAs. The finite element package ABAQUS is used to simulate the problem and compare the obtained results with those experimentally measured. The numerical results demonstrate the pile shaft mode of failure for a relatively short GPA and the bulging mode of failure for a long L/D > 6 GPA.

Keywords Soft clay · Granular pile anchor (GPA) · Pullout resistance · Failure modes · Group efficiency

K. A. Alsaadi (✉) · A. N. Aljorany
Department of Civil Engineering, University of Baghdad, Baghdad, Iraq
e-mail: karrarali2018@gmail.com

A. N. Aljorany
e-mail: alaljorany@gmail.com

M. O. Karkush and D. Choudhury (eds), *Modern Applications of Geotechnical Engineering and Construction*, Lecture Notes in Civil Engineering 112,
https://doi.org/10.1007/978-981-15-9399-4_12

1 Introduction

Many structures are often subjected to uplift loading and usually designed to resist such type of loading. The main sources of the uplift loading are the wind loads on high rise towers, water waves on offshore and onshore structures, and lateral earth pressure that apply overturning moments on the earth retaining structures. Other sources of uplift loading are resulting due to soil swelling and buoyancy forces on underground structures below the water table. The effects of soil swelling and buoyance forces are more pronounced on lightweight structures. Several methods and techniques are used to overcome this problem, such as; ordinary piles, screw piles, and under-reamed piles. The GPA is one of the recently used techniques. In this technique, the granular pile which is commonly used to resist compression loads is modified by adding an anchorage system, to be capable of resisting the pullout loading.

The granular anchor pile footing (GPAF) system consists of a rigid plate (which acts as an anchor) that is placed at the granular pile tip and attached to a steel rod or cable that passes along the pile centerline and connected to the surface footing. Along this rod, the upward loads that are applied on the footing are transferred to the anchor plate at the tip of the pile where the confining pressure is rather high [1–3]. The main features of the GPAF are presented in Fig. 1.

Regarding the load transfer mechanism and modes of failure of GPA under pullout load, many investigators [5–9] have experimentally and numerically dealt with this problem. It is concluded that there are two modes of failure. In the first mode which is relevant to short GPA, the GP is pulled like that of a common (solid) pile. The ultimate pullout load in this mode is the sum of GPA weight plus the generated shaft resistance along the pile. The second mode of failure which occurs when the GP is

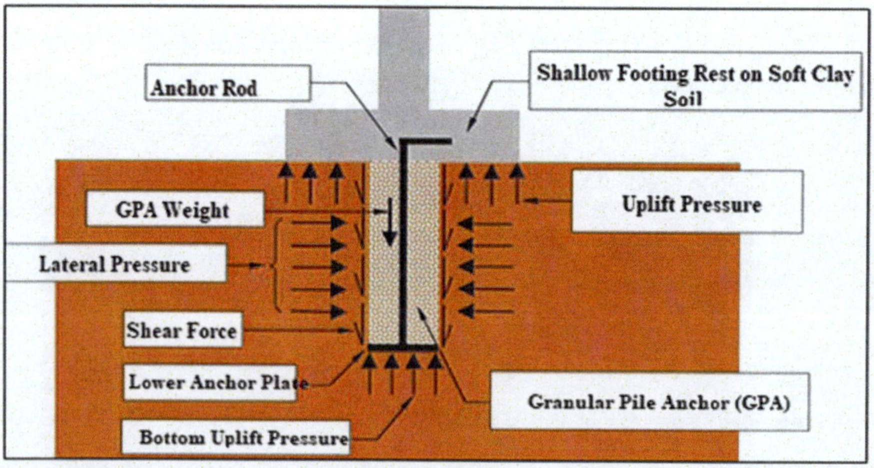

Fig. 1 Components of a granular pile anchor (GPA)

Fig. 2 Schematic graph of failure modes of GPA under uplift loads

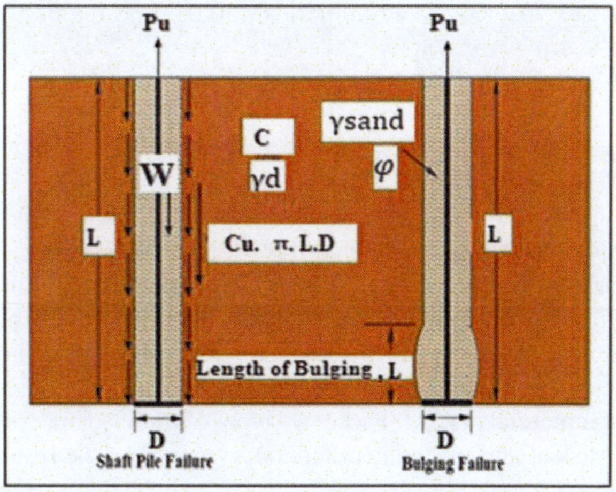

relatively long, is the bulging mode. It is similar to the bearing capacity mode of failure of the stone column under bearing pressure. The only difference is that in GPA, the bulging occurs near the GP tip. The ultimate bearing load of GPA will, therefore, be higher than that of the common stone column because the confining pressure at the pile tip level is much higher than that at the top of the stone column. Figure 2 is shown the schematic graph of failure modes of GPA under uplift loads. The main factor that controls the mode of failure of GPA is the length to diameter ratio (L/D) [4].

Many investigators have concluded that the change of failure mode from solid pile mode to bulging one occurs at L/D ratio from 4 to 8 depending on the strength parameters of the clayey soil and the granular material used in GPA. The following equations are derived [4] to calculate the ultimate pullout load capacity of GPA for both modes of failure:

$$P_{ult}\ (\text{Shaft Pile Failure}) = \gamma g.\pi.\frac{D^2}{4}.L + Cu.\pi.D.L \tag{1}$$

$$P_{ult}\ (\text{Bulging Failure}) = \frac{\pi}{4}\,D^2\,\left(\frac{(1+\sin\Phi)}{(1-\sin\Phi)}\,\left(6vc + \left(1 + \text{Log}\,\frac{Gur}{Cu}\right).Cu\right)\right) \tag{2}$$

where: Gu_r is the undrained shear modulus, Cu_b is the undrained shear strength at the base of GPA, φ is the angle of internal friction of the pile granular material, and σv_c is the lateral confining pressure. From the first equations, it can be noticed that the P_{ult} is in a direct proportion with the GPA length, but for the second equation, this proportion is implemented indirectly through the value of confining pressure. In this study, an experimental model is erected to investigate the pullout behavior of single and group of GPAs. Single GPA with different L/D ratios are tested to figure out the anticipated modes of failure. Also, a numerical analysis using the finite element method is done to compare the experimental results. The details of the experimental and numerical models and the used materials are exposed in the following paragraphs.

2 Experimental Work

Soft Clayey Soil and Granular Material. Disturbed and undisturbed soil samples of saturated soft clay were collected from the region of the Oil Deport at Al-Qadisiyah Province in the middle part of Iraq. The sample was obtained from a depth ranging from 1 to 2 m below the ground surface. The granular material that is used in this research is classified as poorly graded course clean sand that is brought from the commercial market. These two types of the used materials are subjected to the routine physical and mechanical laboratory tests and the results are shown in Tables 1 and 2.

Experimental Set-up. The experimental set-up of this study was designed to investigate the behavior of GPA installed in soft clay under pullout loading. This model consists of three main components:

- Steel box of dimensions 700 x 700 x 500 mm with stiffened sides to contain the soft clay.
- Loading frame to apply the pullout load at a controlled rate of loading of 1mm/min.
- The granular pile and anchorage system.

Table 1 Physical and mechanical properties of the used soft clay

Properties	Values
Liquid limit, LL (%)	33
Plastic limit, PL (%)	18
Plasticity index, PI (%)	15
Specific gravity (Gs)	2.76
Grain size distribution and hydrometer analysis	Sand (1%), Silt (37%), Clay (62%), CL
Standard compaction	γ_d (kN/m^3) (18.04), O.M.C (%) (13)
Undrained cohesion (c_u) (kPa)	18
Unconfined compressive strength (qu) (kPa)	36
Average undrained cohesion (c_u) (kPa)	21.5
Average undrained shear strength (Su) (kPa)	16
Initial void ratio (e_o)	0.83
Compression index (C_c)	0.1993
Swelling index (Cr)	0.0415
Coefficient of consolidation (Cs) (mm^2/sec)	1.907
Coefficient of volume change (avg) (m_v) (m^2/kN)	3.92E-4
Coefficient of permeability (k) (m/day)	2.333E-10

Table 2 Physical properties of the used sand

Properties	Values
Specific of gravity G.s	2.65
Maximum dry unit weight (kN/m^3)	21.5
Dry unit weight (kN/m^3) at RD = 85%	19.5
D_{10} (mm)	0.960
D_{30} (mm)	1.273
D_{60} (mm)	1.742
Coefficient of uniformity (c_u)	1.813
Coefficient of curvature (Cc)	0.894
Peak angle of internal friction (φ_{pk}) (degrees)	40

- Other instrumentations include two dial gauges of accuracy of 0.002 mm with the required holders and load cell of the model S-shape that was used for tension with the maximum load of 200 kg and a digital weighing indicator of the model SI-4010.

Figure 3 displays the schematic drawing of the experimental set-up. Before placing the soft soil in the box, several trial tests were conducted to get a relationship between the undrained shear strength and water content. Sufficient amount of dry clay soil

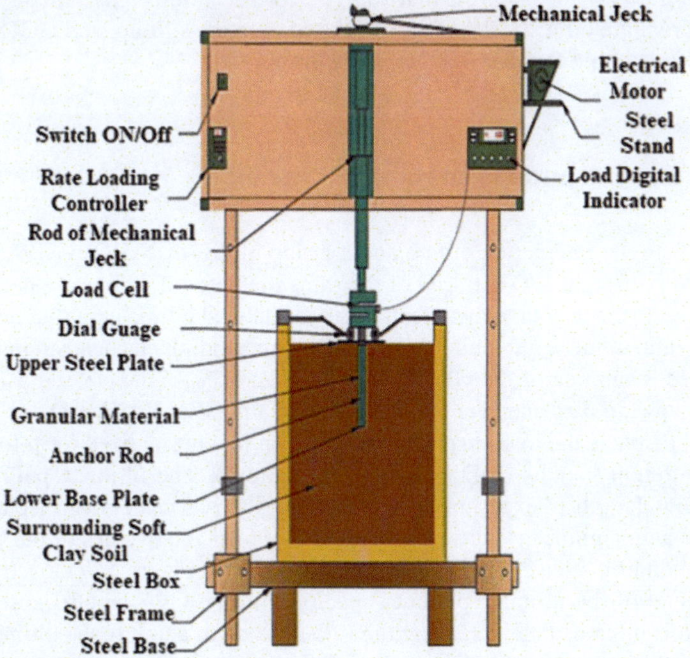

Fig. 3 Sketch of the experimental set-up

that was oven dried for 24 h and then different amounts of water contents were added according to the values of liquid and plastic limits that ranged from 24 to 38%. Six samples were prepared by using standard compaction mold after mixing different amounts of water content and left for 2 days as a curing period. The undrained shear strength was then measured by using two methods (unconfined compressive strength and vane shear apparatus). A relationship between the water content and the undrained shear strength was then obtained. The water content of about 28% that referred to undrained shear strength about 17 kPa by vane shear strength and 20.5 kPa by unconfined compressive strength is selected. Soft clay soil was placed in the box and compacted into five layers. This compaction was done by a specially designed hammer. The moisture content, the moist unit weight, and the undrained shear strength were checked at different locations and depths of the placed soil.

The moist unit weight was about 19.8 kN/m^3 and the dry unit weight was about 15.5 kN/m^3. The average undrained shear strengths were about 25 kPa. Granular material of GPA was installed in the soft clay by making a borehole using a hollow steel cylindrical with a diameter of 30 mm. The steel rod was then used for tamping the granular soil after pouring in the hole among five layers. For single GPA, the selected lengths were 2, 4, 6, 8, and 10 cm with a constant diameter of 3 cm. For a group of two GPA, three spacings between the GPAs were selected 2D, 4D, and 6D, where D is the diameter of GPA. A group of three GPAs and a group of four GPAs were then investigated as well. For the groups of GPAs, the diameter of each GPA was 3 cm with the same length of (18 cm). All the 10 model tests were performed for different L/D ratios. The tests that were performed in this research are denoted as SGPA for single GPA, GGPASD for GPAs of two piles group, and GGPAn for a group of GPAs with n number of piles in the group.

3 Tests Results and Discussion

Ten tests of single, interaction effect, and a group of GPAs (SGPA, GGPASD, and GGPAn) with different L/D and L/H of the same diameter of 3 cm were performed. Table 3 presents the results of these tests. The normalized ultimate pullout resistance (P_{ult}) is the ratio of measured pullout resistance to the undrained shear strength times the cross-sectional area of the GPA. The normalized (P_{ult}) is expressed as (Pu/As. c_u) and the upward displacement is normalized by the GPA diameter D.

Figure 4 displays the load-displacement relation in a normalized form for single GPAs of different lengths (L/D = 2, 4, 6, 8, and 10). The ultimate pullout load for each GPA length is experimentally realized when an excessive displacement occurs with an insignificant increase in the pullout load. The values of the ultimate load corresponding to each length are shown in Table 4. Figure 5 demonstrates the variation trend of the ultimate pullout capacity with the GPA length expressed as L/D ratio. It is obvious that increasing the L/D ratio up to about 6, the ultimate load is almost in a direct proportion of increase. This finding indicates a failure mode of pile shaft rather than bulging failure mode. For values of L/D greater than 6, the rate

Table 3 Description and results of the experimental model testing program.

Test type	Test symbol	L/D ratio	Length (L) (cm)	L/H ratio	S/D ratio	Pullout loading (N) (Experimental model)
Single Granular Pile Anchor (**SGPA**)	SGPA1	2	6	0.13		68
	SGPA2	4	12	0.27		155
	SGPA3	6	18	0.4		283
	SGPA4	8	24	0.53		310
	SGPA5	10	30	0.67		330
Group of the Granular Pile Anchor (The effect of interaction) (**GGPASD**)	GGPA2D	6	18	0.4	2	322
	GGPA4D				4	353
	GGPA6D				6	381
Group of the Granular Pile Anchor (The effect of number of GPA and Patterns) (**GGPAn**) n: No. of the GPAs	GGPA3	6	18	0.4	2	505
	GGPA4				2	757

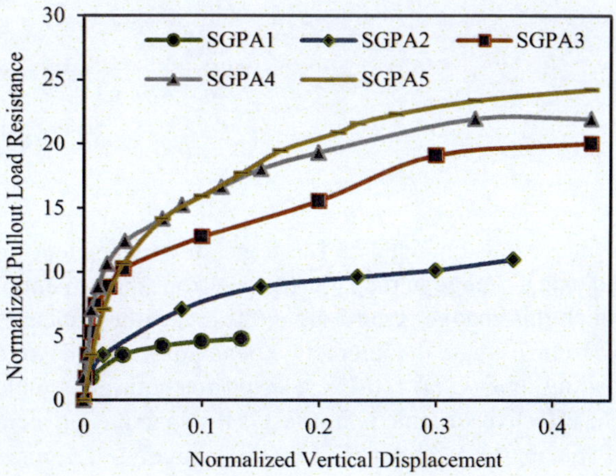

Fig. 4 Normalized pullout resistance versus normalized displacement for single GPA

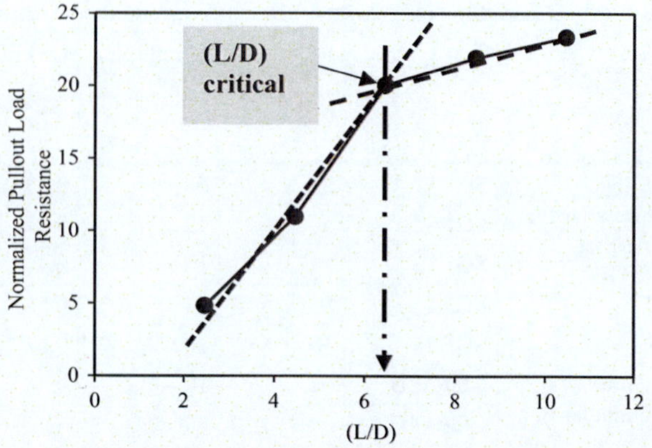

Fig. 5 Relationship between normalized experimental pullout load resistance and (L/D) values for SGPAs

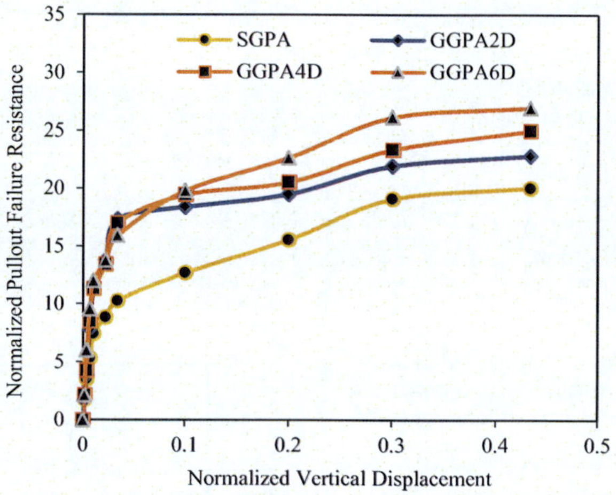

Fig. 6 Normalized pullout resistance versus normalized displacement for all interaction ganular anchor piles (GGPASD) compared with SGPA

of increase is changed and becomes less dependent on the GPA length. This may indicate a change in the failure mode from shaft to bulging. The value of L/D ratio at which this change occurs about 6 is called the critical L/D ratio as shown in Fig. 5.

To investigate the interaction behavior between two adjacent GPAs under pullout loading, one (L/D) ratio is selected for both piles which are the critical L/D ratio. The effective variable in this test is the spacing between the two GPAs. Three values of spacing to diameter (S/D) ratios, namely; 2, 4, and 6 are investigated. Figure 6 displays the relevant load-displacement relation of these three tests in addition to that related to a single GPA of the same L/D ratio for comparison. From this figure and the ultimate load values shown in Table 3, it can be noticed that the group ultimate load is much less than twice that of an individual GPA. By introducing the group efficiency which is defined as the percentage of the group pullout capacity to the sum

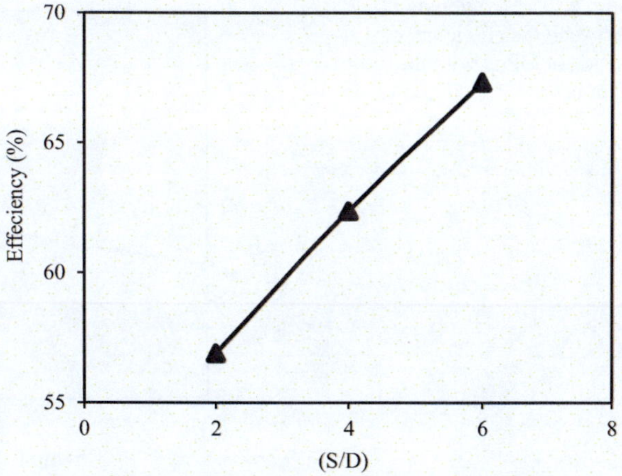

Fig. 7 Variation of the efficiency of group of two GPAs with the spacing between the Piles

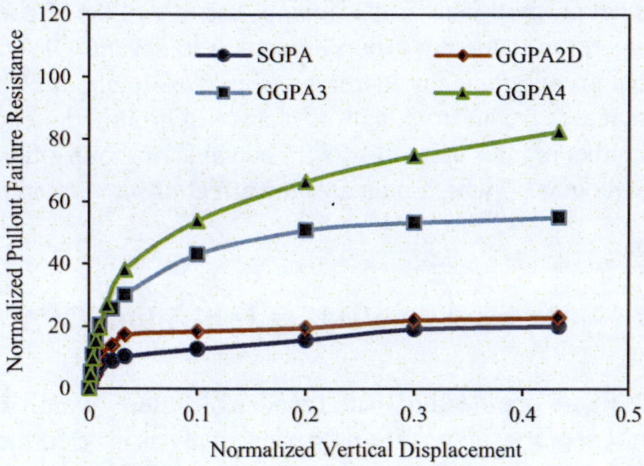

Fig. 8 The relationship between the normalized pullout failure resistance of GGPAN and the normalized vertical displacement

of that related to individual piles in the group, Fig. 7 is introduced. From this figure, it can be noticed that the group efficiency increases as the spacing between GPA becomes greater. It changes from 56% for S/D value of 2 to 67% for S/D equals 6. Since this value is less than 100%, the interaction between the GPA is still effective which implies that the diameter of the zone of influence of each GPA is greater than six times its diameter. In addition to the two GPAs group, a group of three equidistant GPAs in a triangular pattern and a group of four GPAs in a square pattern is tested, as well.

The length of all piles in these groups is 6 times the diameter and the spacing between the GPAs is two times the diameter. Figure 8 displays the load-displacement relation of each group of GPAs in addition to that of a single GPA of the same L/D ratio, for comparison. It is clear that the ultimate pullout capacity of each group

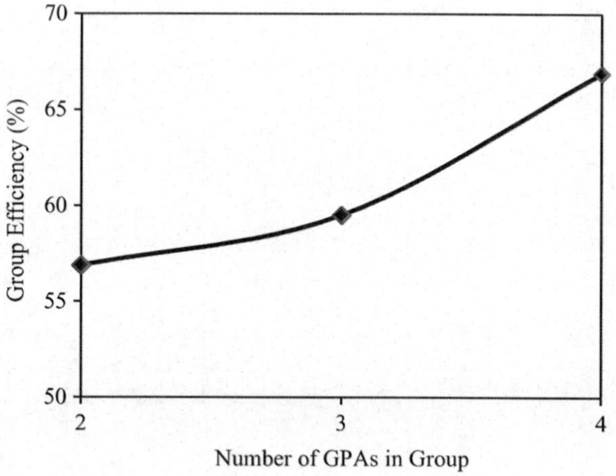

Fig. 9 The relationship between the efficiency of a group of GPAs with the number of GPAs in the group

increases as the number of GPAs in the group becomes greater but the rate of increase is not proportional. This is mainly because of the effect of interaction between piles. Figure 9 relates the group efficiency to several GPAs in the group. It is indicated that group efficiency increases from 57% to about 60% when the number of GPAs increases from 2 to 3, but the change in group efficiency becomes greater when the number of piles increases to 4. The value of group efficiency for the group of 4 GPAs is about 67% which indicates the effect of interaction between piles.

4 Numerical Analysis of Individual GPA

To figure out the anticipated modes of failure of an individual GPA and to compare the experimental results, numerical analysis is performed by using the finite element package ABAQUS. The problem is dealt with as an axisymmetric problem. The soft clay is modeled as an elastic-plastic material obeying the Mohr–Coulomb criterion which is changed to Von Mises criterion when φ becomes zero (to consider the undrained behavior). The granular material of the GP is modeled as elastoplastic materials using Drucker–Prager failure criteria. The dimensions and geometry of the problem are similar to those considered in experimental tests and the same L/D ratio is analyzed. Regarding the finite element mesh, all the used elements are 4-noded axisymmetric stress elements with biased node seeding at edges to get more accurate fine mesh at the regions around and below the base of GPA. The choosing of the boundary conditions was taken similar to that occurred in the experimental testing conditions in terms of loading and displacement.

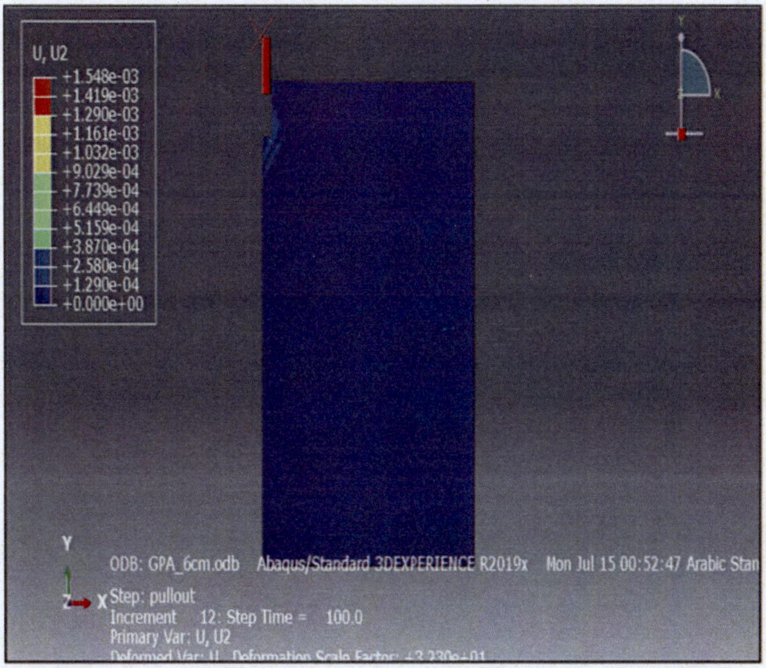

Fig. 10 Contours of the vertical displacements for SGPA1 (Shaft Failure) (Short Pile)

Figure 12 displays the variation of the calculated and measured values of the ultimate pullout capacity with L/D ratio. Both experimental and numerical results indicate the same trend of variation with some differences in the magnitudes. However, the value of a critical L/D ratio of about 6 is realized in both sets of results. For L/D ratio below 6, shaft failure mode is indicated by the FE analysis as shown in Fig. 10, which illustrates the contours of vertical displacement around a GPA with L/D ratio of 2. The contours of vertical stresses around a GPA with L/D ratio of 10 are displayed in Fig. 11. The mode of GPA failure of this long pile is the bulging mode. The extension of the bulging zone above the pile tip is ranging from 1 to 2 times the GPA diameter.

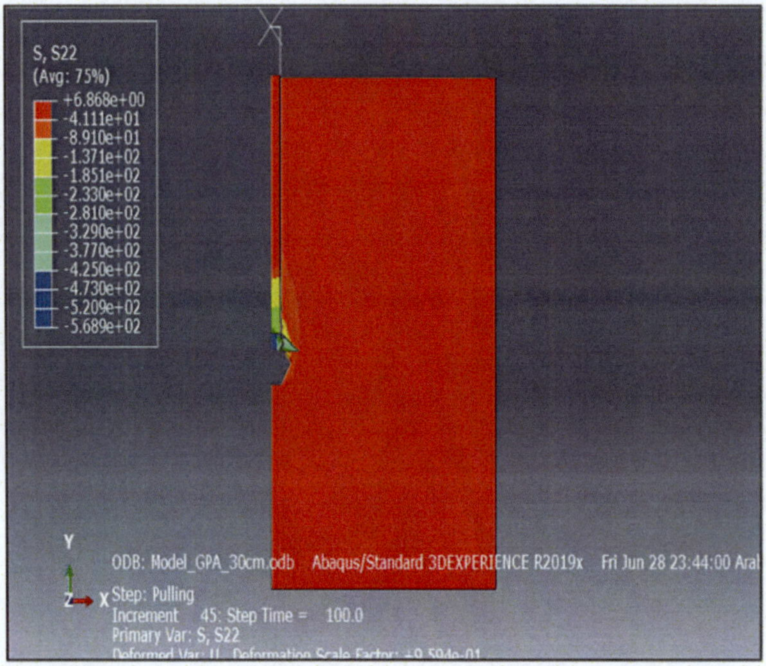

Fig. 11 Contours of the vertical stresses for SGPA5 (Bulging Failure) (Long Pile)

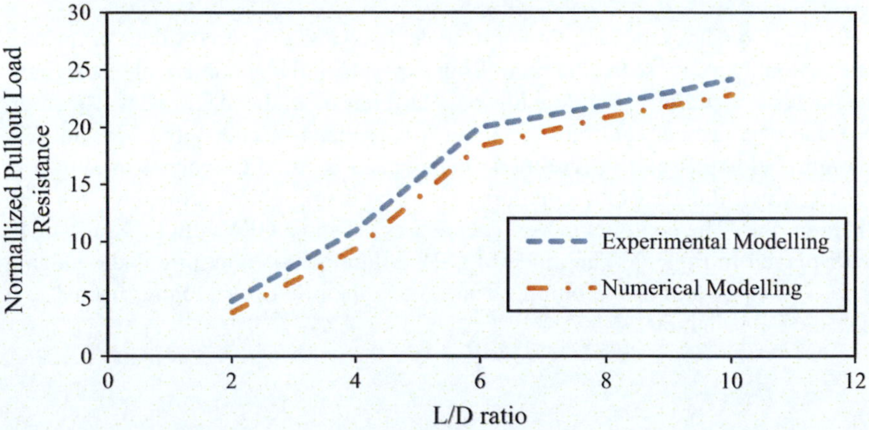

Fig. 12 The Relationship between the normalized pullout failure resistance and (L/D) ratio

5 Conclusion

- The uplift load resistance of GPA increases with its embedment depth but the rate of increase indicates a significant change at a certain depth. The critical depth at which this change occurs is about six times the GPA diameter.
- The diameter of the influence zone around the GPA under a pullout load is greater than six times the pile diameter.
- Within the investigated range spacing between GPAs in a group in this study, the maximum value obtained for group efficiency is about 70%.
- The mode of (Pile shaft) failure is realized by the FE analysis for short GPA (L/D < 6) while (Bulging) mode of failure is recognized for a relatively long GPA. The extension of the bulging zone above the pile tip is (1 to 2) times the GPA diameter.

References

1. Aljorany AN (2012) Modeling and analysis of the pullout behavior of granular pile anchor in expansive soils. In: ISSMGE-TC 211 Int. Symp. Gr. Improv, IS-GI Brussels
2. Phanikumar BR, Muthukumar M (2014) Reducing heave of expansive clay beds through granular pile–anchor groups. Proc. Inst. Civ. Eng. Improv. 167(2):99–107
3. B. R. Phanikumar, A. Srirama Rao, and K. Suresh, Field behaviour of granular pile-anchors in expansive soils, Proc. Inst. Civ. Eng. Improv., 161(4), 2008, 199–206.
4. O'Kelly BC, Brinkgreve RBJ, Sivakumar V (2014) Pullout resistance of granular anchors in clay for undrained condition. Soils and Foundation 54(6):1145–1158
5. B. R. Phanikumar and M. Muthukumar, Reducing heave of expansive clay beds through granular pile – anchor groups, 167(GI2), 2014, 99–108.
6. Muthukumar M, Shukla SK, Muthukumar M, Shukla SK (2016) Swelling behaviour of expansive clay beds reinforced with encased granular pile anchors. Int. J. Geotech. Eng. 6362(November):1–9
7. P. Kumar, M. Kumar, V.K. Chandaluri, and V. A. Sawant, Uplift Capacity of Single And Group of Granular Anchor Pile System, J. Civ. Eng., 9(1), 2018.
8. Abhishek and R. Sharma, A New Foundation Practice for Predicting the Behavior of Granular Pile Anchor Against Uplift Forces in Expansive Soils, in International Conference on Sustainable Waste Management through Design, 2018, 140–147.
9. E.H. Sfoog, A.J.L.M. Siang, W.M. Albadri, N. Naji, S.S. Yi, and N.A.A. Guntor, Finite Element Modeling of Innovative Shallow Raft Foundation with Granular Pile Anchor System for Expansive Clays, in IOP Conference Series: Materials Science and Engineering, 713(1), 2020, 12050.

Proposed Design Charts for Reinforced Concrete Spread Foundations Subjected to Concentric Load

Ali Hussein Ali Al-Ahmed, Methal Qudori Ali, and Hasan Hussein Ali

Abstract The design of reinforced concrete spread foundations mainly depends on soil bearing capacity, loading value, and column size. So for each design case, tiresome calculations and time consumption are needed. In this paper, generalized design charts are presented and plotted according to derivations based on the ACI 318 M-2019 Code. These charts could be used directly by the structural designers to estimate the column size, foundation thickness, and dimensions as well as the foundation reinforcement under a certain given concentric load assuming a uniformly distributed contact pressure underneath the foundation. Of noteworthy, these charts are oriented to deal with square isolated footings with a square concentric column, covering reasonable ranges of applied concentric load, column steel reinforcement ratio, and allowable soil bearing capacity. All derivations included in this research are based on assuming a practical concrete design compressive strength of 21 MPa and steel yielding stress of 420 MPa for both columns and foundations.

Keywords Design charts · Reinforced concrete · Spread foundations · Concentric load

1 Introduction

Foundation is considered one of the most important parts of a structure that transmits the load from superstructure to the soil. It spreads the load over a large area, so the delivered pressure on the soil does not exceed its allowable bearing capacity and

---

A. H. A. Al-Ahmed (✉)
Civil Engineering Department, College of Engineering, University of Baghdad, Baghdad, Iraq
e-mail: dr.ali-alahmed@coeng.uobaghdad.edu.iq

M. Q. Ali · H. H. Ali
Ministry of Higher Education & Scientific Research, Erbil, Iraq
e-mail: methalqudori@yahoo.com

H. H. Ali
e-mail: alhashmyhasan032@gmail.com

M. O. Karkush and D. Choudhury (eds), *Modern Applications of Geotechnical Engineering and Construction*, Lecture Notes in Civil Engineering 112,
https://doi.org/10.1007/978-981-15-9399-4_13

diminishes the settlement of the structure within the permissible limits. Foundations could be classified into shallow and deep foundations. There are substantial differences between them according to geometry, soil nature, and its constructive systems [1, 2]. A shallow foundation is a structural member where the dimensions of the cross-section are large in comparison to the height. Moreover, shallow foundations are provided when the ratio of height to width is less than one [3–7]. Design of shallow foundations in terms of the load as an axial load, an axial load and bending moment in one direction or an axial load and bending moment in two directions [4]. In practice, the spread (isolated) reinforced concrete (RC) foundation is considered one of the most common types of shallow foundations which withstand columns that transmit light to moderate loads from the structure.

The classical model for dimensioning of isolated footings is proposing and utilizing the hypothesis of the bidirectional bending to obtain the stresses acting on the contact surface of the isolated footings (i.e. trial and error), which is meet the condition: the minimum stress should be equal to or greater than zero, (due to the soil is incapable of withstanding tensile stresses and the maximum stress must be equal or less than the allowable capacity that can withstand the soil) [8]. Rojas et al. [9] developed a mathematical model to deal with the real contact soil pressure with the foundation. It was found that the classical model for spread footing design is larger than the proposed model. Abdrabbo et al. [10] conducted a correlation between failure loads predicted by different codes, such as ECP203-11, ACI 318 M-08, BS 8110.1-1997, and EC2-2004 of spread foundations, and the corresponding measured values. It was found that the shear span to foundation depth ratio and the distribution of contact stress at footing–soil interface are the key factors for isolated footing structural design.

Isolated footings were analytically investigated by Backhander and Prasad [11]. Their study was focused on the self-weight of footing concerning safe bearing capacity. They concluded that the increase in the safe bearing capacity led to reduce the self-weight of footing. In practical engineering design, the ACI 318 M-2019 Code [12] allows a rigid behavior for this type of foundation, results in a uniformly distributed contact pressure. Moreover, extensive calculations are needed to calculate the required depth of the footing and the proper reinforcement needed for each design case study. This paper aims to derive generalized equations used to present easy to use and comfortable design charts. These charts deal with square isolated foundations that withstand the square concentric column. So, for a certain given load and allowable soil bearing capacity, column size, footing dimensions, and thickness as well as the footing reinforcement area could be directly obtained. This might be helpful to facilitate the work of the designers and reducing time-consuming.

2 Design Stages for Isolated Footing

Design of RC isolated footing could be passing through several stages. These stages are sequence starting as follow:

1. Calculating axially loaded short column size depending on the given applied load and column steel ratio.
2. Calculating the footing area depending on the applied load and allowable soil bearing capacity.
3. Calculating footing thickness depending on the applied load, bearing capacity of the soil, and footing area which was prepared previously from stage 2.
4. Calculating the required reinforcement area to get proper design requirements.

Calculation of Column Size (c). Square RC column having size dimension c is adopted throughout this study. The ACI 318 M-2019 Code predicts the ultimate axial compression load capacity $\mathbf{P_u}$ as given by Eq. 1.

$$P_u = \varnothing \propto [0.85 f_c^{'} (A_g - A_{st}) + A_{st} f_y] \quad (1)$$

where A_g is the column sectional area ($A_g = c^2$ for square column). A_{st} is the total steel reinforcement area of the column given by Eq. 2.

$$A_{st} = \rho A_g \; \rho (\text{is the column steel reinforcement ratio}) \quad (2)$$

$\varnothing$and $\propto$ are column reduction factors = 0.65 and 0.80 respectively. $f_c^{'}$ and f_y are the specified concrete compressive strength and steel yielding stress, respectively. Practical values for $f_c^{'} = 21$ MPa and $f_y = 420$ MPa have been chosen in this study. Since column size is required, Eq. 1 could be rearranged and given by Eq. 3.

$$c = \sqrt{\frac{P_u}{0.52[0.85 f_c^{'} (1 - \rho) + \rho f_y]}} \quad (3)$$

By applying a reasonable range of ultimate load (0–4000) kN and column steel reinforcement ratio (0.01–0.04) to Eq. 3, Fig. 1 is plotted which represents the relation between the ultimate applied load and the required column size. It is worth to mention that the minimum column size is fixed to 200 mm according to requirements suggested by the ACI 318 M-2019 Code.

Calculation of Footing Dimensions (B). As a uniformly distributed contact pressure is permissible to be assumed according to the ACI Code, the dimension of the square foundation B could be calculated according to Eq. 4. Here, a full-contact pressure is applied and assumed to reach its maximum permissible limit q_a. In this case, the applied load would be the total non-factored load P_n which is satisfied by dividing P_u by 1.4 (this number represents the average factor of safety for dead and live loads).

$$B = \sqrt{\frac{\frac{P_u}{1.4}}{q_a}} \quad (4)$$

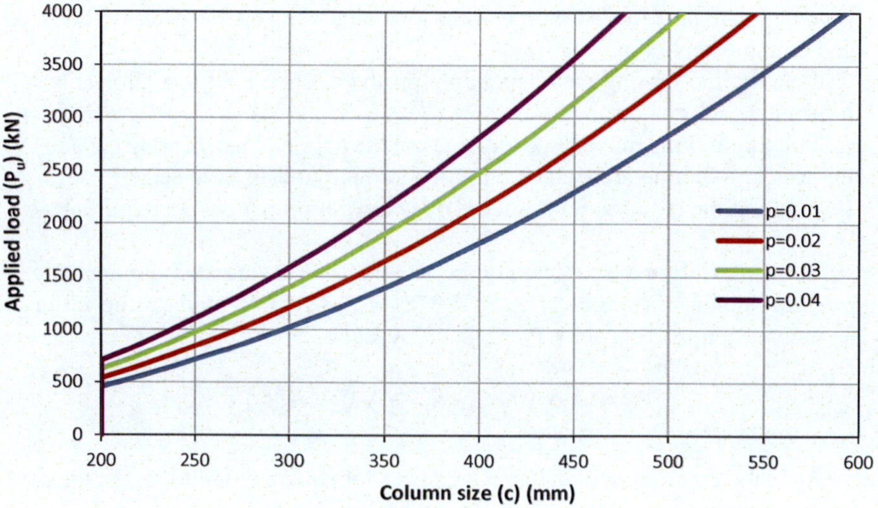

Fig. 1 Ultimate applied load (P_u) versus column size (c)

By applying a reasonable range of ultimate load (0–4000) kN and allowable soil bearing capacity (60, 80, and 100) kPa to Eq. 4, Fig. 2 is plotted which represents the relation between the ultimate applied load and the required footing dimension. It is worth to mention that the minimum dimension for square footings is fixed to 500 mm to be compatible with the minimum column size.

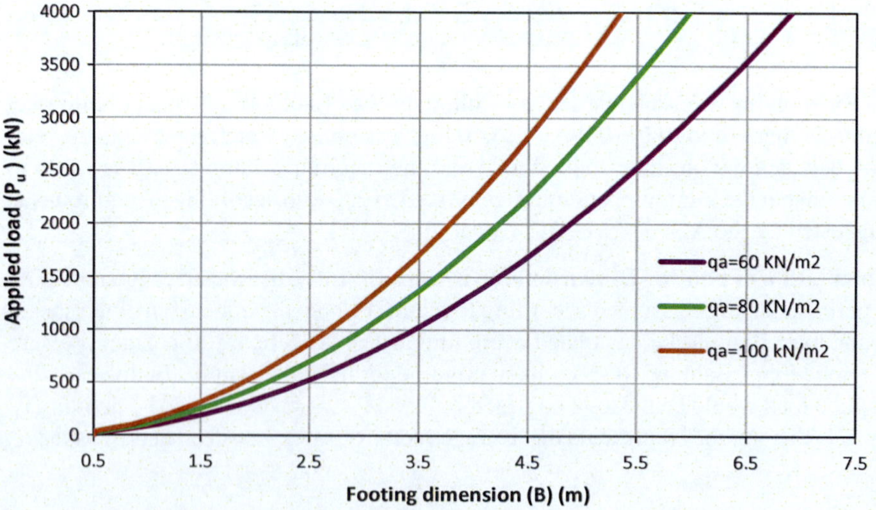

Fig. 2 Ultimate applied load (P_u) versus dimension of square footings (B)

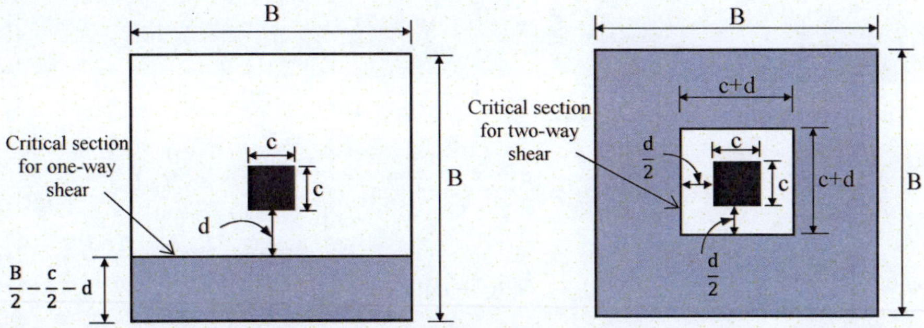

Fig. 3 Types of shear action in square RC isolated foundation

Calculation of Footing Effective Depth (d). The ACI-2019 Code provision [12] considers two types of shear action in RC isolated footing to calculate the required and adequate effective depth d of footing. These actions are one-way shear action and two-way shear action (punching shear). The critical shear force is assumed to occur at distances d and d/2 from the edge of the column for one-way shear and punching shear respectively as shown in Fig. 3. Moreover, the desired depth will be the maximum value obtained from the two types of shear.

One-way shear action according to the ACI 318 M-2019 Code. The critical section for one-way shear action is located at distances d from the edge of the column. In this section, the existing shear force V_c is given by Eq. 5. While the resisting shear force V_{cr} is given by Eq. 6 according to the ACI-2019 Code. Equating the two equations, yielding Eq. 7 which produces the required effective depth of isolated footings that could resist this type of shear action.

$$V_c = q_u B\left(\frac{B}{2} - \frac{c}{2} - d\right) \tag{5}$$

$$V_{cr} = \varnothing_v \frac{1}{6}\sqrt{f_c'}Bd \tag{6}$$

$$d = \frac{q_u\left(\frac{B-c}{2}\right)}{\left(\frac{\sqrt{f_c'}}{8} + q_u\right)} \tag{7}$$

where $\varnothing_v$ is the shear reduction factor equal to 0.75 and q_u represents the ultimate contact pressure which equals the ultimate load P_u divided by the footing area B^2.

Two-way shear action according to the ACI 318 M-2019 Code. The critical section for two-way shear action is located at distances d/2 around column edges. In this section, the existing shear force V_c is given by Eq. 8. While, the resisting shear force V_{cr} is given by Eqs. 9, 10, and 11 proposed by the ACI 318 M-2019 Code.

$$V_c = q_u[B^2 - (c + d)^2] \tag{8}$$

$$V_{cr1} = \varnothing_v \frac{4}{3}\sqrt{f'_c}(c + d)d \tag{9}$$

$$V_{cr2} = \varnothing_v \frac{4}{6}\sqrt{f'_c}(c + d)(1 + \frac{2}{\beta})d \tag{10}$$

$$V_{cr3} = \varnothing_v \frac{4}{12}\sqrt{f'_c}(c + d)[2 + \frac{\alpha_s d}{4(c + d)}]d \tag{11}$$

where β is column shape factor (equals 1 for square column) and α_s is column location factor (taken 40 for concentric column). According to the ACI-2019 Code, the lowest value of the resisting shear force would be taken and equating with the existing shear force. For the case of an isolated square foundation with a square concentric column, Eq. 9 always achieves the lowest value. Thus, Eq. 12 is written to give the required effective depth for isolated square footings to resist this type of shear action.

$$d = \frac{\sqrt{c^2(2q_u + \sqrt{f'_c})^2 - 4q_u(c^2 - B^2)(q_u + \sqrt{f'_c})} - c(2q_u + \sqrt{f'_c})}{2(q_u + \sqrt{f'_c})} \tag{12}$$

Square Spread Foundation Thickness (t). The total thickness of isolated square footing t could be obtained by adding about 100 mm (concrete cover plus steel reinforcing diameter) to the effective depth performed from the maximum value obtained from one-way and two-way shear action. (i.e. Equation 7 and Eq. 12). Figures 4, 5, and 6 show the variation of footing thickness with the ultimate applied load for different values of allowable soil bearing capacity and different ranges of column steel ratio. Moreover, the minimum footing thickness is fixed to 200 mm to fit with other footing dimensions.

Calculation of Foundation Steel Reinforcement Area (A_{sf}). Figure 7 shows the critical section for bending moment M_u in isolated foundation which occurs at the face of the RC column. This moment per meter width could be calculated according to Eq. 13. Moreover, the required steel reinforcement A_{sf} is given by Eq. 14. While minimum steel reinforcement is limited according to Eq. 15. The ACI 318 M-2019 Code states that the maximum value from the two equations is reliable.

$$M_u = \frac{q_u(\frac{B-c}{2})^2}{2} \tag{13}$$

$$A_{sf} = \left(\frac{1 - \sqrt{1 - \frac{2.36\frac{M_u}{\varnothing_b}}{f'_c bd^2}}}{1.18\frac{f_y}{f'_c}}\right)bd \tag{14}$$

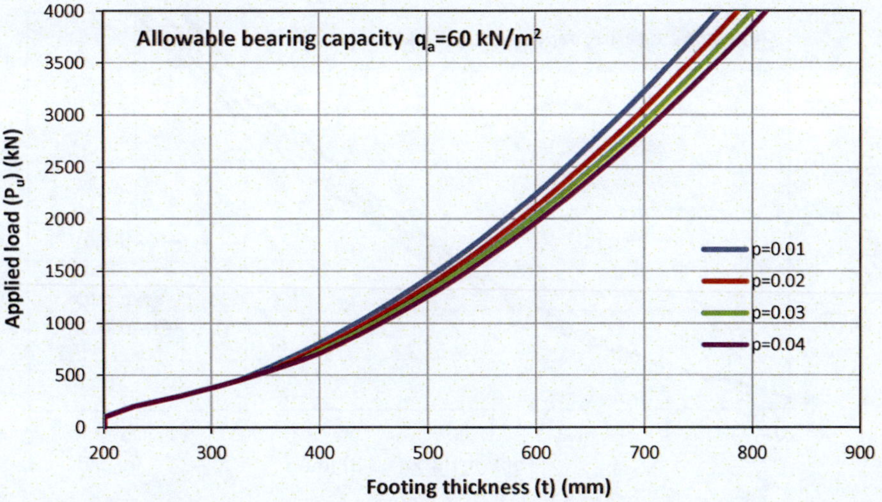

Fig. 4 Ultimate applied load (P_u) versus square footing thickness (t) for $q_a = 60$ kPa

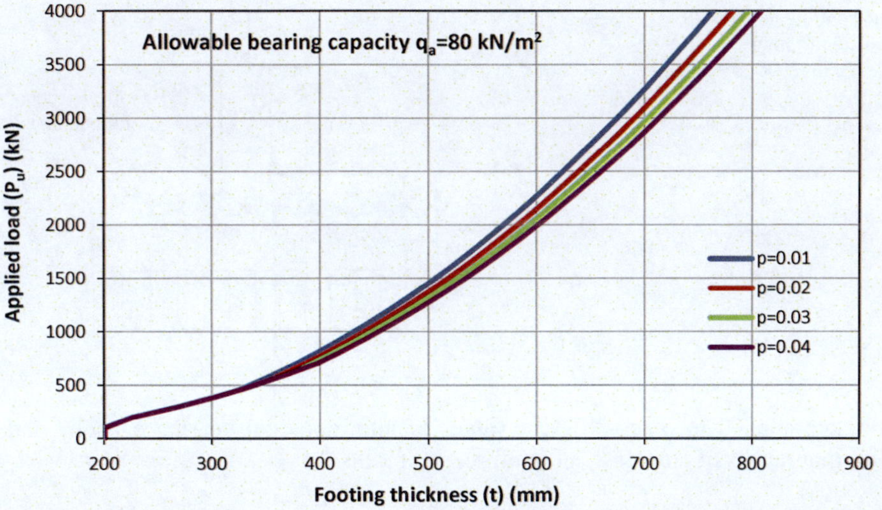

Fig. 5 Ultimate applied load (P_u) versus square footing thickness (t) for $q_a = 80$ kPa

$$A_{s,min} = 0.0018bt \tag{15}$$

where $\varnothing_b$ is the bending reduction factor taken 0.9 and b is one-unit width of footing. Figures 8, 9, and 10 show the required footing reinforcement corresponding to the

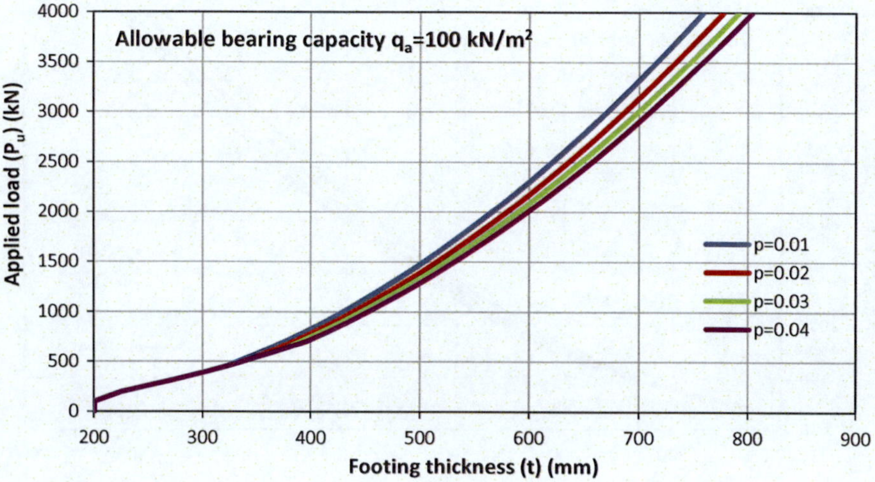

Fig. 6 Ultimate applied load (P_u) versus square footing thickness (t) for $q_a = 100$ kPa

Fig. 7 Critical section for the bending moment in isolated footing

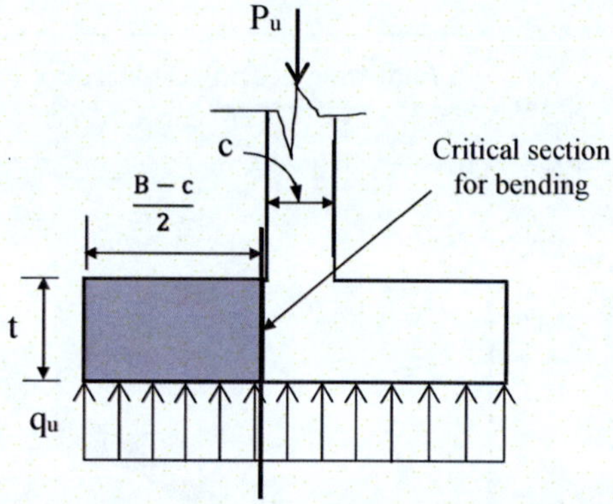

ultimate applied load for different values of allowable soil bearing capacity and different ranges of column steel reinforcement ratio.

3 Conclusions

From the obtained results presented in this study which are mainly depended on the proposed limitations for square isolated foundation under concentric loading which is ranged from 0 to 4000 kN, main conclusions could be drawn as follow:

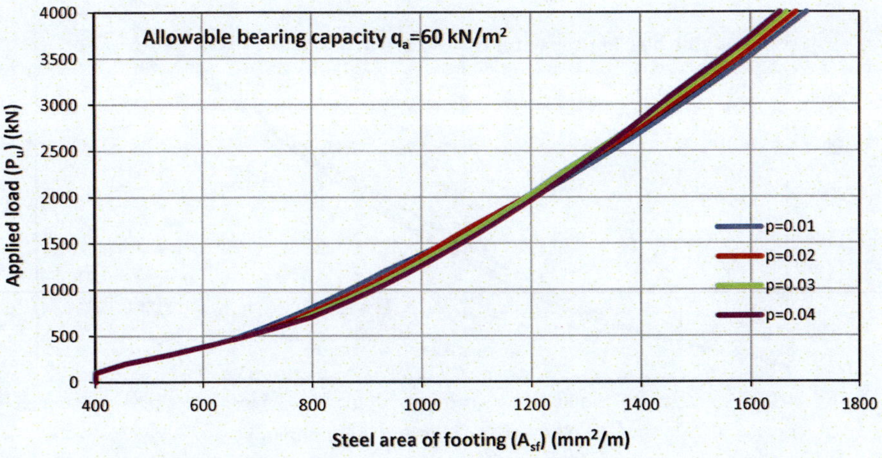

Fig. 8 Ultimate applied load (P_u) versus footing reinforcement (A_{sf}) for $q_a = 60$ kPa

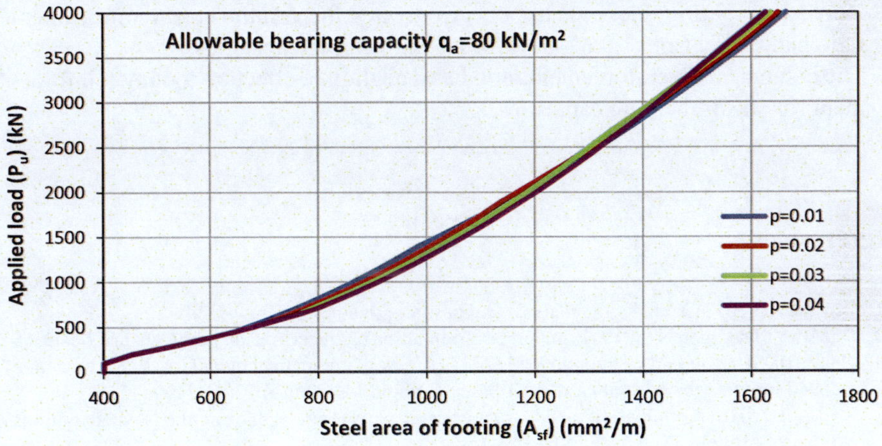

Fig. 9 Ultimate applied load (P_u) versus footing reinforcement (A_{sf}) for $q_a = 80$ kPa

- In general, as the column steel reinforcement ratio is increased, a significant decrease in column size occurs. Increasing this ratio from 1.0% to 4.0% led to reduce the column size by about 20%.
- In general, increasing allowable soil bearing capacity led to reduce footing dimension. When soil bearing capacity is increased from 60 to 100 kPa, about 23% decreasing in the footing dimension is maintained.
- The variation of column steel reinforcement ratio had a slight effect on footing thickness. In general footing, thickness is increased with increasing column steel reinforcement ratio. For concentric load up to 500 kN, a negligible change in footing thickness has occurred. While, for more than 500 kN load, about a 6%

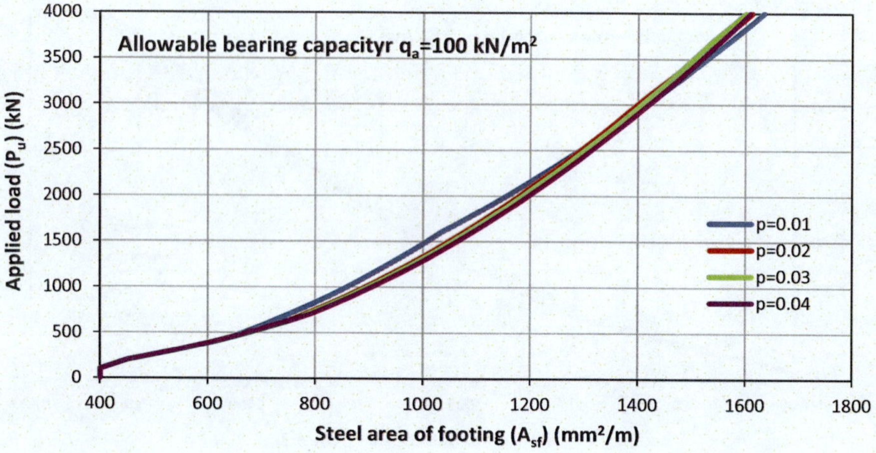

Fig. 10 Ultimate applied load (P_u) versus footing reinforcement (A_{sf}) for $q_a = 100$ kPa

increase in footing thickness has occurred when the column steel reinforcement ratio had varied from 1.0 to 4.0% for all soil bearing capacities.

- The column steel reinforcement ratio has a negligible effect on footing reinforcement for all soil bearing capacities.

References

1. Bowles JE (2017) Foundation analysis and design, 5th edn. McGraw-Hill
2. Das BM, Sivakugan N (2019) Principles of foundation engineering, 9th edn. Cengage Learning
3. Chavarrìa SL, Rojas AL, Elizondo MM (2017) A new mathematical model for design of square isolated footings for general case. Int J Innov Comput Inf Control 13(4): 1149–1168
4. Rojas AL (2016) A comparative study for the design of rectangular and circular isolated footings using new Models. DYNA 83(196):149–158
5. Magade SB, Ingle RK (2019) Numerical method for analysis and design of isolated square footing under concentric loading. Int J Adv Struct Eng 11(1):9–20
6. Magade SB, Ingle RK (2019) Comparative study of moments with plate and solid elements for an isolated footing under axial load. In: International conference on innovation in concrete for infrastructure challanges
7. Rojas AL (2014) Design of isolated footings of circular form using a new model. Struct Eng Mech 52(4):767–786
8. Tabsh SW, Al-Shawa AR (2005) Effect of spread footing flexibility on structural response. Pract Period Struct Des Constr 10:109–114
9. Rojas AL, Herrera JG, Vallejo RA, Alvarez MA (2013) Design of isolated footings of rectangular form using a new model. Int J Innov Comput Inf Control 9(10):4001–4021
10. Abdrabbo F, Mahmoud ZI, Ebrahim M (2016) Structural design of isolated column footings. Alex Eng J

11. Balachandar S, Prasad DN (2017) Analysis and design of various types of isolated footings. Int J Innov Res Sci Eng Technol 6(3):3980–3986
12. ACI Committee 318M (2019) Building code requirements for structural concrete (ACI 318M-19) and commentary. American Concrete Institute, Farmington Hills, 623

Three-Dimensional Modeling of Laterally Loaded Pile Embedded in Unsaturated Sandy Soil

Maha H. Abood, Mahmood R. Mahmood, and Nahla M. Salim

Abstract A finite element formulation for the static, cyclic, and dynamic analyses of the laterally loaded single pile model embedded in sandy soil was carried out. The sandy soil was brought from Baghdad; the state of soil changes from dry, saturated, and unsaturated conditions. The analysis was performed by using "ABAQUS 3D V.6.14". The program validity was checked by using experimental results. The soil follows Drucker–Prager failure criteria. The pile head displacement, deflection, and soil resistance for different cases were chosen for comparison. The finite element results of the pile under static load and embedded in dry, saturated, and unsaturated sandy soil are compatible with the experimental results. Furthermore, in the case of cyclic load, the results are compatible with experimental results for saturated and unsaturated conditions while there is a divergence in the results between experimental and finite element results when the pile is embedded in dry soil. A parametric study was carried out by increasing the frequency of the cyclic load from 0.5, 1.0, and 1.5 Hz (only 0.05, 0.1, and 0.2 used in the experimental study). The results indicated that as the frequency of the load increased, the pile deflection and soil resistance increased and an increment in the deflection approaches 12% for each increment in frequency value.

Keywords Abaqus · Lateral loading · Unsaturated soil · Deflection · Mobilizes soil stresses

M. H. Abood (✉) · M. R. Mahmood · N. M. Salim
Civil Engineering Department, University of Technology, Baghdad, Iraq
e-mail: civil_eng001@yahoo.com

M. R. Mahmood
e-mail: mahmoudal_qaissy@yahoo.com

N. M. Salim
e-mail: Nahla_salim2007@yahoo.com

M. O. Karkush and D. Choudhury (eds), *Modern Applications of Geotechnical Engineering and Construction*, Lecture Notes in Civil Engineering 112,
https://doi.org/10.1007/978-981-15-9399-4_14

1 Introduction

Pile foundations are used to support the heavy structures by carrying the applied load to deeper or stronger layers and also reinforcing the soil. Piles are also subjected to lateral loads in case of foundations of bridges, transmission towers, offshore structures, and other types of huge structures. The pile foundations' lateral load resistance is important in the design of structures under loading from earthquakes, soil movement, waves, etc. [1]. The maximum deflection of the pile is the major criterion on the design according to [2]. The common theory of soil mechanics treats soil as either dry or fully saturated. The recognition of the difference between saturated and unsaturated soils becomes necessary due to basic differences in their nature and engineering behavior. A lot of factors that influence the soil condition such as climate, which considered one of these factors that play an important role in whether a soil is saturated or unsaturated [3].

For two-way cyclic loading, the results show an increase in the accumulated deformation due to the changing loading direction. The accumulated displacement is not limited to the main loading direction but includes transverse movement as well [4]. Three-dimensional linear and nonlinear finite element analyses are gradually becoming commonplace for geotechnical problem simulation. The available computational power continues to provide a more realistic representation of the actual involved geometric scenarios. In the present paper, nonlinear three-dimensional finite element analysis is carried out to investigate the behavior of a single pile subjected to different lateral loads.

Abaqus built-in features and Abaqus extensions can be applied to solve a wide range of boundary value problems in the field of soil mechanics by the use of subroutines and coupling with external programs. This offers the researchers a tool to examine some of the open problems in geomechanics like the influence of construction processes for geotechnical constructions, the optimization of geotechnical constructions regarding deformations, or the optimization of construction processes [5]. The soil used in this study is fine sandy soil with unit weight $\gamma = 15.5$ kN/m^3 for medium relative density (Dr = 50%) and the angle of internal friction was Ø = 34°.

An aluminum single pile model of 16 mm diameter and 500 mm length was used to conduct all the tests. A steel container of 600 mm diameter and 600 mm depth with four valves fixed at different levels to achieve partial saturation of the soil was used, and the pile model was embedded to a depth of 500 mm within the soil. The unsaturated state in this study was obtained by supplying water to the soil in the container from a valve in the bottom of the container as shown in Fig. 1 and was left to saturated for 24 h until reaching the state of fully saturated soil, and the other valves were used for lowering the water table to a specific level, to obtain unsaturated conditions with different matric suction values by the use of the tensometers [6]; the suction was10 kPa according to the tensometer reading in the laboratory. The two-way cyclic lateral loading is applied by a digital weighing indicator used to display the lateral loading. The load was applied to the single pile model by using two rods

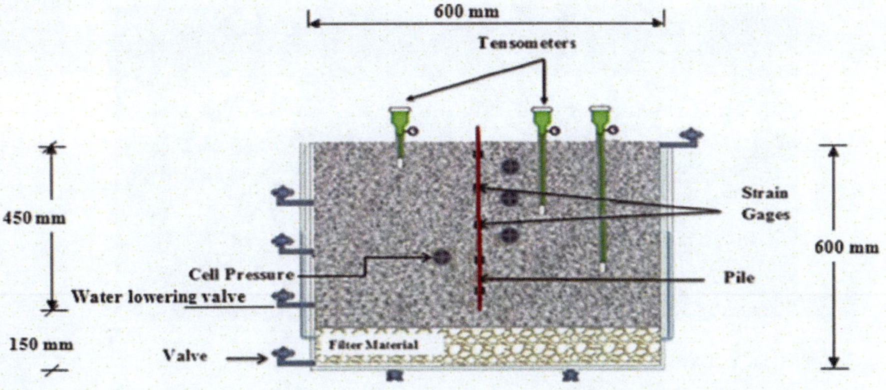

Fig. 1 Sketch shows the model test

of (19 mm) diameter with rubber pads attached ends of rods, both rods are connected to 150 kg load cell capacity.

2 Numerical Simulation

The experimental laboratory model was simulated by the Abaqus program to verify the FEM models' results with the experimental models' results in addition to making a comparison between the FEM models' and experimental models' behavior. 3D analysis was used in the numerical simulation analyses; three cases were selected to investigate and simulate the experiment by the finite element model which are.

- Single pile model subjected to static lateral load of 80 N, 120 N, and embedded in dry, saturated, and unsaturated sandy soil.
- Single pile model subjected to cyclic lateral load of 60 N and embedded in dry, saturated, and unsaturated sandy soil with different frequencies (0.05, 0.1, and 0.2 Hz).
- Single pile model subjected to the dynamic lateral load applied as El Centro 1940 NS with three different accelerations, 0.05 g, 0.15 g, and 0.32 g.

The numerical model has the same dimensions of the experimental laboratory model as shown in Fig. 2. The soil is presumed to follow a Drucker–Prager failure criterion. Material properties utilized in the numerical analysis are listed in Table 1.

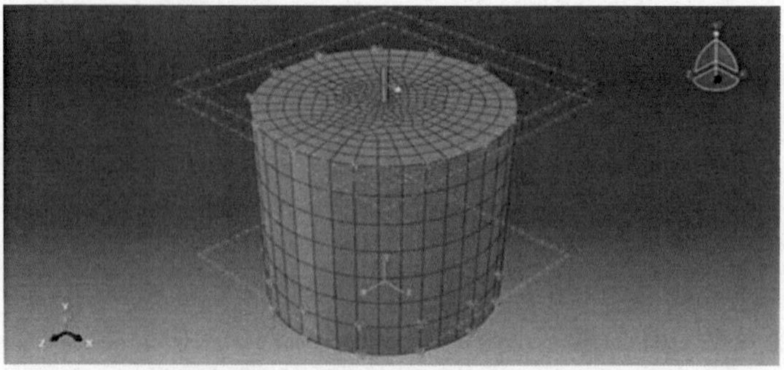

Fig. 2 Mesh and parts assembly of simulated laboratory numerical analysis model

Table 1 The utilized numerical analysis of material properties

Parameter	Dry sand	Saturated sand	Unsaturated sand	Pile
Material model	Drucker–Prager			Linear elastic
Unit weight (kN/m^3)	15.45	19.53	17.44	27
Modulus of elasticity (MN/m^2)	20			67 * 10^3
Friction angle (ϕ)	34°	30°	39° (10 kPa suction)	–
Dilatancy angle (ψ)	4	0	9	–
Poisson's ratio (ν)	0.3	0.3	0.3	0.3
Initial void ratio ($e_{initial}$)	0.58	0.58	0.58	–

2.1 *Results and Discussion of the Numerical Analysis of Simulated Laboratory Model*

To show the results of the numerical analysis of the simulated laboratory models with the experimental laboratory results, the same measurement positions in the experimental models were used. The pile head displacement, deflection, and soil resistance for selected cases were chosen for comparison. Figure 3 shows the results of the experimental measurement and the numerical evaluation of the pile head displacement for the single pile model under lateral static load and embedded in (dry, saturated, and unsaturated) sandy soil. Figure 4 shows experimental and numerical results of deflection for the single pile model subjected to static lateral load embedded in (dry, saturated, and unsaturated) sandy soil, subjected to lateral static load (a) 80 N and (b) 120 N.

Figure 5 shows the experimental and numerical results of mobilized soil resistance for the single pile model subjected to static lateral load embedded in (dry, saturated, and unsaturated) sandy soil. It can be noted that the pile head displacement and soil stresses in experimental laboratory models have lower values than the

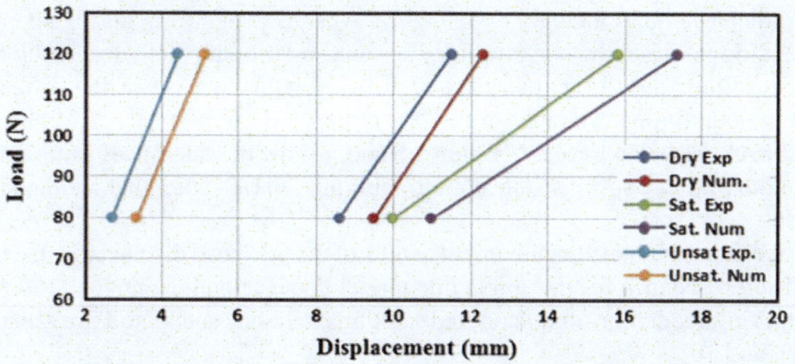

Fig. 3 Experimental and numerical result evaluated pile head displacement for the single pile model under lateral static load (80 N and 120 N) embedded in sandy soil

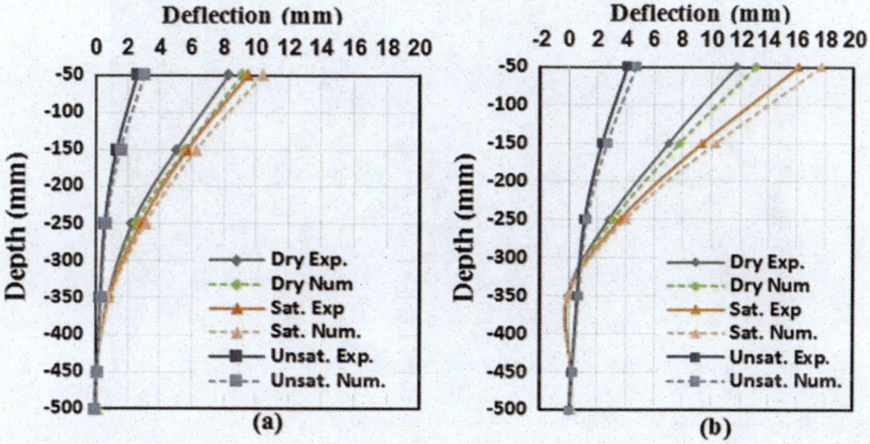

Fig. 4 Experimental and numerical results of deflection for the single pile model subjected to static lateral load embedded in (dry, saturated, and unsaturated) sandy soil subjected to lateral static load **a** 80 N and **b** 120 N

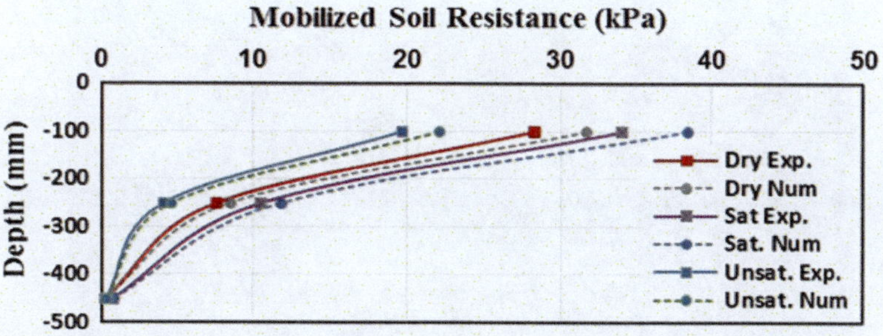

Fig. 5 Experimental and numerical results of mobilized soil resistance for the single pile model under lateral static load and embedded in (dry, saturated, and unsaturated) sandy soil

predicted displacement obtained by using the Abaqus program in all the soil saturation conditions. Also, the soil stresses measured by pressure cells located at three different positions are differ than measured by Abaqus because of the difference between the conditions available in the laboratory and the perfect condition taken by the Abaqus as shown in Fig. 5. Figures 6 and 7 show the lateral deformation that occurs in the pile and soil resistance after applying lateral static load for the single pile model.

Also, the experimental and numerical results of the pile head displacement, deflection, and soil resistance for the single pile model under lateral cyclic load (60 N for 100 cycles) embedded in (dry, saturated, and unsaturated) sandy soil are shown in Figs. 8, 9, and 10.

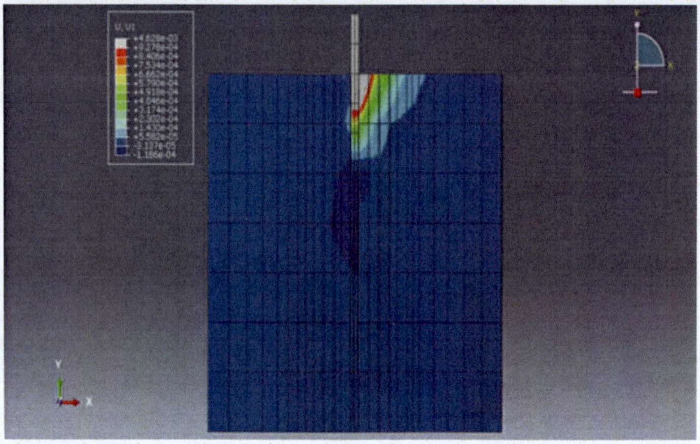

Fig. 6 The resistance of soil by the effect of lateral static load

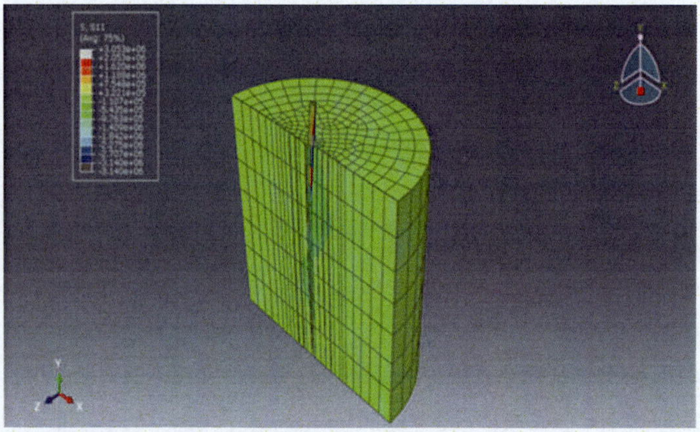

Fig. 7 The deformation of the pile under the effect of lateral static load

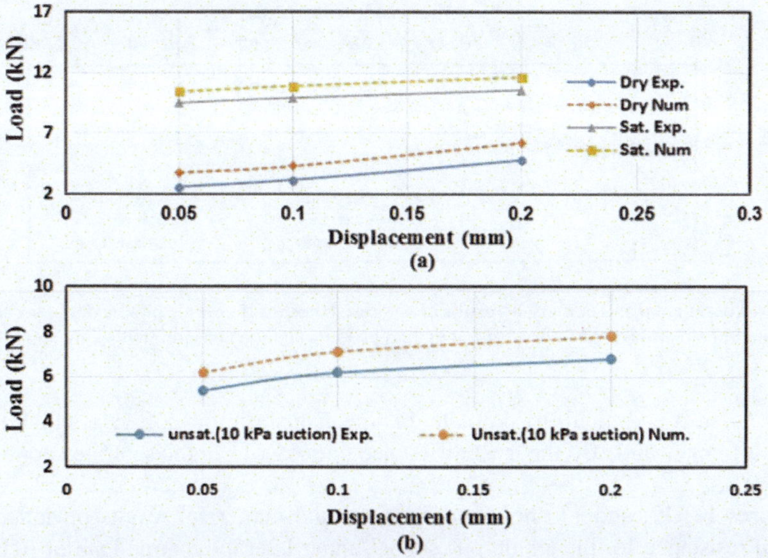

Fig. 8 Experimental and numerical results of pile head displacement for the single pile model under lateral cyclic load (60 N) with (0.05, 0.1, and 0.2 Hz) frequency embedded in **a** dry and saturated sandy soil **b** unsaturated sandy soil

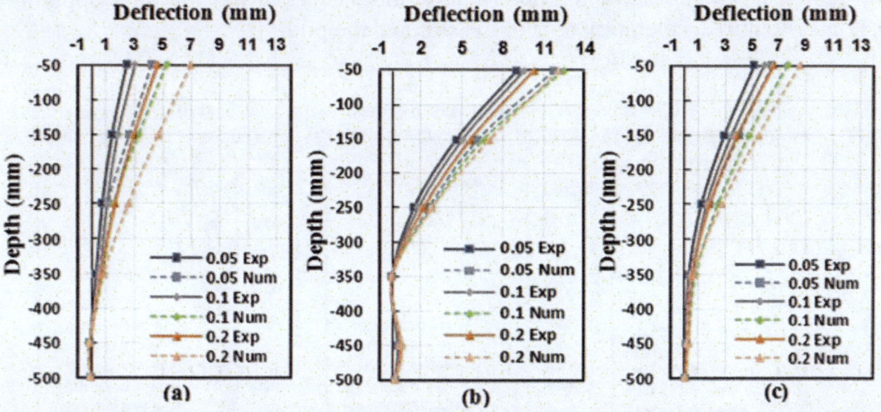

Fig. 9 Experimental and numerical results of deflection for the single pile model under lateral cyclic load (60 N) with (0.05, 0.1, and 0.2 Hz) frequency and embedded in **a** dry, **b** saturated, and **c** unsaturated soil with 10 kPa suction

The results showed that numerical results have the same behavior but higher than that of the experimental results. For the pile under cyclic lateral loading, the higher difference in soil resistance was noticed in the dry condition about 40%. This difference occurs in the laboratory when the soil is dry because the displacement of the pile will decrease with increasing the number of loading cycles due to soil

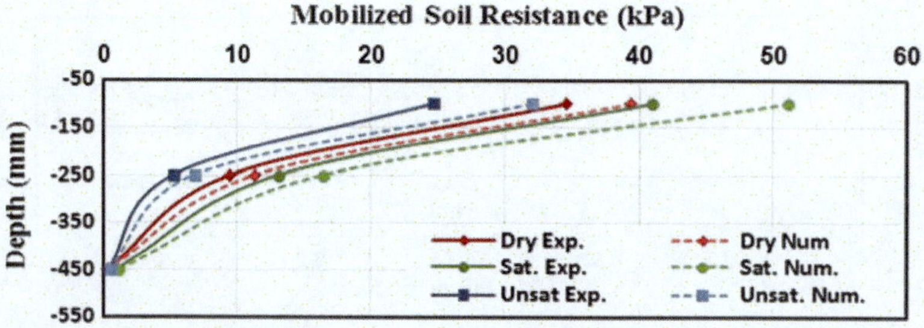

Fig. 10 Experimental and numerical results of soil resistance for the single pile model under lateral cyclic lateral load embedded in dry, saturated, and (10 kPa suction) unsaturated sandy soil

densification. It is difficult to simulate this mechanism in the software which leads to this difference and also the least difference noticed was in the saturated condition about 20%.

Figures 11, 12, and 13 show experimental and numerical results of deflection and soil resistance for the single pile model under lateral dynamic load of (0.05 g, 0.15 g, and 0.32 g El Centro acceleration record) and embedded in (dry, saturated, and unsaturated) sandy soil. The results show that, similar to the static and cyclic lateral loading, the numerically evaluated and experimental results have the same behavior but are higher for the numerical results than the experimental results.

Figures 14 and 15 show the deformation that occurs in the soil after applying 0.32 g El Centro acceleration data to the soil and the pile.

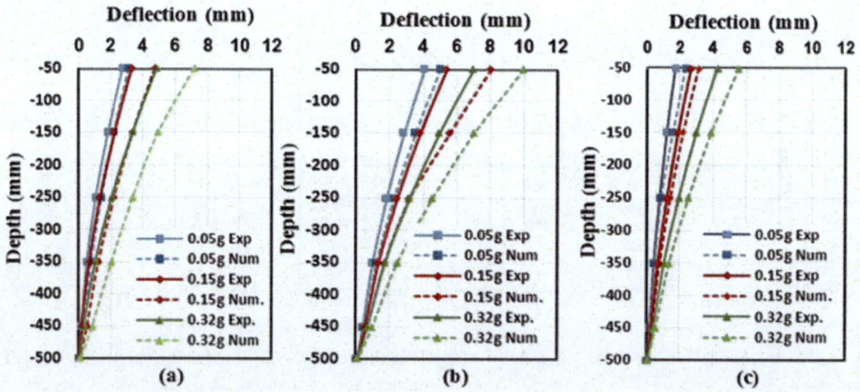

Fig. 11 Experimental and numerical results of deflection for the single pile model embedded in (dry, saturated, and unsaturated) sandy soil subjected to lateral dynamic load **a** dry, **b** saturated, and **c**) unsaturated (10 kPa suction)

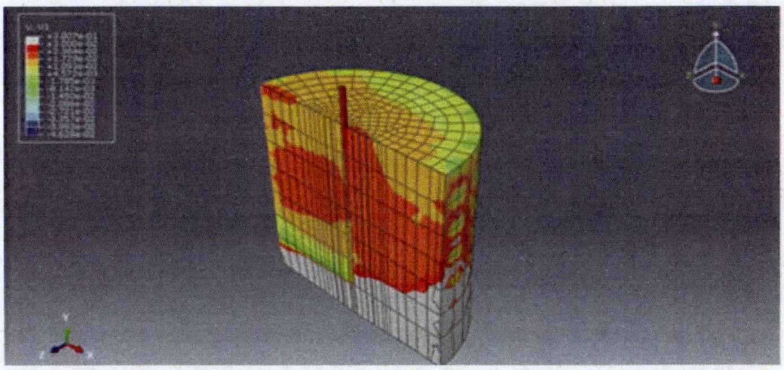

Fig. 12 Pile deformation and mobilized soil resistance by the effect of lateral Dynamic load

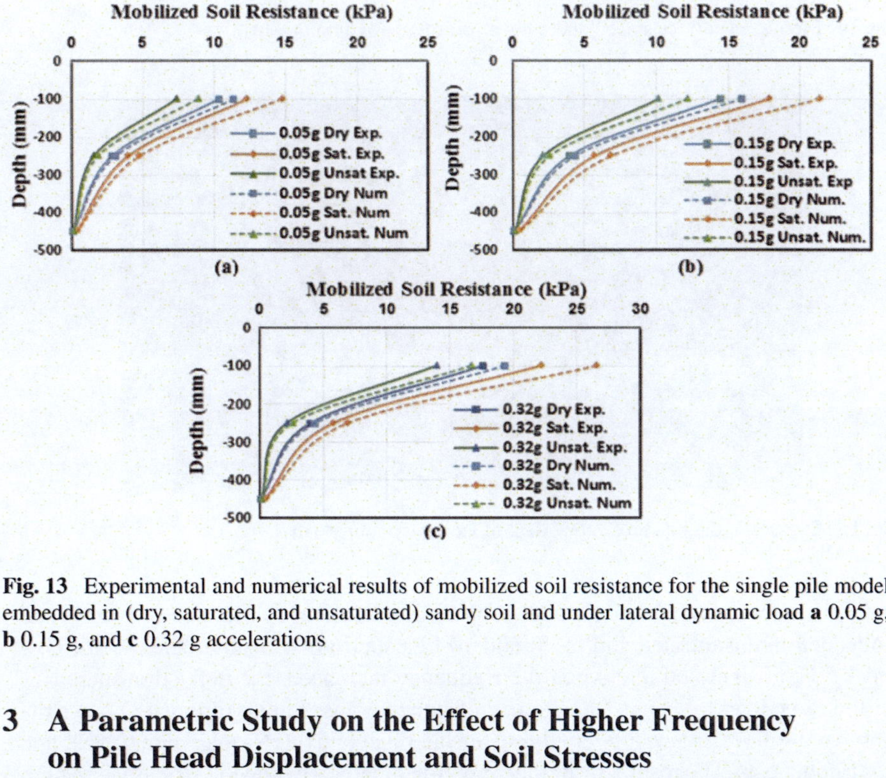

Fig. 13 Experimental and numerical results of mobilized soil resistance for the single pile model embedded in (dry, saturated, and unsaturated) sandy soil and under lateral dynamic load **a** 0.05 g, **b** 0.15 g, and **c** 0.32 g accelerations

3 A Parametric Study on the Effect of Higher Frequency on Pile Head Displacement and Soil Stresses

The same model geometry (600 mm diameter and 600 mm depth) and the same pile (16 mm diameter and 50 mm high) and sandy soil properties with a higher frequency (0.5, 1, and 1.5 Hz) are simulated by Abaqus V6.14. Figure 16 shows the lateral deformation for the pile and soil resistance after applying lateral cyclic load with different frequencies of 0.5, 1, and 1.5 Hz. Figure 17 shows the deflection

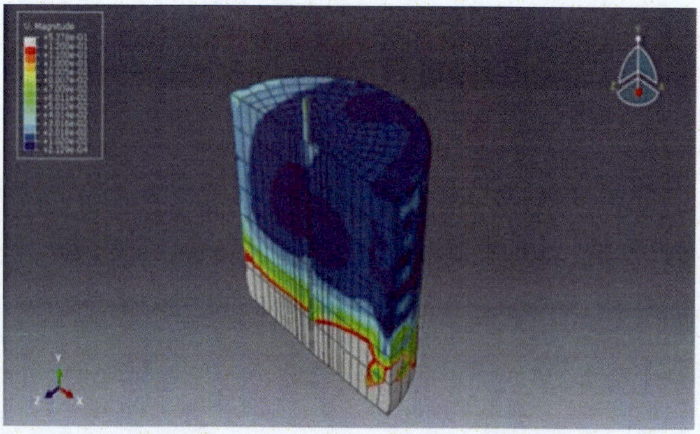

Fig. 14 Mobilized soil resistance under the effect of lateral dynamic load

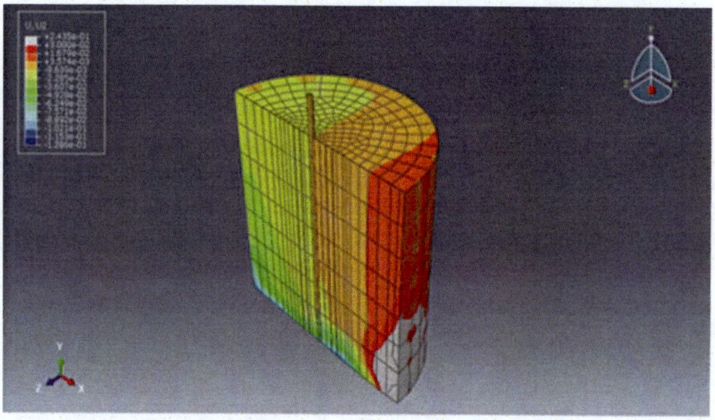

Fig. 15 Stresses in the pile under the effect of lateral dynamic load

that occurs for the single pile model subjected to a lateral cyclic load of (60 N) and embedded in unsaturated soil (5, 7, and 10 kPa suction) with frequencies of 0.5, 1 and 1.5 Hz. It is noted that when the frequency increased, the deflection increased for the same load by about 12% for each frequency increment. Figure 18 shows the mobilized soil resistance after applying the lateral cyclic (60 N) load with (0.5, 1, and 1.5) frequencies. The results show that applying of lateral cyclic load (60 N) with high frequency (0.5, 1 and 1.5 Hz) on the single pile model embedded in (dry, saturated, and unsaturated soil) will cause an increase in the pile deflection and mobilized soil resistance with the increase of the frequency for all the soil conditions.

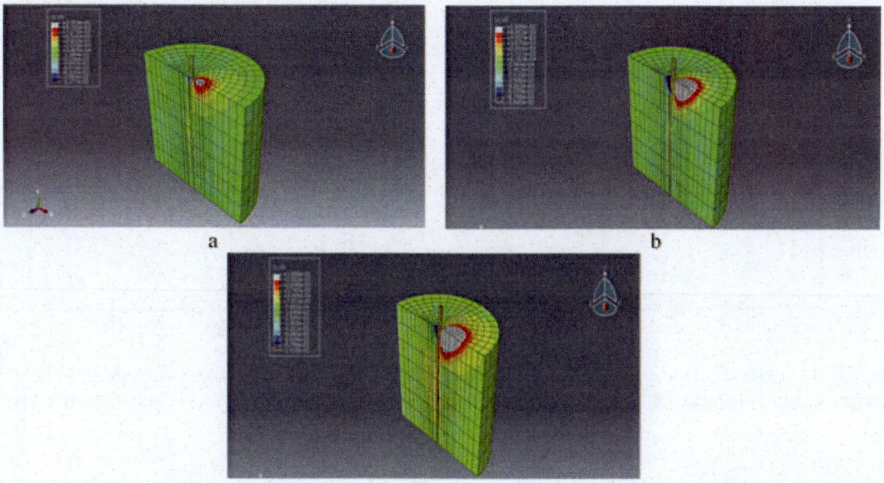

Fig. 16 The deformation of the pile and mobilized soil resistance by the effect of lateral cyclic load with **a** 0.5 Hz **b** 1 Hz, and **c** 1.5 Hz frequencies

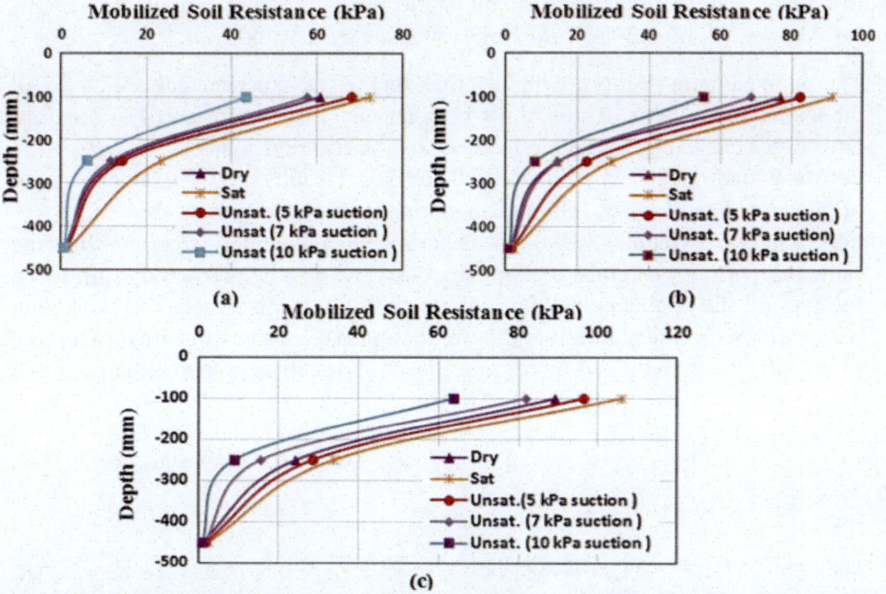

Fig. 17 Numerical deflection of the single pile model subjected to lateral cyclic load with different frequencies (0.5, 1, and 1.5) embedded in sandy soil **a** dry, **b** saturated, and **c** Unsaturated soil with 10 kPa suction

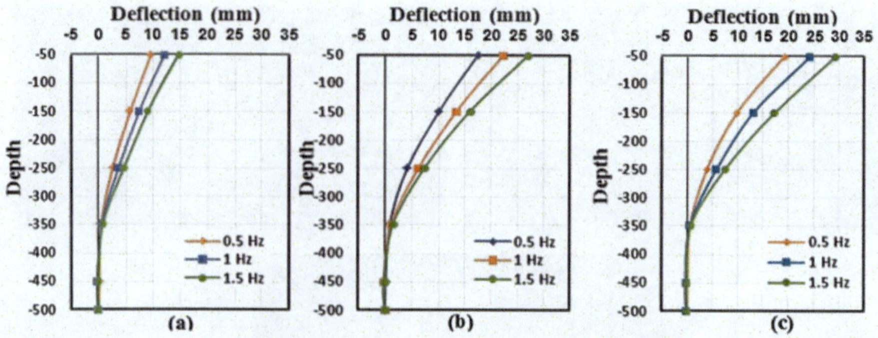

Fig. 18 Numerical results of mobilized soil resistance for the single pile model embedded in (dry, saturated, and unsaturated) sandy soil and under lateral cyclic load (60 N) with **a** 0.5 Hz **b** 1 Hz **c** 1.5 Hz accelerations

4 Conclusion

The displacement and deflection of the pile and soil resistance are investigated for the single pile model subjected to lateral (static, cyclic, and dynamic) loading by the use of Abaqus V6.14 and the following conclusions are withdrawn:

- The finite element results are highest than that of the experimental results for all the loading conditions. It should not be concealed that the presented simulations have limits caused by modeling features and hardware resources.
- For static loading, the experimental and numerical results show the same behavior of pile embedded in (dry, saturated, and unsaturated) sandy soil.
- For cyclic and Dynamic loading, the experimental and numerical results show the same behavior of pile embedded in (saturated and unsaturated) sandy soil, while for dry soil the results show high deference in the displacement and deflection because the experimental results show reduction in displacement with the increase of several cycles and it is difficult to achieve this mechanism numerically.

References

1. Ahmed MY, Neelima SD (2009) Numerical analysis of a pile subjected to lateral loads, IGC 2009, Guntur, India, pp 669–673
2. Poulos HG, Davis EH (1980) Pile foundation analysis and design. Wiley, New York, NY
3. Esmat ZA, Fadhil AI (2016) Effect of relative density on the matric suction and its contribution to shear strength of unsaturated sandy soil. Appl Res J 2(4):134–142
4. Rudolph C, Bienen B, Grabe J (2014) Effect of variation of the loading direction on the displacement accumulation of large-diameter piles under cyclic lateral loading in sand. Can Geotech J 51:1–11

5. Hügel HM, Henke S, Kinzler S (2008) High-performance Abaqus simulations in soil mechanics. In: 2008 Abaqus users' conference
6. Mahmood MR, Salim NM, Abood MH (2018) The behavior of laterally loaded single pile model in unsaturated cohesionless soil. In: International conference on materials engineering and science. IOP Conference

Hydrodynamic Pressure Effect of the Tank Wall on Soil–Structure Interaction

Ahmed A. Hussein, Mohammed A. Al-Neami, and Falah H. Rahil

Abstract In the traditional analysis of fluid tanks, the base of the tank is assumed to be fixed, and the flexibility of the soil is not taken into account in the analysis. However, it is known that the actual behavior of the tank does not only depend on the stiffness of the tank, but it also depends on the stiffness of the foundation and supporting the soil system. During this study, the tank resting on the medium and hard soils with the relative density of 55 and 65%, respectively, are analyzed considering the impact of soil when subjected to seismic excitation. Two different cases have been evaluated; the first is the tank with hydrodynamic pressure and, the second is the tank without hydrodynamic pressure. The tank, fluid, and soil are modeled using FLAC3D software. Two types of soil with different flexibilities are considered to analyze the soil–structure interaction impact. It has been concluded that the soil properties have an extreme effect on the base displacement and stresses developed in the tank. The hydrodynamic pressure led to a great increase in the settlement and two-dimensional lateral displacement of the tank base. Also, the frequency domain and its magnitude are considerably affected by the tank displacements.

Keywords Cylindrical tank · Hydrodynamic pressure · Fluid–soil–structure interaction · Seismic

A. A. Hussein (✉) · M. A. Al-Neami · F. H. Rahil
Civil Engineering Department, University of Technology, Baghdad, Iraq
e-mail: 42080@student.uotechnology.edu.iq

M. A. Al-Neami
e-mail: 40008@uotechnology.edu.iq

F. H. Rahil
e-mail: 40029@uotechnology.edu.iq

M. O. Karkush and D. Choudhury (eds), *Modern Applications of Geotechnical Engineering and Construction*, Lecture Notes in Civil Engineering 112,
https://doi.org/10.1007/978-981-15-9399-4_15

1 Introduction

A ground supporting the circular tanks is concedes the important structures in activities and is used to store a multiplicity of liquids, e.g. water for drinking and fire-fighting, chemicals, petroleum, and liquefied natural gas. In the latest years, the significance of fluid–structure–soil interaction of the dynamic response of vital structures such as silos, storage tanks, and also other types of structures has been offered in several studies [1]. Particular problems associated with fluid–structure–soil interaction should be taken into account: the dynamic interaction between fluid and tank, and the dynamic interaction between tank and soil, i.e. soil–structure interaction. Understanding the fluid impacts, acting on the walls and bottoms of the containers during the earthquake, is a vital part of the reliability and durability of the earthquake-resistant structure [2–4].

Housner's mass-spring model [5] is the major model for the estimation of the seismic response of rigid cylindrical liquid storage tanks. In the model of Housner, the whole liquid is divided into two portions; the portion of the liquid that excites individually from the tank wall near the free surface is called Convective and the other portion of the liquid that excites with the tank wall is called Impulsive. Housner's equation splits the hydrodynamic pressure into two parts. The impulsive portion is the part of the fluid that is moved in accordance with the tank, while the convective portion is the part of the fluid in the tank. In addition to the equation made by Hoskins and Jacobsen (1934), Housner supposed that the fluid was held between the vertical membranes and that these displacements were small. He also proposed the following equation to determine the impulsive pressure.

$$p_i(z,t) = a(t)\rho H\sqrt{3}\left[\frac{z}{H} - \frac{1}{2}\left(\frac{z}{H}\right)^2\right]\tanh\left[\sqrt{3}\frac{L}{H}\right] \tag{1}$$

A break of 2% of the tank height is sufficient between the free surface of the fluid and the bottom surface of the top plate of the tank for free oscillation of the fluid. The following equations are suggested by Housner for convective pressure after the kinetic and potential energy expressions of the fluid are replaced by the Hamilton principle.

$$p_c(z,t) = 0.527\rho L^2 {\omega_n}^2 \Phi \frac{\cosh\left[1.581\frac{H-z}{L}\right]}{\sinh\left[1.581\frac{H}{L}\right]}\sin(\omega_n t) \tag{2}$$

where z is the height of the liquid in the tank, t is the time of acceleration, a(t) is acceleration response, ρ is the density of the liquid, H is the height of the tank, L is the half-width of the tank in the direction of the seismic action, ωn is the angular frequency of the fluid, and Φ is equal to S_a/g, where S_a is the acceleration of spectrum and g is the acceleration due to gravity. The Housner approach is commonly used in the dynamic study of the tanks and implemented into the applicable design codes. Epstein (1976) enhanced Housner's work and introduced design curves for estimating the bending and overturning moment affected by hydrodynamic pressure for rectangular and cylindrical rigid tanks [6]. Each of the structure–soil interaction and structure–fluid interaction is distinctly a complex apparent for structures. Although there were many researches on the fluid–structure interaction in liquid storage tanks, most of them are concerned with cylindrical tanks. Chen and Huang [7] have been using a time-independent finite difference technique to solve a fully nonlinear sloshing fluid in a tank mounted on a structure. The interaction between the dynamic response and the sloshing fluid of the structure has also been investigated. The fourth-order Runge–Kutta method was used to calculate the dynamic response of the structure, and the response of the tank-structure system was reported to be sensitive to the frequency of excitation and the depth of the water in the fluid tank.

Chougule [8] investigated the seismic analysis of the ground support water tank rest on soft soil containing a mass of tank wall, the mass of roof, the mass of base slab, and mass of water. Also, a parametric study was carried out considering the spring-mass model, a period in impulsive and convective modes, horizontal seismic design coefficients, hydrodynamic pressure and base shear due to the convective and the impulsive mass of water; they observed that the D/h ratio increases the decrement of the impulsive period is smaller and considerably convective period increase, also, total base shear reduces with increments d/h ratio. Sadek et al. [9] constructed a 3D Finite Element Model of a shallow rectangular rested on soil Reinforced Concrete. The vertical and horizontal components of three ground motion records are used in the study with different frequency domains. It is concluded that the dynamic behavior of the rectangular tank system is sensitive to the frequency domain of the ground motion. Also, the FSSI has great effects on the seismic behavior of such types of tanks, which should be taken into account in the design.

This research work aims to attain a complete apperception of the seismic conduct of the deep cylindrical ground rested tanks during earthquakes; accordingly, the influence of the Fluid–Soil–Structure Interaction (FSSI) on the seismic behavior of deep cylindrical tanks during earthquakes has been investigated. Moreover, the effect of the soil properties and the frequency domain of the ground motion on the tank's seismic behavior have been investigated. To get such understanding, the well-known FLAC3D software has been used to construct a finite difference model for a deep cylindrical tank and subsoil conditions. A rectangular modal analysis has been carried out firstly for the tank with a fixed base to verify the basic dynamic properties of the tank against the obtained analytical ones from other published research and design codes worldwide. Then, a parametric study using nonlinear dynamic analysis of the time history using two ground motion records representing a range of frequency domain is analyzed. Furthermore, two different subsoil cases have been considered, the first one is the tank exposures to the hydrodynamic pressure. And the other case is the tank in the rest (no hydrodynamic pressure) supported by hard and medium soils.

2 Description of the Model

To deal with such a complex problem, which is the fluid–soil–structure interaction (FSSI), resting on the soil, a deep cylindrical tank with dimensions 15.0 m diameter and 10.0 m wall height (H_w) has been used in this study. The wall thickness of the tank (t_w) is 0.25 m and the height of the liquid inside the tank (H_L) is 8.0 m. A schematic configuration of the tank is shown in Fig. 1. The material properties of the concrete and the liquid are listed in Table 1.

Furthermore, to investigate the impact of soil–structure interaction (SSI) on the dynamic response of the liquid storage tank, two types of soil have been used. The first soil represents a hard soil with the relative density of (65%), while the second one represents a medium soil with the relative density of (55%). The nonlinear parameters of the studied types of soil are listed in Table 2.

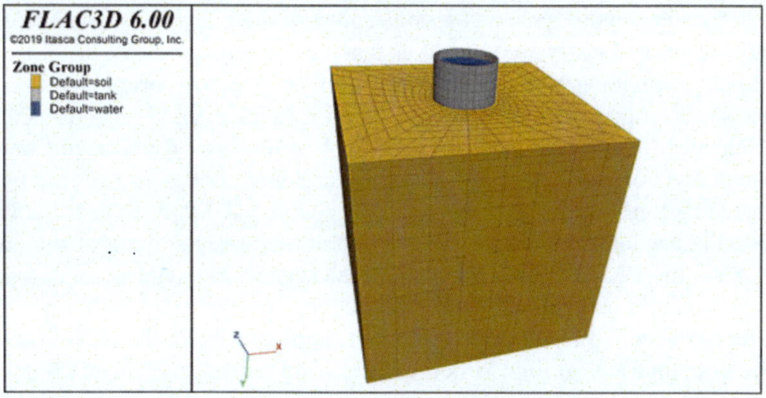

Fig. 1 Finite difference model of considered cylindrical concrete tank

Table 1 Material properties of liquid and tank for a concrete rectangular tank

Concrete			Liquid	
Young's modulus (Pa)	Density (kg/m^3)	Poisson's ratio	Density (kg/m^3)	Bulk modulus (Pa)
35×10^9	2400	0.17	1000	2×10^9

Table 2 Characteristics of soil used in the analysis

Soil type	Density (kg/m^3)	Poisson's ratio	Young's modulus (Pa)	Friction angle (°)
Hard	1700	0.28	26×10^7	40
Medium	1600	0.33	9×10^7	30

3 Verification of the Model

Before going into an investigation and a complex parametric study, a verification process has been carried out to assess the reliability of the model and assumptions comparing its results to the numerical findings that are available in the previously published researches and design code [6]. The verification process includes an assessment of the suitability of the model to simulate the hydrodynamic pressure and Fluid–Structure Interaction (FSI) behavior by carrying out a modal analysis for a fixed-base tank which is fully anchored to the rigid ground. The verification process includes comparing the obtained results to analytical results provided by (Dogangun and Ramazan) [6] and, comparing the results to Housner's equation.

A rectangular storage tank shall be considered as shown in Fig. 2. This tank is selected as a similar tank considered by Koh et al. to match the results [10]. The bulk of the module is 2×10^9 N/mm^2 and the density of the fluid is taken to be 1000 kg/m^3. It is assumed that the tank is being subjected to the N-S component of the El Centro Earthquake parallel to the short sidewalls and fixed to the ground. The damping ratio of the fluid is used to be $\xi = 0.5\%$ as recommended in Eurocode 8. For this analysis, the unit width of the tank model vertical to the direction of the motion is considered to be a comparison of the results of analytical and finite difference methods. In this model, the fluid is free to move in the direction of motion along the walls and the horizontal direction at the bottom. For this purpose, the well-known finite difference software FLAC3D has been employed to simulate the behavior of the rectangular tank [6]. The tank structure has been modeled using eight nodes with three degrees

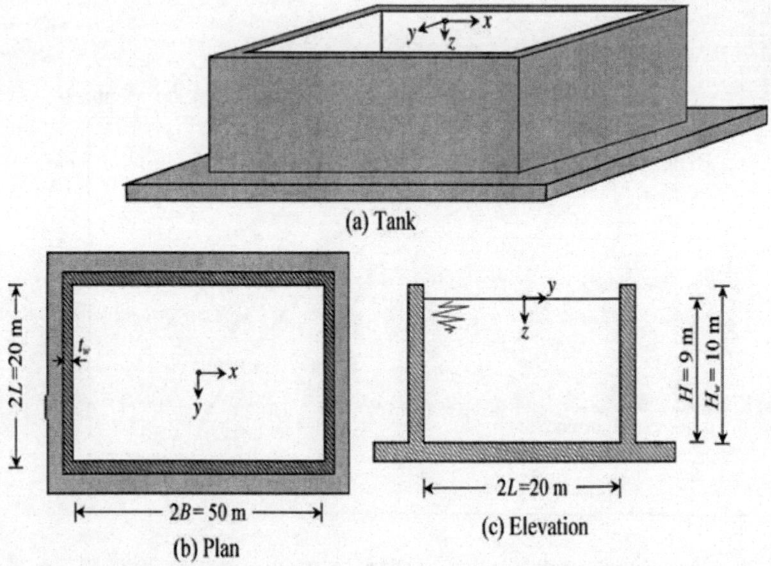

Fig. 2 Elevation and plan of the rectangular tank sample

of freedom in each node. The finite element idealization of the tank with a fixed base is shown in Fig. 3.

The distribution of hydrodynamic pressures by the Housner method, (Dogangun and Ramazan) [6] and our study is shown in Fig. 4. It can be seen from this figure that the distribution of the hydrodynamic pressure obtained in these three different studies is normally given in good agreements. Hydrodynamic pressures from this

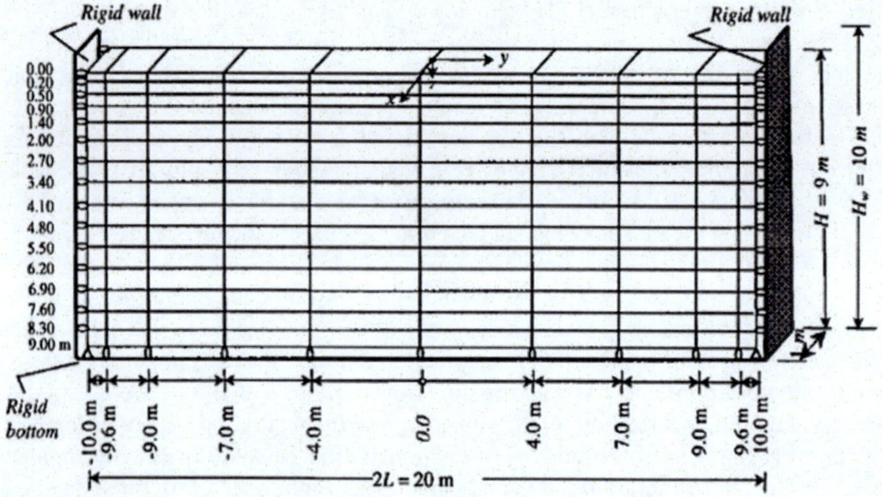

Fig. 3 Finite element mesh is considered to be the unit width of the tank in the rigid solution

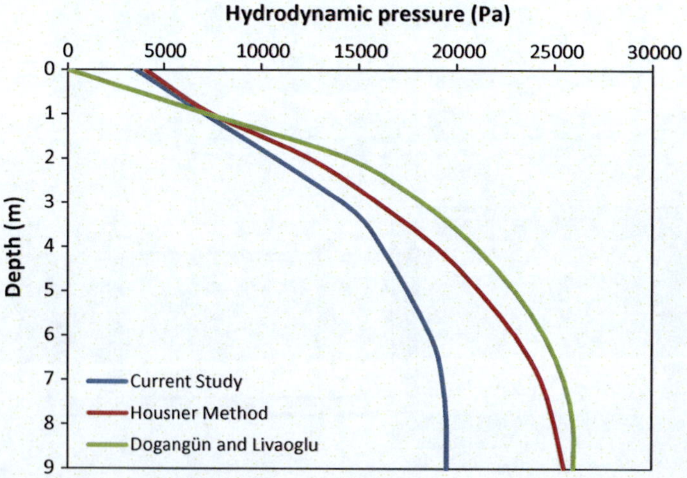

Fig. 4 Hydrodynamic pressure distributions obtained by assuming that the walls are rigid

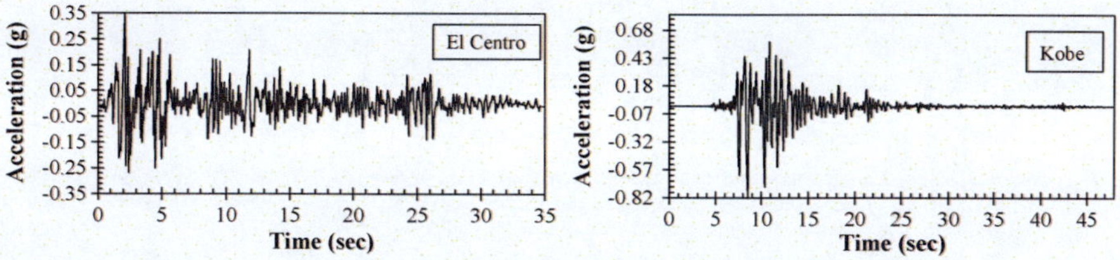

Fig. 5 Horizontal components of the selected ground motion records

study are generally lower than those obtained using the other methods. So the results approved the ability of software **FLAC3D** to analyze these types of problems.

4 Parametric Study

After verifying the ability of the finite difference model to simulate the rectangular tank and the Fluid–Structure Interaction, the seismic response of the partially filled cylindrical deep RC tank underground the motion excitations in horizontal directions has been studied. The cylindrical concrete tank dimensions used in the parametric study are mentioned before and the material properties are listed in Table 1. Furthermore, to account for the soil–structure interaction, two types of soil have been used and modeled considering the effect of soil–structure interaction (SSI) on the dynamic response of the liquid storage tank (Fig. 1). The first soil has been chosen to represent hard soils and the second one represents medium soils. The ground accelerations for the horizontal components are shown in Fig. 5. The seismic excitation has been applied at the base of the soil as an acceleration time history in horizontal directions and the structural responses have been obtained at each time step. The displacement values have been detected at the center of the tank and it is found that the tank structure variation has a great concern in the design of cylindrical tanks.

5 Results and Discussion

The time histories of base displacement due to different ground motion records and different subsoil cases have been obtained at the middle base of the tank as shown in Figs. 6, 7, 8, 9, 10, 11, 12 and 13. To encompass the expected range of ground motions, two ground motion records with different frequency domains have been selected to perform the dynamic time history analysis; the two records are Kobe (1995) and El Centro (1940).

The responses of the maximum settlement represented in y-displacement in the hard soil of the tank range 8.8×10^{-3} to 9.56×10^{-3} m as shown in Fig. 6 for the El Centro earthquake and Fig. 7 range 6.16×10^{-2} to 1.42×10^{-1} m for El

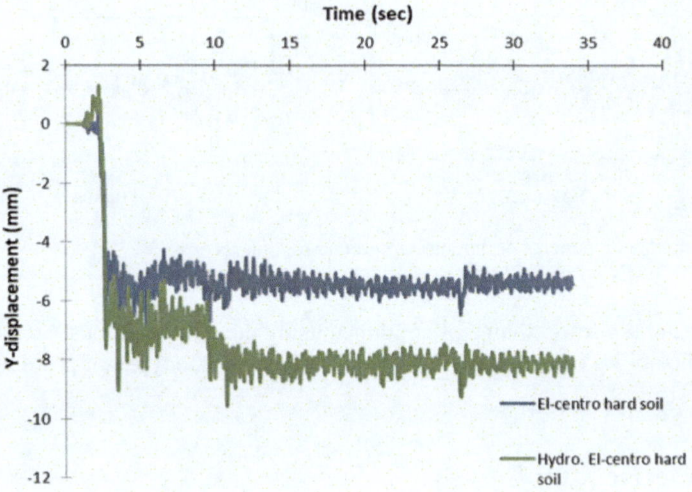

Fig. 6 Time history of settlement in y-displacement of the tank under El Centro excitation for hard soil with and without hydrodynamic pressure

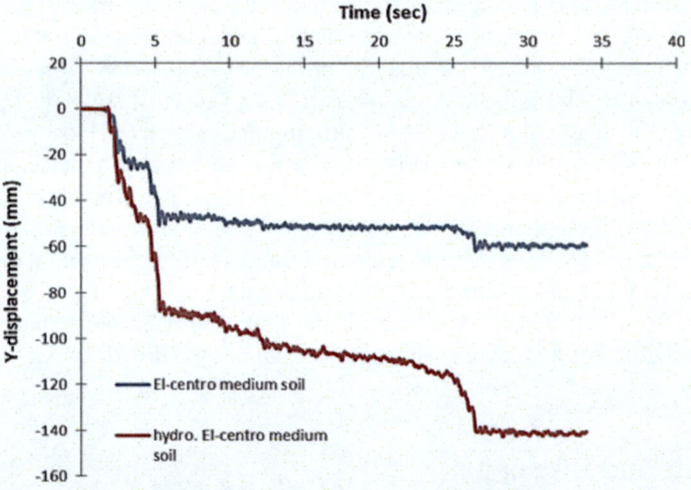

Fig. 7 Time history of settlement in y-displacement of the tank under El Centro excitation for medium soil with and without hydrodynamic pressure

Centro earthquake considering medium soil, while they range 6.18×10^{-2} to 9.91×10^{-2} m as shown in Fig. 8 for Kobe earthquake considering hard soil. In Fig. 9, the maximum displacement ranges 8.36×10^{-1} to 9.35×10^{-1} m for the Kobe earthquake considering the medium soil. Figures 6, 7, 8 and 9 clarify the effect of hydrodynamic pressure on the vertical settlement in all the y-directions. Figures 10 and 11 show the effect of the lateral displacement in the x-direction of the tank with

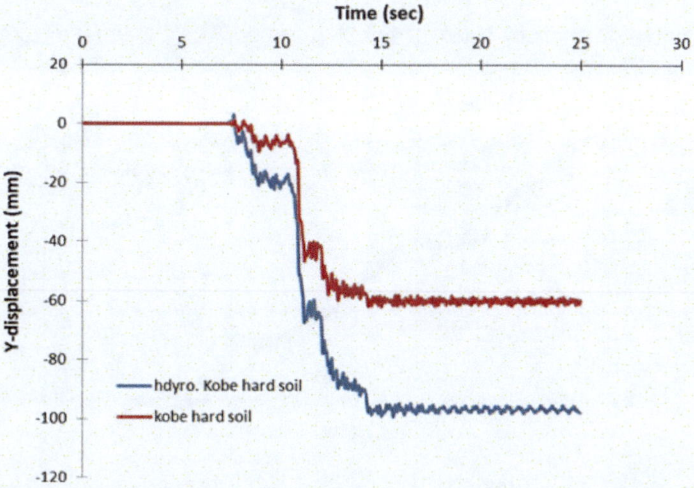

Fig. 8 Time history of settlement in y-displacement of the tank under Kobe excitation for hard soil with and without hydrodynamic pressure

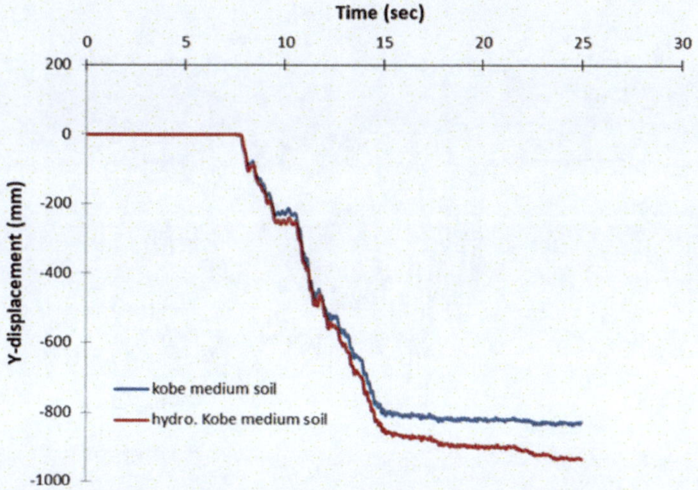

Fig. 9 Time history of settlement in y-displacement of the tank under Kobe excitation for medium soil with and without hydrodynamic pressure

a maximum range of 5.02×10^{-2} to 5.77×10^{-2} m under El Centro excitation in hard and medium soils, while it ranges 8.7×10^{-3} to 1.19×10^{-1} m in Figs. 12 and 13 for the lateral displacement in the z-direction in the Kobe earthquake considering hard and medium soils. The influence of the earthquake frequency domain was noticeable in the x-displacement between the Kobe and El Centro records and was significantly influenced by the soil properties as shown in the x-displacement

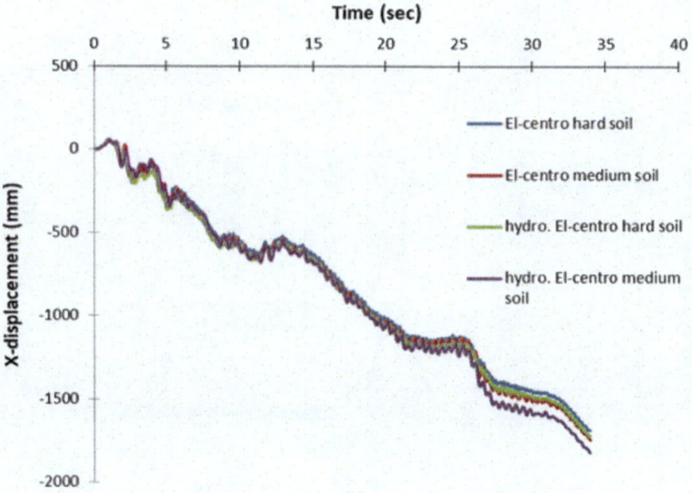

Fig. 10 Time history of lateral displacement in the x-direction of the tank under El Centro excitation for hard and medium soil with and without hydrodynamic pressure

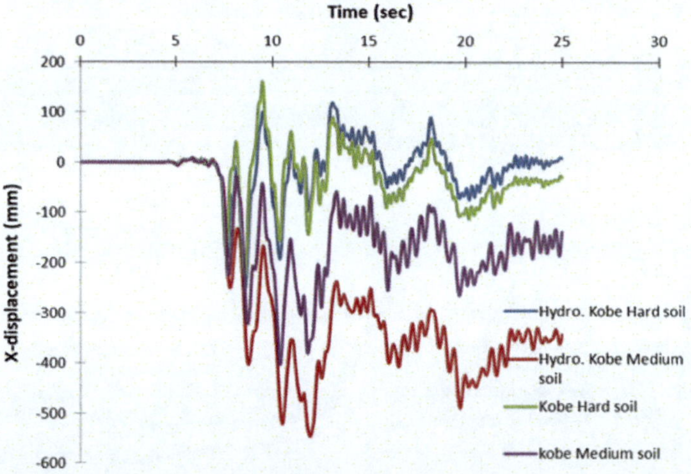

Fig. 11 Time history of lateral displacement in the x-direction of the tank under Kobe excitation for hard and medium soil with and without hydrodynamic pressure

behavior (Figs. 10 and 11). Also, the z-displacements as shown in Figs. 12 and 13 were marginally affected in all cases due to the perpendicularity of the excitation axis, despite the change in behavior through differences in soil and earthquakes. The soil–structure interaction was affected by soil properties and the behavior of the natural frequency and amplitude of the earthquake. However, it is worth to mention that the

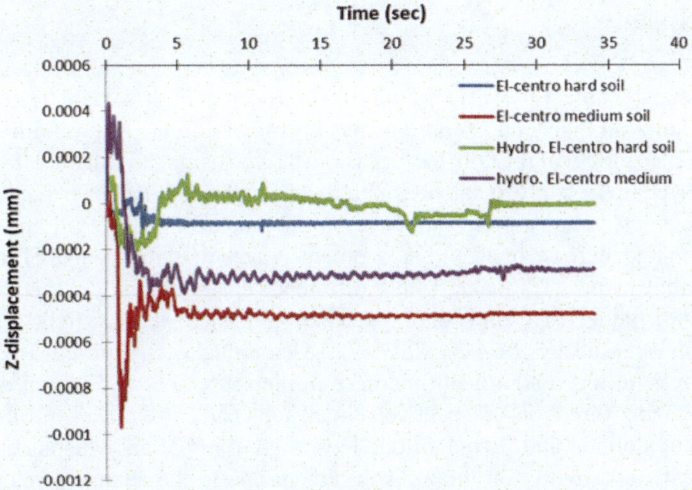

Fig. 12 Time history of lateral displacement in the z-direction of the tank under El Centro excitation for Hard and Medium soil with and without hydrodynamic pressure

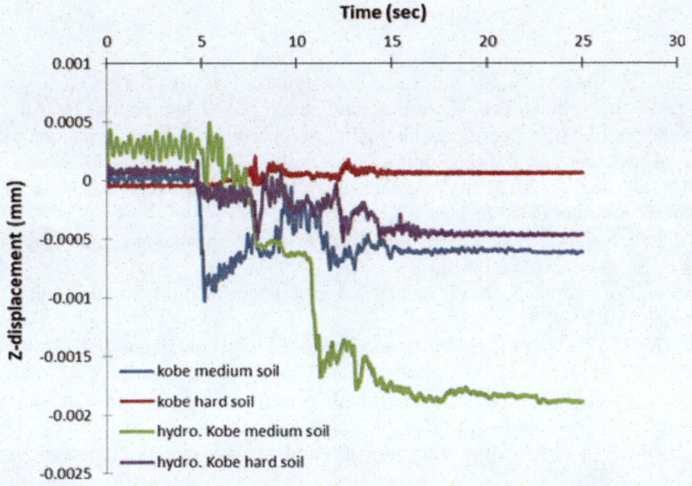

Fig. 13 Time history of lateral displacement in the z-direction of the tank under Kobe excitation for hard and medium soil with and without hydrodynamic pressure

frequency domain of ground motion is the ratio between the normalized Peak Ground-Acceleration (PGA) and the normalized Peak Ground-Velocity (PGV). Hence, the El Centro earthquake record and Kobe are representing the Low-frequency ground motion and Medium-frequency ground motion, respectively.

6 Conclusions

In this study, a dynamic time history analysis for a deep cylindrical tank rested on the soil under horizontal ground excitations has been carried out. A three-dimensional finite difference model of the tank has been constructed using the software, FLAC3D, to be employed in the study. It can be concluded from the results above that the structure–soil interaction has a great influence on the displacement of soil and stresses response. A huge increase in the y-displacement as a result of the change of soil properties under the same earthquake, 14.6% difference change in the x-direction due to change of soil under El Centro earthquake, despite there it is a slight difference in z-direction caused by the normality of the earthquake direction. Also, hydrodynamic pressure has a significant increase in the settlement and lateral displacement y-, x-, and z-directions, respectively, in all soils and earthquakes. Furthermore, the frequency domain and its magnitude have a great effect on y-displacement, x-displacement, and slightly on the z-displacement due to the change of earthquake excitation.

References

1. Kotrasova K, Harabinova S, Panulinova E, Kormanikova E (2016) Seismic analysis of cylindrical liquid storage tanks considering of fluid-structure-soil interaction. In: Advances and trends in engineering sciences and technologies—proceedings of the international conference on engineering sciences and technologies, ESaT, pp 87–92
2. Livaoğlu R, Durmuş A (2015) Investigation of wall flexibility effects on seismic behavior of cylindrical silos. Struct Eng Mech Int J 53(1):159–172
3. Seitl S, Merta I, Kersner Z (2015) Wedge splitting test of foam concrete specimens: calibration curves. Key Eng Mater 627:281–284
4. Kotrasová K, Harabinová S (2017) Numerical experiment of fluid-structure-soil interaction. Procedia Eng 190:291–295
5. Housner GW (1957) Dynamic pressures on accelerated fluid containers. Bull Seismol Soc Am 47:15–35
6. Dogangun A, Livaoglu R (2004) Hydrodynamic pressures acting on the walls of rectangular fluid containers. Struct Eng Mech 17(2):203–214
7. Chen B, Huang S (2015) A numerical study of liquid tank and structure interaction. J Mar Sci Technol Taiwan: 781–791
8. Chougule AC, Patil PA (2017) Study of seismic analysis of water tank at ground level. Int Res J Eng Technol 4(7):2895–2900
9. Aly SS, Gomaa MS, El Ghazaly HA (2019) Seismic behavior of ground rested rectangular RC tank considering fluid-structure-soil interaction. IOSR J Mech Civil Eng 16(1):01–15
10. Koh HM, Kim JK, Park JH (1998) Fluid-structure interaction analysis of 3-D rectangular tanks by a variationally coupled BEM-FEM and comparison with test results. Earthq Eng Struct Dyn 27:109–124

Use of Lightweight Deflectometer for Evaluating in Situ Strength Characteristics of Pavement Foundations in Kerbala

Rawq Al-Fattehallah, Alaa M. Shaban, and Shakir Al-Busaltan

Abstract Pavement performance depends mainly on the strength characteristics of subgrades. California Bearing Ratio (CBR) plays a vital role in the evaluation of subgrade strength in pavement design. Determination of the field bearing resistance of subgrades requires an equipment, labors, and is time consuming, therefore, several researchers have attempted to predict CBR value of a subgrade in the field from other nondestructive testing devices. One of the most versatile, rapid, and in situ nondestructive testing device is a lightweight deflectometer (LWD) which is used in estimating subgrade properties in various pavement projects. Developing correlations between CBR and LWD parameters including surface deflection, dynamic modulus, and degree of compatibility makes LWD an important alternative test used in the performance evaluation of pavement layers. In this paper, an attempt has been made to evaluate in situ strength by selecting two under construction street sites located at Karbala: Al-Takahe district and Al-Emam Ali district. The subgrade soils et al. -Takahe district and Al-Emam Ali district were classified according to AASHTO as A-1-b and A-3, respectively. The initial results showed the ability to predict bearing resistance of subgrades based on their basic physical properties and dynamic measurement obtained by an LWD test device.

Keywords Lightweight deflectometer (LWD) · California bearing ratio (CBR) · Subgrade · Field density · Moisture content

R. Al-Fattehallah (✉) · A. M. Shaban · S. Al-Busaltan
Department of Civil Engineering, University of Kerbala, Kerbala, Iraq
e-mail: rawq.m@s.uokerbala.edu.iq

A. M. Shaban
e-mail: alaa.shaban@uokerbala.edu.iq

S. Al-Busaltan
e-mail: s.f.al-busaltan@uokerbala.edu.iq

M. O. Karkush and D. Choudhury (eds), *Modern Applications of Geotechnical Engineering and Construction*, Lecture Notes in Civil Engineering 112,
https://doi.org/10.1007/978-981-15-9399-4_16

1 Introduction

During the 1930s, an empirical testing method was developed by California Department of Transportation to evaluate the strength of subgrade soils and unbound base materials designated as the California Bearing Ratio (CBR) [1]. This test method was widely utilized by highway agencies which considered an essential design parameter that needs to be considered in the design of highway and airport pavements is the characteristic of the subgrade where the pavement is placed on. Subgrade materials are typically characterized by their resistance to deformation under load, which can be either a measure of their strength or stiffness. In general, subgrade resistance to deformation which can be expressed in terms of stiffness, modulus, density, and California Bearing Ratio (CBR) could reflect its load-carrying capacity before reaching a critical deformation value. However, many researchers developed correlative models to relate CBR with other test results to estimate the strength and stiffness of subgrades using different tests such as Dynamic Cone Penetration (DCP), Portable Light Weight Deflectometer (PFWD), Cyclic Triaxial (CT), Clegg Impact Values (CIV), and Miniaturized Pressuremeter Test (MPMT).

A lightweight deflectometer (LWD) is a portable device used to measure the in situ elastic modulus of pavement layers and fill embankments. The LWD is primarily used for soil subgrade evaluation before pavement design (mechanistic approach) and verification [2]. Although LWD is widely utilized for compaction quality control of soil layers in highway construction projects by determining the deflection (δ) of flexible pavement layers, it can also be used to determine the stiffness of surface soils (E_{LWD}). The method of calculation of LWD modulus (E_{LWD}) was based on Boussinesq elastic half-space theory [3]. LWD is gaining acceptance and has been widely used over the years as the pavement and transportation geotechnics community is currently moving toward more mechanistic-based design and quality control evaluation of pavement layers and embankment fills. The developments in testing methods such as LWD display more precise, ease of use, ease of portability, fast, non-distractive, and modern methods, in addition to the ability of determining the surface deflection and surface stiffness of compacted soil [4].

The main purpose of this research is to investigate the validity and reliability of two statistical models developed to predict the bearing resistance of local subgrade soil properties depending on either their physical properties or their dynamic measurement determined by an LWD test device. The first model was developed using LWD parameters which include surface deflection S_d, dynamic modulus E_d, degree of compatibility D_c as independent variables, and the second model was represented by considering Wc, and γd_f as independent variables while the dependent variable is represented by the CBR value. This verification process could facilitate the understanding of the role of LWD parameters in determining the in situ bearing resistance of a subgrade.

2 Site Location and Layout of the Tests

In this study, two roadway projects in Karbala were selected to evaluate the strength of their subgrade soils: the first site project was Al-Takahe district, and the second was Al-Emam Ali district. The Al-Takahe zone is located at the west of Karbala which includes an arterial two-lane roadway with length 2000 m, and width 40 m, while Al-Emam Ali district is located at southwestern Karbala, which consists of four local streets with length 150 m and width 12 m and a collector with length 200 m and width 12 m; see Fig. (2). A construction project was conducted in Al-Takahe district, and 20 test points were selected for every 100 m on the center for constructed roadway, which starts from station 00 + 00 to station 20 + 00. Two testing techniques, including field density by core cutter method (CCM) and the LWD test, were performed, while eight test points were conducted for Al-Emam Ali district which starts from station 00 + 100 and to station 00 + 800. The LWD test was conducted in the same location of the field density test at both sites, as clarified in Figs. 1 and 2. After completing the field tests, a disturbed-soil sample has been collected and tested in a laboratory for identifying their basic properties. Standard laboratory tests according to ASTM and AASHTO procedures were conducted on each soil type to define the basic soil properties including sieve analysis, Atterberg limits, modified Proctor tests, and laboratory California bearing ratio test (CBR). Also, chemical tests were included to investigate the content of deleterious materials in subgrade soils (i.e., SO_3, gypsum content, and total soluble salts).

Fig. 1 Schematic diagram of field test points—Al-Takahe district

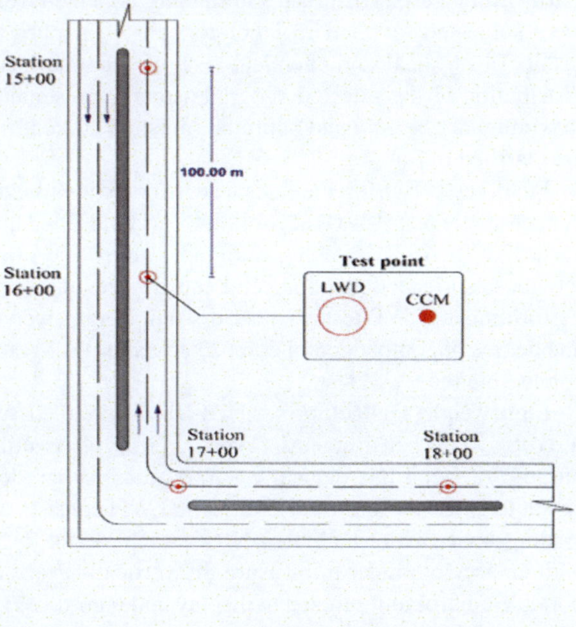

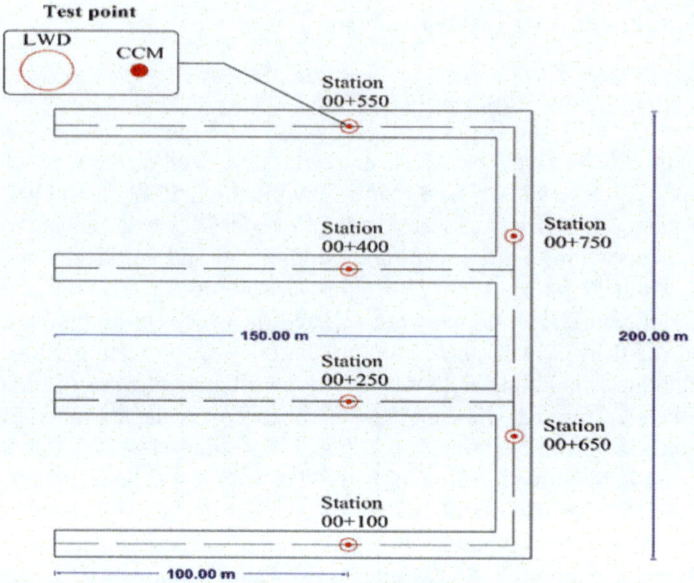

Fig. 2 Schematic diagram of field test points—Al-Emam Ali district

3 Experimental Program

Laboratory Tests. Disturbed soil samples collected from the natural subgrade layers were subjected to standard laboratory tests for purposes of characterization. The results of physical and chemical tests are summarized in Table 1. The grain size distribution of the selected subgrade soils is presented in Fig. 3. Additionally, the maximum dry density and optimum moisture content from the modified Proctor test are clarified in Fig. 4.

Field Tests. Two test methods were adopted in this experimental work: lightweight deflectometer test (LWD) according to ASTM E2583 and core cutter method (CCM) according to ASTM D2937. Subgrades' dynamic properties in terms of dynamic modulus, surface deflection, and the degree of compatibility were obtained by performing the LWD test whereas, dry unit weight, wet unit weight, moisture content, and degree of compaction of compacted natural subgrade layers were determined by conducting the CCM test.

Lightweight Deflectometer (LWD). A portable LWD device which is manufactured by ZORN instruments (Type ZFG 3.0) was utilized in the work. The main purpose of using this device was to evaluate subgrades' dynamic and compaction characteristics. As shown in Fig. 5, the LWD consists of three main components: 1) loading device which includes a (10 kg) drop weight, 2) 30-cm loading plate equipped with an accelerometer to measure the vertical surface deflection of a soil, and 3) a portable control unit utilized to display and record LWD testing measurements [12].

Table 1 Physical and chemical properties of selected types of subgrade soils

Property	Subgrade soil type		Specification
	AL-Meelad	AL-Fares	
Site coordination	408,595.011, 3,605,065.385	404,229.144, 3,611,462.039	/
USCS classification	Poorly graded sand with silt (SP-SM)	Poorly graded sand (SP)	ASTM D2487 [5]
AASHTO classification	**A-1-b**	**A-3**	AASHTO M145 [6]
Dry unit weight	21.28 kN/m^3	20.89 kN/m^3	ASTM D1557 [7]
O.M.C	8%	10.4%	ASTM D4643 [8]
D_{10}, D_{30}, D_{60}	0.14,0.29, 0.53	0.17, 0.29, 0.402	ASTM D2487 [8]
Uniformity coefficient, C_u	3.79	2.36	ASTM D2487 [8]
Curvature coefficient, C_c	1.13	1.23	ASTM D2487 [5]
Gravel fraction, GF	4.60%	1.52%	ASTM D2487 [5]
Fine content	5.30%	4.01%	ASTM D2487 [5]
Lab. CBR—unsoaked	57%	38%	ASTM D1883 [9]
Lab. CBR—soaked	47%	26%	ASTM D1883 [9]
Liquid limit	0	0	ASTM D4318 [10]
Plasticity index	N. P	N. P	ASTM D4318 [10]
Sulfate (SO_3)	1.27%	1.04%	BS1377-3:1990 [11]
gypsum content ($CaSO_4$ $2H_2O$)	3.73%	1.77%	BS1377-3:1990 [11]
Total soluble salts	1.22%	3.80%	BS1377-3:1990 [11]

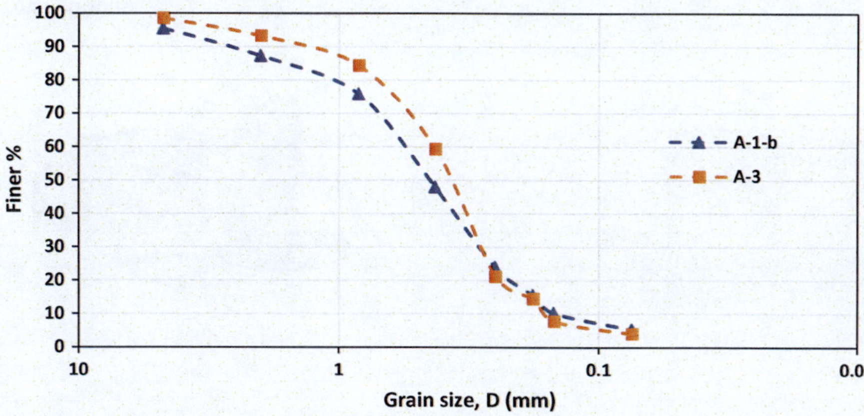

Fig. 3 Grain size distribution for selected types of subgrades

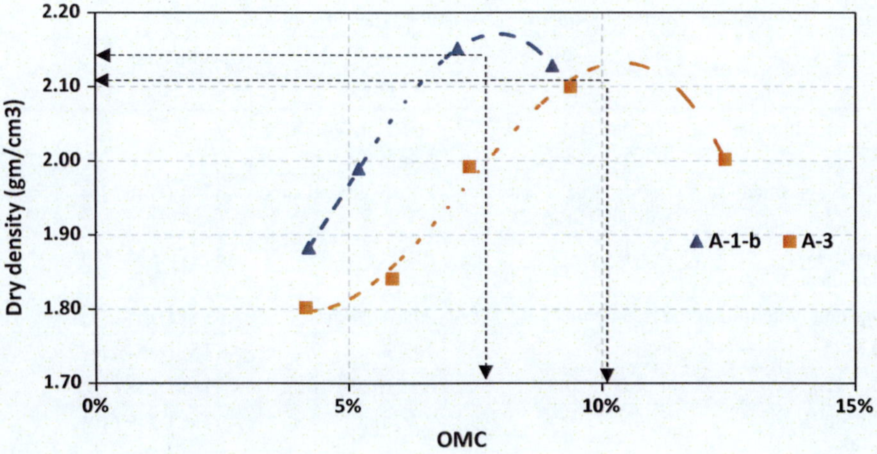

Fig. 4 Modified Proctor test for selected types of subgrades

Fig. 5 Components of LWD field test equipment

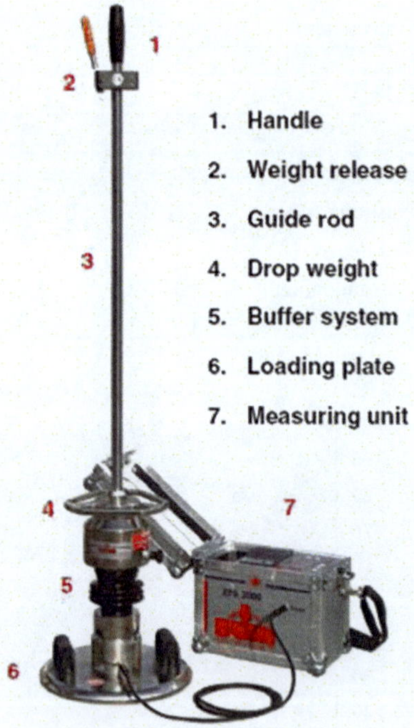

Analysis of Lightweight Deflectometer (LWD) Data. LWD is defined as a nondestructive testing device used to determine in situ stiffness properties of pavement materials under the effect of dynamic impact loads at in situ conditions. This device provides a single dynamic stiffness back-calculated based on actual wave velocity propagated inside a pavement layer [12]. The influence depth of LWD pulse is at range 1.5–2.0 times the diameter of the loading plate; for this reason, the LWD device is considered as not a suitable device to evaluate in situ stiffness for depth that is often greater than 20 inches (50.8 mm).

The LWD parameters measured during this test are listed below:

- Surface Deflection (S_d): it is derived by double integrating the acceleration data concerning time of pulse waves. The acceleration data are recorded using an accelerometer subjected inside the loading plate.
- Dynamic Modulus (E_d): it is back-calculated from S_d depending on Boussinesq elastic half-space theory that can be presented by the following expression [13].

$$E_d = \frac{(1 - \upsilon^2)\sigma_o a}{\delta}.f \quad (1)$$

where is the E_d (MPa), σ_o is peak applied stress (MPa), a is the radius of loading plate (mm), and υ is the Poisson's ratio, δ is peak vertical deflection (mm), and f = plate rigidity factor which is typically assumed (f = 2) for LWD ZFG model.

- Degree of Compatibility (D_c): a soil parameter that gives an indication about compaction characteristics of the pavement layer and can be determined by dividing the mean surface deflection by the mean pulse velocity of dynamic impact load applied in the surface layer. Generally, the compaction effort is good if the degree of compatibility is less than or equal to 3.5, while further compaction effort is recommended if the Dc is greater than 3.5 [14].

Determination of In situ Density by Core Cutter Method (CCM). Dry unit weight (gm/cm^3), wet unit weight (gm/cm^3), moisture content (%), and degree of compaction (%) can be determined using a core cutter test method according to ASTM D2937. This method can be used successfully whenever soil conditions permit pushing of cutter for sampling and taking it out in the laboratory without much disturbance. AASHTO restricted CCM for determining the in-place density in soils containing particles not larger than 50 mm (2 in.) in diameter.

4 Results and Discussion

Table 2 summarizes the LWD measurements of soil samples collected from the designated sections. The average of surface deflection indicates that the data extracted

Table 2 Summary of LWD results for selected test points

Test No	Station No	Surface deflection (mm)				E_d (MPa)	D_c (ms)
		S_1	S_2	S_3	Mean		
Al-Takahe district (A-1-b subgrade soil)							
1	00 + 100	0.284	0.279	0.280	0.281	80.07	3.110
2	00 + 200	0.244	0.234	0.264	0.247	91.09	2.354
3	00 + 300	0.376	0.391	0.395	0.387	58.14	2.252
4	00 + 400	0.409	0.388	0.405	0.401	56.11	2.250
5	00 + 500	0.576	0.584	0.566	0.575	39.13	2.982
6	00 + 600	0.427	0.423	0.414	0.421	53.44	2.487
7	00 + 700	0.209	0.192	0.206	0.202	111.39	2.567
8	00 + 800	0.429	0.431	0.430	0.430	52.33	2.473
9	00 + 900	0.188	0.186	0.177	0.184	122.28	2.840
10	01 + 00	0.214	0.207	0.208	0.21	107.14	2.750
11	01 + 100	0.359	0.363	0.351	0.358	62.85	2.503
12	01 + 200	0.500	0.502	0.504	0.502	44.82	2.760
13	01 + 300	0.285	0.294	0.299	0.293	76.79	2.780
14	01 + 400	0.165	0.178	0.171	0.171	131.58	2.118
15	01 + 500	0.212	0.226	0.215	0.218	103.21	2.530
16	01 + 600	0.374	0.382	0.383	0.380	59.21	2.465
17	01 + 700	0.203	0.204	0.199	0.202	111.39	2.088
18	01 + 800	0.166	0.156	0.163	0.162	138.89	2.242
19	01 + 900	0.193	0.195	0.193	0.194	115.98	2.285
20	02 + 00	0.205	0.208	0.206	0.206	109.22	2.163
Average		**0.301**	**0.301**	**0.301**	**0.301**	**86.25**	**2.500**
Al-Emam Ali district (A-3 subgrade soil)							
1	00 + 100	0.32	0.319	0.274	0.304	74.01	2.548
2	00 + 250	0.377	0.381	0.37	0.376	59.84	2.436
3	00 + 400	0.223	0.229	0.223	0.225	100.00	3.285
4	00 + 550	0.398	0.392	0.389	0.393	57.25	2.652
5	00 + 650	0.295	0.297	0.29	0.294	76.53	2.734
6	00 + 750	0.303	0.302	0.304	0.303	74.26	2.465
Average		**0.319**	**0.320**	**0.308**	**0.316**	**73.648**	**2.687**

from the integration process for 20 test points performed was recoded as 0.301 mm for A-1-b and 0.316 mm for A-3 subgrade soil, while the values of average dynamic modulus for A-1-b was 86.25 Mpa and 73.648 Mpa for A-3. Lastly, the average value of the degree of compatibility of subgrade soil was 2.50 ms and 2.687 ms for A-1-b and A-3, respectively.

Table 3 Summary of field density results for selected test points

Test No	Station No	Wet unit weight (gm/cm^3)	Moisture content %	Dry unit weight (gm/cm^3)	Degree of compaction %
Al-Takahe district (A-1-b subgrade soil)					
1	00 + 100	1.952	8.91	1.792	82.58
2	00 + 200	1.990	6.76	1.864	85.90
3	00 + 300	2.010	8.00	1.861	85.76
4	00 + 400	1.883	6.90	1.761	81.15
5	00 + 500	1.988	6.40	1.868	86.08
6	00 + 600	1.951	5.80	1.844	84.98
7	00 + 700	1.910	7.00	1.785	82.26
8	00 + 800	1.927	4.80	1.839	84.75
9	00 + 900	1.837	4.70	1.755	80.88
10	01 + 00	1.844	3.50	1.782	82.12
11	01 + 100	2.010	8.00	1.861	85.76
12	01 + 200	1.951	8.80	1.793	82.63
13	01 + 300	1.854	2.40	1.811	83.46
14	01 + 400	1.977	2.60	1.927	88.80
15	01 + 500	1.854	0.50	1.845	85.02
16	01 + 600	1.854	1.50	1.827	84.19
17	01 + 700	1.829	0.90	1.813	83.55
18	01 + 800	1.951	3.50	1.885	86.87
19	01 + 900	1.922	5.60	1.820	83.87
20	02 + 00	1.893	1.67	1.862	85.81
Average		**1.919**	**4.912**	**1.830**	**84.32**
Al-Emam Ali district (A-3 subgrade soil)					
1	00 + 100	2.198	8.91	2.018	95.60
2	00 + 250	2.141	6.76	2.005	95.00
3	00 + 400	2.175	8.00	2.014	95.50
4	00 + 550	2.182	6.90	2.041	96.70
5	00 + 650	2.178	8.00	2.017	95.60
6	00 + 750	2.189	8.80	2.012	95.30
Average		**2.177**	**7.895**	**2.018**	**95.62**

The data of moisture and density of soil samples from the pavement layers at the designated sections are presented in Table 3. The results of A-1-b indicate that the wet unit weight average was 1.919 gm/cm^3 and A-3 average results were 2.177 gm/cm^3, while the average moisture content was 4.912% for A-1-b and 7.895% for

A-3. Lastly, the average of dry unit weight for A-1-b and A-3 was 1.927 gm/cm^3 and 2.018 gm/cm^3, respectively.

To evaluate a bearing resistance of subgrade soils, a comparison analysis was performed between two developed statistical models developed in the previous work by Al-Fatteh Allah et al. (2019) [14]. Experimental results and regression analyses by using SPSS software was employed to develop two models: the first model was represented by considering LWD parameters (E_d, S_d, D_c) as independent variables; see Eq. (1), while the second model was represented by considering Wc and γd_f as independent variables to predict CBR value as a dependent variable for subgrade soils as shown in Eq. (2):

$$CBR\ predicted\ = 17.225 + 0.051 * E_d + 0.023 * D_c^2 - 0.25 * D_c \tag{2}$$

$$CBR\ predicted\ = 776.99 - 38 * W_c + 275 * {}^{2}_{\gamma df} - 912 * {}_{\gamma df} \tag{3}$$

The bivariate Pearson correlation between variables for statistical analysis was present a correlation between dependent variables CBR and independent variables (LWD parameters), that caused to absent the effect of surface deflection (S_d) and illustrate E_d has the most significant correlation to bearing soil capacity, then S_d, and lastly D_c [15]. A comparison between two predicted models was conducted; the bearing resistance of the subgrade layer A-1-b depending on LWD parameters ranged 23.78 to 28.71% with an average of 26.22%, and for A-3 it ranged 26.62 to 27.56% with an average of 26.98%, while the bearing resistance of subgrade depending on physical properties ranged 20.59 to 57.75% with an average of 32.92% for A-1-b and 50.61–52.70% with an average of 52.7 for A-3. See Fig. 6.

Comparison and verification between results for two predicted models to determine the bearing resistance of subgrade according to LWD parameters once and to their basic physical properties otherwise were conducted and they show that the

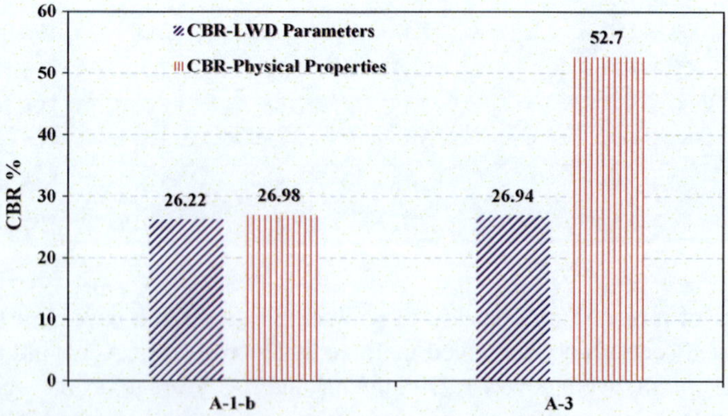

Fig. 6 CBR results for statistical models

in situ CBR value is lower than that of the laboratory results, this decrease in results is due to the difference in environmental conditions, such as a change in moisture content including the temperature and evaporation effect, in addition to the effect of boundary restricted and applied compaction effort which is an effect in the degree of compactions, which affect during field tests and hence will change the particle size distribution at remolded CBR specimens of the laboratory tests. Also, the results indicate that the ratio between two developed models (CBR predicted according to LWD parameters/CBR predicted according to the physical properties) were 1.03; this result indicates that the developed models were acceptable to predict bearing resistance for A-1-b subgrade soils, while the difference between predicted CBR value for A-3 subgrade soil is equal to 1.96 due to the effect of gypsum content of subgrade which employed to build the models and hence caused to increase CBR value for the developed model that agreed completely with Ahmed [16] reported that the compressive strength values for poorly graded sandy soil samples stabilized with recycled gypsum increased from 14.42 kPa to 25.43, 81.99 and 331.18 kPa due to adding 5, 10, and 20% content of recycled gypsum, respectively. This can be explained by the addition of recycled gypsum to the soil causing cementation or hardening of soil particles; thus, cohesion strength between soil particles is developed for this reason and this developed model is considered unreliable to predict the CBR value.

5 Conclusion

From experimental works of this research study, the following can be concluded:

- Modeling the CBR value for A-1-b subgrade soil concerning the dynamic measurement of LWD parameter inputs is achievable and satisfactory in terms of predicting the bearing resistance of subgrade soils.
- Modeling the CBR value for A-3 subgrade soil regarding their basic physical properties is unsatisfactory to the prediction of the bearing resistance of subgrade soils.
- The experimental works show that the LWD parameters have the most significant correlation than the field dry density and moisture content to predict bearing resistance of subgrade soils.

References

1. Shaban AM, Cosentino PJ (2017) Characterizing structural performance of unbound pavement materials using california bearing ratio tests. J Testing Evaluation. ASTM, 3–45
2. Jitareekul P, Sawangsuriya A, Singhatiraj P (2017) Integration of pavement layer evaluation using LWD for road safety control. Procedia Engineering. (May), 111–117

3. Fleming P, Frost M, Lambert J (2009) Lightweight deflectometers for quality assurance in road construction, Bearing Capacity of Roads; Railways and Airfields: Proceeding of the 8th International Conference, 809–818
4. Kongkitkul W, Saisawang T, Thitithavoranan P, Kaewluan P, Posribink T (2014) Correlations between the surface stiffness evaluated by light-weight deflectometer and degree of compaction. In Tunneling and Underground Construction, 65–75
5. ASTM D2487 (2011) Standard Practice for Classification of Soils for Engineering Purposes (Unified Soil Classification System)
6. AASHTO M145 (2012) American association of state and highway transportation officials. Standard Specification for Classification of Soils and Soil-Aggregate Mixtures for Highway Construction Purposes
7. ASTM D1557 (2012) Standard Test Methods for Laboratory Compaction Characteristics of Soil Using Modified Effort
8. ASTM D4643 (2008) Standard Test Method for Determination of Water (Moisture) Content of Soil by Microwave Oven Heating 1
9. ASTM standard D1883 (2014) Standard Test Method for CBR (California Bearing Ratio) of Laboratory-Compacted Soils. ASTM Int West Conshohocken, PA
10. BS1377–3 (2010) Standard Test Methods for Liquid Limit, Plastic Limit, and Plasticity Index of Soils 1. 1990 (March)
11. BS1377–3 (1990) Soils for civil engineering purposes—Chemical and electro-chemical tests (1)
12. Ryden N, Mooney MA (2009) Analysis of surface waves from the light weight deflectometer
13. Shaban AM, Cosentino PJ (2016) Comparative analyses of granular pavement moduli measured from lightweight deflectometer and miniaturized pressuremeter tests. Transportation Research Boar
14. Shaban AM (2016) Evaluation of unbound pavement layers moduli using the miniaturized pressuremeter test
15. Al-Fatteh Allah R, Al-Busaltan SMA (2019) Determining field bearing resistance of subgrade soils using physical characteristics. Fourth Scientific Conference for Engineering and Post-graduate Research, 8–14
16. Ahmed KI (2013) Effect of gypsum on the hydro-mechanical characteristics of partially saturated sandy soil [Ph.D. thesis]. Geoenvironmental Research Centre, Cardiff School of Engineering, UK: Cardiff University

Stability of Gypseous Soil Slopes Using the Upper Bound Theorem of Limit Analysis

Akram H. Abd and Manar M. Abdullah

Abstract The upper bound theorem of limit analysis is utilized to study the stability of gypseous soil slopes. A simplified equation is proposed to estimate the effect of gypsum content on the strength parameters of saturated soil. This equation is found to be less conservative than the equation proposed by other researchers in the literature, hence, it may lead to a more economical design. The modified strength parameters were implemented by instability analysis software called OptumGE assuming dry and saturated soil with different gypsum contents and different slope inclinations. The effect of gypsum content on the stability of gentle slope is found to be more damaging than that for steep slopes. For saturated gentle soil slopes, the reduction in factor of safety can reach up to 90 or 70% depending on the equation used to estimate the soil strength parameters. The aim of the study to implement the upper bound theorem to analyze gypseous slope stability. Results are formed as charts that can be used by practitioners to quickly estimate the reduction in factor of safety due to the presence of gypsum.

Keywords Gypseous soil · Gypsum · Slope stability · Upper bound · Limit analysis

1 Introduction

In geotechnical engineering, the evolution of slope stability analysis has followed the developments in soil and rocks mechanics. The need for developing such analysis is raised mainly because of the rapid expansion of cities and extending highways via natural geotechnical difficulties. However, some types of soil need special attention when dealing with stability problems as they can be metastable.

A. H. Abd (✉) · M. M. Abdullah
Civil Engineering Department, College of Engineering, Tikrit University, Tikrit, Iraq
e-mail: ahabd@tu.edu.iq

M. M. Abdullah
e-mail: engineera.m95@gmail.com

M. O. Karkush and D. Choudhury (eds), *Modern Applications of Geotechnical Engineering and Construction*, Lecture Notes in Civil Engineering 112,
https://doi.org/10.1007/978-981-15-9399-4_17

Gypseous soils cover large arid and semiarid areas around the world. The main problem with these soils is that although they seem very hard or stiff when dry, they may exhibit large deformations upon wetting. This has put these soils under the classification of collapsible or metastable soils. Collapsible soils are defined by Clemence and Finbarr [1] as "any unsaturated soil that goes through a radical rearrangement of particles and great loss of volume upon wetting". Gypsum is the mineral salt represented in hydrated calcium sulphurate ($CaSO_4.2H_2O$) formed from 20.9% combined with water (H2O), 46.6% sulfur trioxide (SO_3), and 32.5% calcium oxide (CaO) [2].

In Iraq, gypseous soils cover 12.2% of the total area [3]. Another researcher Ismael [4] wrote that gypsum soil percentage equals 31.7% of the total soil of Iraq. These differences in percentage might be attributed to the limit of gypsum content at which the soil is considered gypseous.

The upper bound of limit analysis is being used since the mid-twentieth century to analyze slope stability. The very first implementation of this method was by Drucker and Prager [5] for structural analysis. Chen et al. [6] applied the general theory for the perfect plasticity for stability problems in the mechanism of the soil. The basic assumption is the soil is to deform according to the flow rule which said that the relationships for the soil are ideal and the soil is perfect plastic. Michalowski [7] used the upper bound theorem to consider the effect of pore water pressure. Utili [8] investigated the stability of cohesive frictional slopes with cracks employing the kinematical approach of limit analysis. The upper bound theorem of the limit analysis method was merged with the finite element method by Lyamin and Sloan [9]; further work was done by Krabbenhoft et al. [10]. These methods, known as finite element limit analysis, do not require assumptions to be made regarding the failure mode [11].

Quantifying the reduction in shear strength parameters of gypseous soils is essential for the calculations of slope stability. This problem arises especially when using a computer software because current software cannot simulate the effect of the presence of gypsum on the stability. Al-Mufty [12] has come up with equations based on stochastic analysis to estimate the reduction in soil strength parameters due to the presence of gypsum in both dry and saturated soils. However, his equation for the saturated soil case seems to be overconservative. Hence, an equation is proposed in this study, and results are compared to those obtained based on Al-Mufty equations.

2 Al-Mufty's Equations and Parameters Used

The following equations are suggested by Al-Mufty [12], for dry case:

$$c'_r = \chi^2 c'_g + (2\chi - 2\chi^2)c'_{gs} + (1-\chi)^2 c'_s \tag{1}$$

$$\tan\phi'_r = \chi^2 \tan\phi'_g + (2\chi - 2\chi^2)\tan\phi'_{gs} + (1-\chi)^2 \tan\phi'_s \tag{2}$$

For saturated case:

$$c'_r = (1 - S)\left[\chi^2 c'_g + \left(2\chi - 2\chi^2\right)c'_{gs}\right] + (1 - \chi)^2 c'_g \tag{3}$$

with c'_r being the reduced value for the effective soil cohesion corresponding to χ gypsum content and c' being the effective soil cohesion when soil has zero gypsum content. c'_s, c'_g, and c'_{gs} are the effective cohesion for soil, gypsum, and between gypsum and soil, respectively, similarly ϕ'_s, ϕ'_g, and ϕ'_{gs} are the effective angle of shearing resistance for soil, gypsum, and between gypsum and soil, respectively, and S is the degree of saturation. It should be noted that there are no given values for c'_g, and c'_{gs} and $\tan\phi'_g$, and $\tan\phi'_{gs}$ but according to Al-Mufty [12], they are equal in the dry case who recommended ranges for these factors. In the current study for dry case, c'_g, and c'_{gs} and $\tan\phi'_g$ and $\tan\phi'_{gs}$ are both taken equal to 0.75 of c' and $\tan\phi'$. This would be within the range suggested by Al-Mufty [12]. The initial and reduced values of c'_g and $\tan\phi'_g$ were entered to thesoftware (initially c'_s=20 kPa and $\phi'_s = 30°$); as it is sandy soil, these values are chosen arbitrarily just to show the effect of gypsum on the soil strength.

3 Proposed Equation

Estimating the strength parameters for soil with a certain amount of gypsum content can be manifold. Hence, assumptions can be made to simplify the problem and achieve an acceptable estimation. For a slope stability problem, the shear strength along the shear band (or the potential failure surface) mainly comes from soil particles contacting each other.

For the case of dry soil with zero gypsum content, let us assume that the summation of contact points among soil particles along this potential failure surface is equal to P. Now if the soil contains, for example, 10% gypsum, and this gypsum is uniformly distributed within the soil then one can presume that there is 0.1 P contact points is either gypsum–gypsum or gypsum–soil contacts (see Fig. 1). It might be argued that the number of contact points may change during the process of failure which is most likely to be true. However, it is assumed that there are a constant number of contact points erring on the side of safety.

For saturated soil, another assumption is made according to which all contact points along the shear band that involve gypsum are to be eliminated. In other words, all gypsum particles are assumed to have zero contribution to soil resistance when the soil is saturated. This is another conservative assumption because not all of the gypsum is to be dissolved during saturation. Furthermore, even if all gypsum in the soil is dissolved, the soil particles will be reoriented to provide the highest possible shear strength. The above assumption leads to the following mathematical expressions:

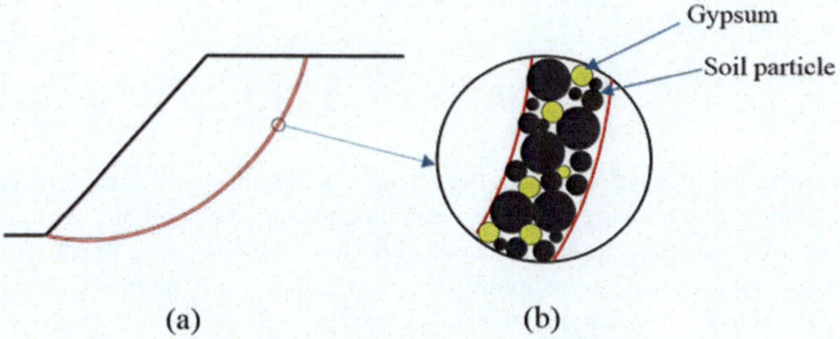

Fig. 1 **a** Slope failure mechanism. **b** Imaginary enlargement of shear failure surface

$$c'_r = (1 - \chi)c' \tag{4}$$

$$\tan\phi'_r = (1 - \chi)\tan\phi' \tag{5}$$

with c'_r being the reduced value for the effective soil cohesion corresponding to χ gypsum content and c' being the effective soil cohesion when soil has zero gypsum content. The above equations are found to be less conservative than those obtained by [12]. This has been illustrated in the results section.

4 Software Used

OptumGE is used in this study to perform different types of analysis among which is a finite element limit analysis. The gravity multiplier, which is here considered as a factor of safety, is presented instead of the critical slope height which is usually sought in the limit analysis of slopes. The highest of the slope is chosen to be 3 m to cover most cases of this study. The software can deal with many methods of analysis and different types of soil structures, by drawing geometry and entering the soil properties, analysis method, and the slope state and then analysis, and the program will give the results. Figure 2 shows the geometry of the problem in the current study.

5 Results and Discussion

The kinematic approach of limit analysis is used to investigate the effect of gypsum on the stability of soil slopes. The reduced strength parameters were implemented by instability analysis software called OptumGE using dry and saturated soil with different gypsum contents and different slope inclinations.

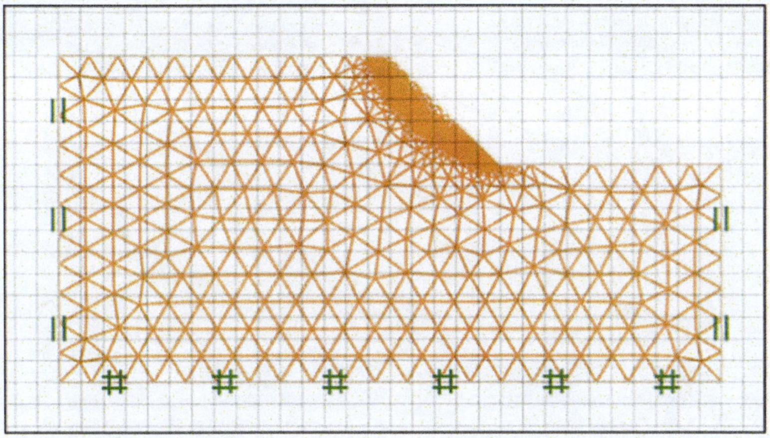

Fig. 2 Geometry, boundaries, and mesh from OptumGE software

The relationship between gypsum content and the reduction percentage in the factor of safety was presented as charts (see Fig. 3). The factor of safety (herein the same as the gravity multiplier) sought for slopes with inclination varies, 45, 60, 75, and 90° with a percentage of gypsum content equal to 0, 10, 20, 30, 40, and 50%.

As can be seen in Fig. 3, the factor of safety decreases as the content of gypsum increases. This holds for the three curves presented in Fig. 3 with the dry case of least reduction in the factor of safety. For saturated gentle soil slopes, the reduction in the factor of safety can reach up to 90% or 70% (see Fig. 3a) depending on the equation used to estimate the soil strength parameters.

The effect of gypsum content on the stability of gentle slopes is found to be more than that for steep slopes. This is related to the angle of shearing resistance which contributes more in the stability of gentle slopes than steep ones because the normal force acting on the failure surface would be larger in gentle slopes than that for steep slopes. So, when reducing the angle of internal friction, the gentle slopes are affected more than the steep ones.

6 Conclusions

The gypsum content can decrease the factor of safety up to 70% or 90% based on the equation used to estimate the reduced shear strength parameters and based on the gypsum content. The proposed equation to estimate the reduction in soil strength parameters for saturated soils was found to be less conservative than Al-Mufty's [12]. The reduction estimated by the proposed equation can be up to 20% less than that calculated based on Al-Mufty [12]. The produced charts can be used by practitioners for a rough estimation of the effect of gypsum content on slope stability.

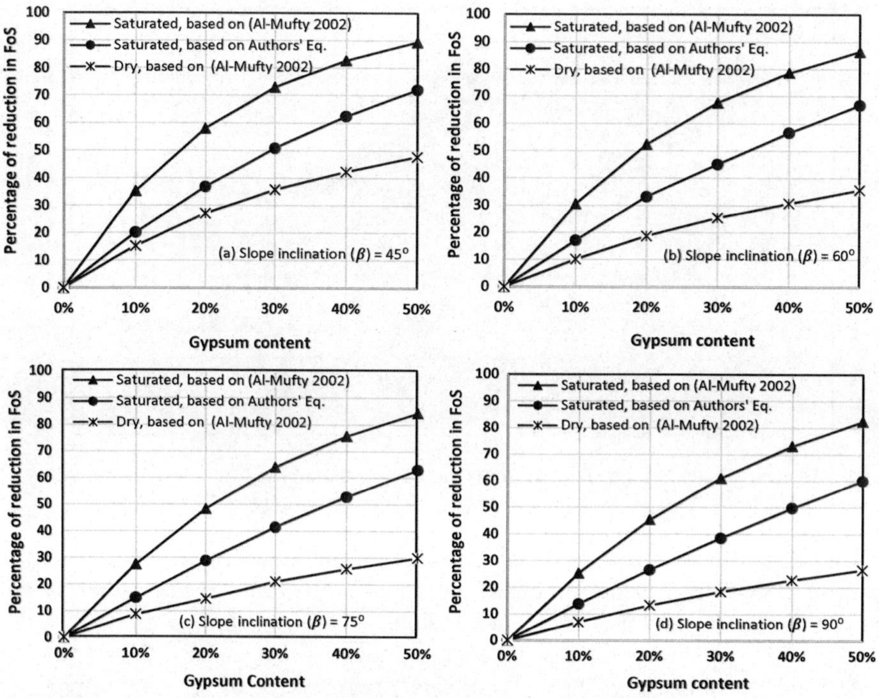

Fig. 3 Effect of gypsum content on the factor of safety comparing results of the current proposed equation and those based on Al-Mufty 2002

References

1. Clemence SP, Finbarr AO (1981) Design considerations for collapsible soils. J Geotech Geoenviron Eng 107:305–317
2. Nashat IH (1990) Engineering Characteristics of Some Gypseous Soils in Iraq, Ph.D. thesis, University of Baghdad
3. Barazanji AF (1973) Gypsiferous Soils of Iraq, Ph.D. Thesis, State of University of Ghent, Belgium
4. Ismael HN (1994) The Use of Gypseous Soils, Symposium o the Gypseous Soils and its Effect on Structures. N.C.C.L, Baghdad
5. Drucker DC, Prager W (1952) Soil mechanics and plastic analysis for limit design. Q Appl Math 10(2):157–165
6. Chen WF, Giger MW, Fang HY (1969) On the limit analysis of stability of slopes. Soils and Found. 9(4):23–32
7. Michalowski RL (1995) Slope stability analysis: a kinematical approach. Geotechnique 45(2):283–293
8. Utili S (2013) Investigation by Limit Analysis on the Stability of Slopes with Cracks. Geotechnique 63(2):140–154
9. Lyamin AV, Sloan SW (2002) Upper bound limit analysis using linear finite elements and non-linear programming. Int J Numer Anal Meth Geomech 26(2):181–216
10. Krabbenhoft K, Lyamin AV, Hjiaj M, Sloan SW (2005) A new discontinuous upper bound limit analysis formulation . Int J Numer Methods Eng 63(7):1069–1088

11. Sloan SW (2013) Geotechnical stability analysis. Géotechnique 63(7):531–571
12. Al-Mufty AA (2002) A probabilistic approach to the variation of shear strength parameters of gypseous soils. Iraqi J Civil Eng 1(3)

Studying the Effect of Water Level for a River on the Shallow Foundation Near Riverbank

Noor Salim Atia, Asma Thamir Ibraheem, and Qassun S. Mohammed Shafiqu

Abstract Erosion of a Riverbank is an inevitable and natural phenomenon and can be defined as a natural dynamic process which includes the removal of materials from the river banks. This phenomenon occurs when the flowing water magnitude exceeds the shearing or strength resistance of sediments or soils on the basal side of a riverbank. In this study, the interpretation of results obtained from laboratory experiments was carried out on a comparatively fine (silty sand) riverbank model to investigate the possible instability factors and compare the bearing capacity and settlement under the shallow foundation (15 × 15 cm) near the riverbank at different river-water levels. The container has dimensions of 2 × 1 m and 1 m height. Also, investigation is carried out on the influence of reinforcing the soil with geogrids. More than 24 experiments were carried out in a steel tank with one side of the glass, the bank model was of angle 90° and a height of 70 cm, 50 cm and 30 cm, and the soil type is SP-SM according to (ASTM 2009) (D2487-98). During the experiments, the shallow foundation resting on the bank was subject to gradual loading and measurement was made for the settlement. The instability of the bank was considered associated with a superimposition of primary (geometrical) factors. It was observed that a geogrid is capable to interpret a notable different response in terms of failure and stability mechanisms. The rise of the river-water level (in flood season) is intended to increase the settlement. And the enhancement of the bearing capacity (strengthening the riverbank by geogrid) reduces the settlement. The bearing capacity increased by 7.7 and 7.2 times without geogrids for Netlon CE121 and Tensar SS2, respectively.

Keywords Riverbank · Silty sand · Water level · Shallow foundation · Geogrid

N. S. Atia (✉) · A. T. Ibraheem · Q. S. M. Shafiqu
Department of Civil Engineering, Al-Nahrain University, Baghdad, Iraq
e-mail: snoor0017@gmail.com

A. T. Ibraheem
e-mail: drasma2005@eng.nahrainuniv.edu.iq

Q. S. M. Shafiqu
e-mail: qassun.almohammed@eng.nahrainuniv.edu.iq

M. O. Karkush and D. Choudhury (eds), *Modern Applications of Geotechnical Engineering and Construction*, Lecture Notes in Civil Engineering 112,
https://doi.org/10.1007/978-981-15-9399-4_18

1 Introduction

Bank failure event is considered one of the major processes of meandering channels, and it is considered a challenging issue for advances in the interpretation and prediction of the planimetric and altimetric evolution for meanders of the river. Most studies dealing with bank failure modeling are concentrated on fine-grained (clay, sand, and silt) and cohesive banks. These banks retreating often occurs along the tidal channels and sandy rivers furthermore, these banks have been usually relating to mass wasting and instability processes [1]. In the geotechnical engineering field of study, the foundation bearing strength is very important. The foundation's design on a level surface ground is based on the soil's mechanical properties such as unit weight, shear strength, and others; also, it is based on the foundation's physical properties such as shape, depth, and width. It is only these two points that must be taken into account in deciding the allowable-bearing pressures for different types of shallow foundations; the settlements under allowable-bearing pressures should not exceed tolerable limits and the factor of safety against shear ultimate failure must be sufficient.

Several methods exist which can highly anticipate the foundations' bearing capacity located in or on level grounds [2–6]. These methods are based on in situ or laboratory test results. Examples include buildings constructed near rivers, foundations that are set up on electrical transmission towers, and bridge piers supported on approach embankments. The bearing capacity and the stability of a foundation built close to the edge of a slope are significant factors that influence the structure performance constructed near a slope. As a foundation built near a river and sloping ground, it could be expected that one of the foundation sides is exposed to the sloping surface and the foundation influence by the river level of water. Accordingly, as the foundation soil approaches the limit state, the plastic area of failure is very specific and it highly affects the bearing capacity of the foundation and the mechanical stability of the slope [7].

Thus, the foundation bearing capacity on a flat ground is more than that sit on the riverbank and near a filled slope. Subsequently, in many cases, the shallow foundation's traditional types are insufficient despite being much easier to build and economical. Base on that, in recent years many researchers have given extra attention to studying the settlement behavior and bearing capacity of footings near rivers and slopes. The main methods used to determine the footing bearing capacity in or on horizontal grounds are consider finely developed. Nerveless, the footing bearing capacity on or near riverbanks and slopes still needs to be investigated. The theoretical methods used to predict footing bearing capacity on a slope [3, 5, 6, 8] are usually developed depending on the footing bearing capacity formula suggested by Terzakihi [9] and are adequate for a specific limit of footing location. Meyerhof suggested a theoretical solution to determine the shallow foundation's ultimate bearing capacity located on the face of a slope [3].

Nardi [1] showed results and interpretation from a series of tests in a glass-walled tank on sandy gravel soil; a bank model was built inside it. The bank was to hydrograph the water level associated with static oscillation. The bank instability was related to geometric factors and a gradual reduction in apparent cohesion during the rising water stage. Dapport [10] described failure mechanisms in two parts of the riverbank along the Arno River (Central Italy) based on field observations. The major objective of this investigation is to study the ultimate square footing bearing capacity placed on top of a riverbank by performing laboratory tests. In the investigation, the relation between the variable parameters and the footing response involving the distance between the crest slope and the footing of the riverbank, the height of riverbank. Studying the effect of water level of a river on footing bearing capacity, and improve the critical states by using two types of geogrids Netlon CE121 and Tensar SS2.

2 Experimental Work

Model Container. The experimental soil box with dimensions of 2.0 × 1.0 m and a height of 0.90 m was manufactured from steel except one of the lateral walls was made of glass to have a clear observation of the detail of the failure mechanism. The soil box is shown in Fig. 1. A pipe was used to supply water to the box, and in the bottom, a drainpipe with a filter was installed to prevent the flow of fine particles. Cork plates with a sharp edge are installed to absorb waves caused due to bank failure. The steel frame is used to support the static load on the footing. A bank model was constructed within the soil tank. The bank model is designed for imitative geometry. The cross-section of the geogrid reinforcement used in the model test is shown in Fig. 2.

Selected Soil. The tested silty sand soil was extracted from the region of Abo-Noaass located in Baghdad, Iraq. The dry unit weight of the tested silty sand ranges 13.9 to 15.0 kN/m^3, and from the grain size distribution shown in Fig. 3, the silty sand has a uniformity coefficient equal to 1.167 and curvature coefficient equal to 1.204.

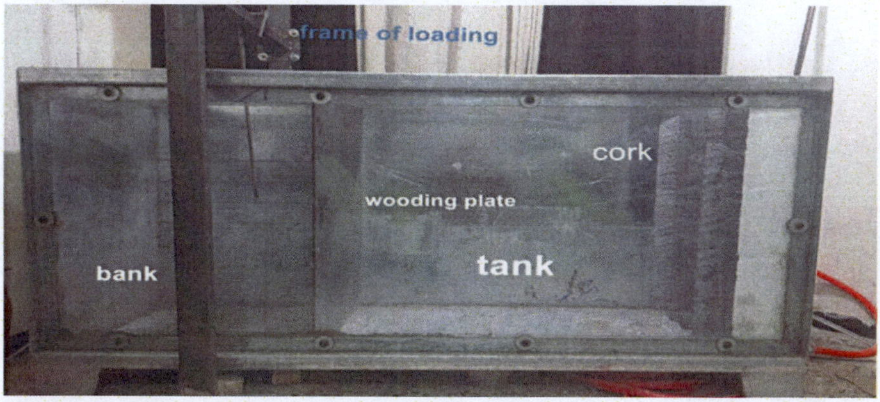

Fig. 1 The soil box: (container and frame of loading)

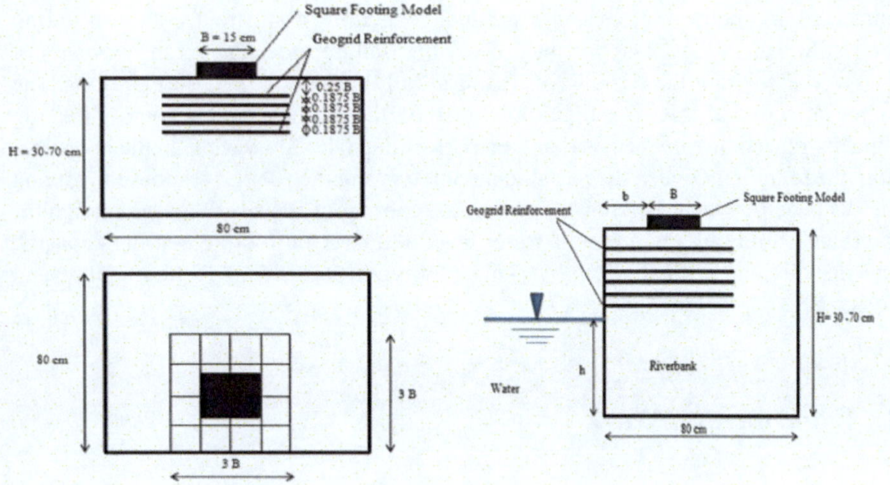

Fig. 2 Geometry and top lane of the geogrid reinforcement and square footing model

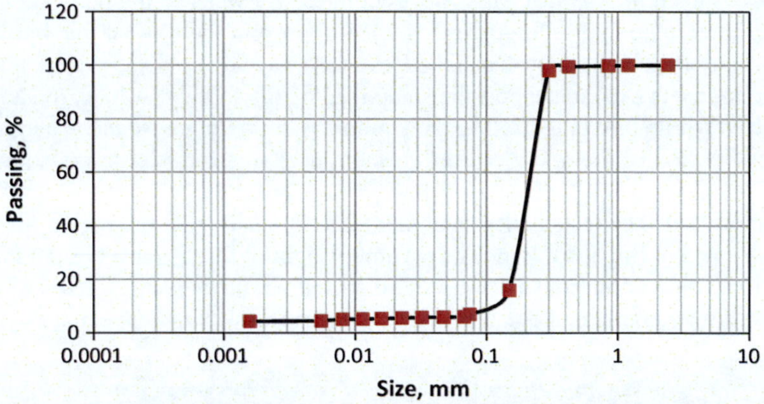

Fig. 3 Grain size distribution for the silty sand soil used

The saturated unit weight of the tested silty sand varies from 17.34 to 18.33 kN/m^3. For all performed tests, the soil was dried up to zero moisture content and then used in a saturated state. The reinforced soil construction models were accurately controlled during the setup of an experimental model to assure a uniform soil density, which corresponds to a relative density approximately around 75.0%, and the saturated unit weight of 18.20 kN/m^3 was obtained. The compacted silty sand internal friction angle measured was approximately 34.50°, at a relative density equal to 75.0%, from a group of direct-shear tests. Other geotechnical parameters of the tested sand are illustrated in Table 1.

Table 1 The geotechnical properties of the used soil

Parameters	Value
Cohesion, c (kPa)	0.00
Internal friction angle of (ϕ)	34.50
Maximum Dry unit weight (kN/m^3)	15.0
Minimum Dry unit weight (kN/m^3)	13.9
Maximum and minimum void ratio (e_{max}., e_{min})	0.900, 0.760
Maximum saturated unit weight (kN/m^3)	18.33
Minimum saturated unit weight (kN/m^3)	17.34
D60, D30, D10 (mm)	0.20, 0.170, 0.120
Uniformity Coefficient, Cu	1.167
Curvature Coefficient, Cc	1.204
Specific gravity Gs	2.66
SO_3 (%)	0.052
Gypsum (%)	0.112
Organic matter	None
Classification (USCS)	SP-SM

Preparation of Sand Slope. A model silty sand slope with a slope angle (β) of 90.0° was prepared by using layers of 100.0 mm thick that were compacted effectively. The test box inner surfaces were marked at 100.0 mm intervals to simplify the preparation process of putting the soil bed in layers and the plate of wood is inserted vertically between the soil (riverbank) and water. The rigid box was half-filled by soil following the steps of pouring with the compaction technique of thin sub-layers of 10 cm thickness formed by homogeneous compacted soil, and the other was half-filled with water that represents the river. Before compacting the soil, a wooden plate is placed inside the container between the soil and water. This plate is slipped slightly before the test began to permit water to flow inside the soil. The process includes putting each layer in the steel box until the target height of the slope was accomplished. Then, the wooden plate is dragged to permit the water contact and flow through the soil. Bank heights of 30, 50, and 70 cm were selected, while the slope angle is 90°.

Reinforcements Methods. The two reinforcing geogrid (Netlon CE121 and Tensar SS2) properties are illustrated in Table 2 and their shapes are shown in Fig. 4 [11]. The dimensions of the geogrids are 0.45 length and 0.45 width and the layers of the geogrids used were 5.

Table 2 The geogrids (Netlon CE121 and Tensar SS2) properties [11]

	Netlon CE121	Tensar SS2
Mesh type	Diamond	Square
Type of polymer	PP	PP
color	Black	Black
Elastic modules	0.39 GPa	0.57/0.99 GPa
Tensile strength	9 MPa	24/30.7 MPa
tensile Peak resistance	6.4 kN/m	14.4/28.2 kN/m
yield strength (Upper)	5 MPa	1/3 MPa
yield strength (Lower)	5 MPa	1/3 MPa

a b

Fig. 4 The two types of geogrids used; a: Netlon CE121 b: Tensar SS2

3 Testing Procedure

The experimental work includes performing a series of 24 bearing load tests on the square footing model located at various locations relative to the slope of the riverbank and different heights of the riverbank. The following variables are taken: the height of the riverbank, the distance of footing from crest slope (dimensions report b/B), the level of river's water, and the type of reinforcement (geogrid). The test model preparation process included compacting the silty sand in layers, each layer ranging 100 mm thick up to 300, 500, and 700 mm heights. The compacting of sand was performed manually using a rammer with length, width, and thickness equal to 150.0, 150.0, and 5.0 mm, respectively, at a relative density equal to 75.0%. The details of the testing procedure are illustrated in Table 3. The formed sand top surface was leveled using a sharpened steel straight plate, and then the footing model was adjusted on the sand surface that was already compacted. The footing model was set on the topsoil surface with the dial gauges which were located on the footing. The applied load was increased from zero at a small rate. Finally, the footing settlement values were recorded using the dial gauges at the end of each increment until failure occurred.

Table 3 The experiment testing program depended on

Square footing mm	Height of riverbank (H) mm	Dimensions report b/B	Level of river water	Reinforcement geogrid (Netlon and Tensar)
150 × 150	300	0	0	-
			1/2 H	Netlon
				Tensar
		1	0	-
			1/2 H	-
		2	0	-
			1/2 H	-
	500	0	0	-
			1/2 H	Netlon
				Tensar
		1	0	-
			1/2 H	-
		2	0	-
			1/2 H	-
	700	0	0	-
			1/2 H	Netlon
				Tensar
		1	0	-
			1/2 H	-
		2	0	-
			1/2 H	-

4 Results and Discussion

For investigating the influence of river-water levels and the footing location to the crest of the slope and the height of the riverbank, a series of 24 tests have been performed on a square footing model resting on a dense silty sand slope (riverbank). The ultimate bearing capacity was determined by dividing the load on the footing area. Figures 5, 6, and 7 represent bearing pressure-settlement variations for riverbank heights of 2B, 3B, and 5B, respectively, and for the two levels of river water 0 and 1/2 H, where H represents the height of riverbank, and the variations by using two types of geogrids.

In Fig. 8, for the height of riverbank H/B = 2, the bearing pressure with the presence of geogrids increased by 4.1 and 3.8 times that without geogrids for Netlon CE121 and Tensar SS2, respectively. And the increase becomes 7 and 6.7 times that without geogrids for the height of riverbank H/B = 3. While for the height of riverbank H/B = 5, the bearing pressure increased by 7.7 and 7.2 times that without

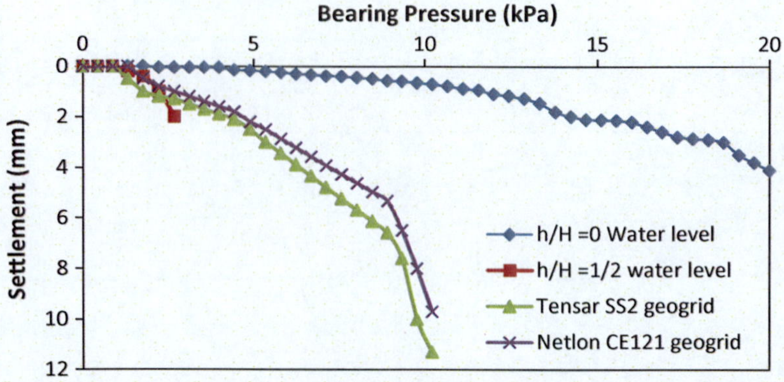

Fig. 5 Variation of bearing pressure with settlement for height of riverbank = 2B

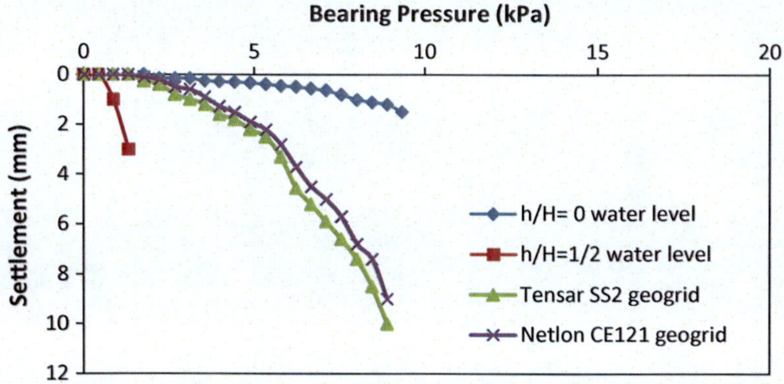

Fig. 6 Variation of bearing pressure with settlement for height of riverbank = 3B

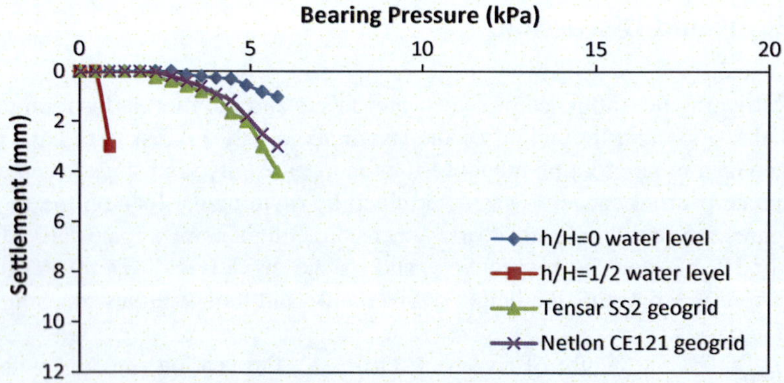

Fig. 7 Variation of bearing pressure with settlement for height of riverbank = 5B

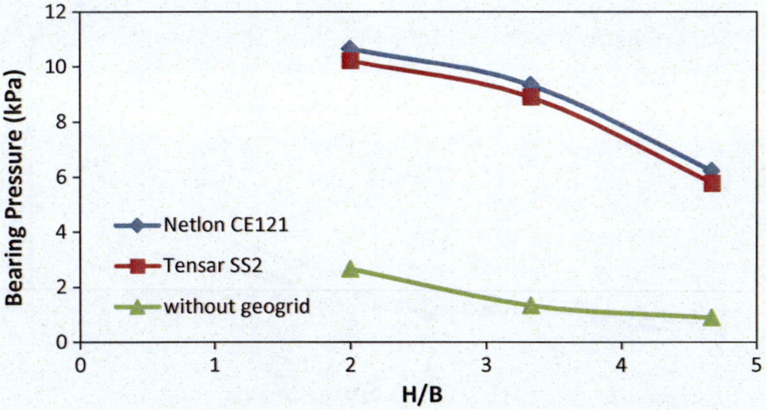

Fig. 8 Comparison of bearing pressure height of a riverbank with and without geogrid

geogrids for Netlon CE121 and Tensar SS2, respectively. To evaluate the effect of the distance between crest and footing (b/B) on the soil bearing pressure, Figs. 9 and 10 have been presented.

Figure 9 presents the results for no water level which shows that enlarging the distances to b/B = 1 increases the bearing pressure by about 3, 1.5, and 1.6 times for H/B 2, 3, and 5, respectively. Increasing the distance to twice the width of the footing will result in higher bearing pressure by about 2, 2.7, and 3.5 times for H/B = 2, 3, and 5, respectively. Figure 10 presents the results for water level h = 1/2 H which shows that increasing the distances to be increased the bearing pressure 2.3, 3.8, and 4.9 times for H/B 2, 3, and 5, respectively. And increasing the distance to twice the width of the footing will result in higher bearing pressure by about 6.4,

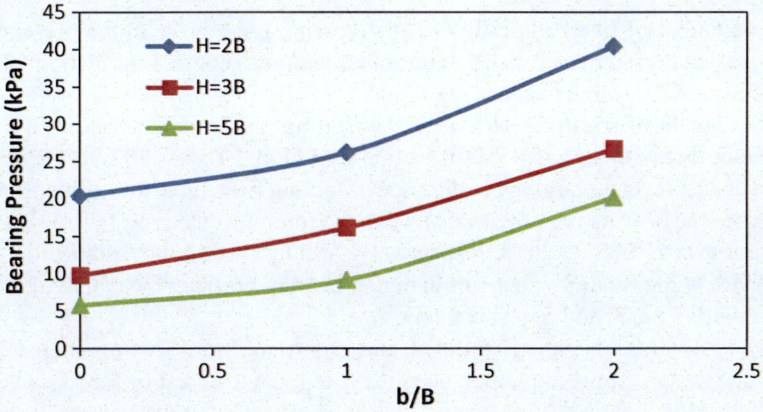

Fig. 9 Variation of bearing pressure distance from crest b/B for different heights of riverbank in river-water level = 0

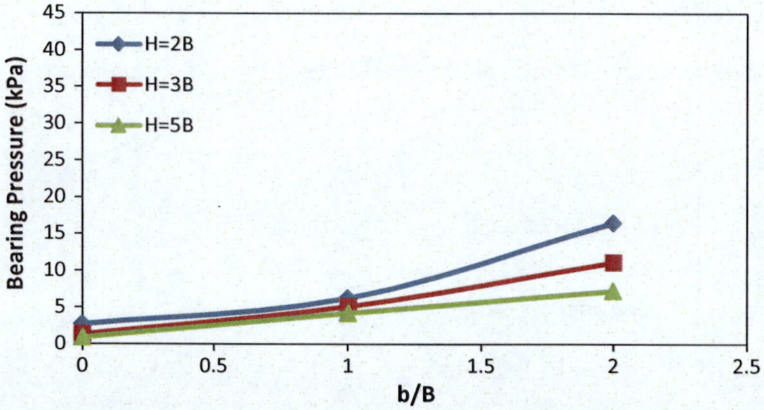

Fig. 10 Variation of bearing pressure distance from crest b/B for different heights of riverbank in river-water level = 1/2 H

8, and 8.3 times for H/B = 2, 3, and 5, respectively. From Figs. 9 and 10 it can be observed that, when the level of water rised to 0.5 h and the distance of footing to the crest increased to b/B = 1, the bearing pressure decreased by 73.3%, 68.5%, and 55.3% for H/B of 2, 3, and 5, respectively. However, by increasing the distance to twice the width of footing, the bearing pressure decreased by 59.4, 58.4, and 57.6% for H/B of 2, 3, and 5, respectively.

5 Conclusions

From the study, the following conclusion may be drawn:

- For the height of riverbank H/B = 2, the bearing pressure with the presence of geogrids increased by 4.1 and 3.8 times that without geogrids for Netlon CE121 and Tensar SS2, respectively.
- For the height of riverbank H/B = 5, the bearing pressure increased by 7.7 and 7.2 times that without geogrids for Netlon CE121 and Tensar SS2, respectively.
- For no water level, enlarging the distance of footing from crest to width of footing increases the bearing pressure to that when the footing at the crest by 1.3, 1.5, and 1.6 times for H/B 2, 3, and 5, respectively. And increasing the distance to twice the width of the footing will result in higher bearing pressure by about 2, 2.7, and 3.5 for H/B = 2, 3, and 5, respectively.
- For water level 0.5 H enlarging the distance to the width of footing, increase the bearing pressure concerning that when the footing at the crest by 2.3, 3.8 and 4.9 times for H/B 2, 3, and 5, respectively. And increasing the distance to twice the width of the footing will result in higher bearing pressure by about 6.4, 8, and 8.3 times for H/B = 2, 3, and 5, respectively.

- When the level of water rised to 0.5 H and the distance of footing to the crest increased to b/B = 1, the bearing pressure decreased concerning that when there is no water level by 73.3%, 68.5%, and 55.3% for H/B of 2, 3 and 5, respectively.

References

1. Nardi L, Rinaldi M, Solari L (2012) An experimental investigation on mass failures occurring in a riverbank composed of sandy gravel. Geomorphology 163:56–69
2. Abdi A, Abbeche K, Athmania D, Bouassida M (2019) Effective width rule in the analysis of footing on reinforced sand slope. Studia Geotechnica Et Mechanica 41(1):42–55
3. Meyerhof GG (1957) The ultimate bearing capacity of foundations on slopes, Proc. 4th Int. Conf. on Soil Mechanics and Foundation Engineering, vol. 1
4. Meyerhof GG (1963) Some recent research on the bearing capacity of foundations. Can Geotech J 1(1):16–26
5. Hansen JB (1970) A revised and extended formula for bearing capacity
6. Vesic AS (1973) Analysis of ultimate loads of shallow foundations, Journal of Soil Mechanics & Foundations Div 99(sm1)
7. Keskin MS, Laman M (2013) Model studies of bearing capacity of strip footing on sand slope . KSCE J Civil Eng 17(4):699–711
8. Saran S, Sud VK, Handa SC (1989) Bearing capacity of footings adjacent to slopes. J Geotech Eng 115(4):553–573
9. Terzaghi K (1943) Theoretical soil mechanics. JohnWiley & Sons, New York
10. Dapporto S, Rinaldi M, Casagli N (2001) Failure mechanisms and pore water pressure conditions: analysis of a riverbank along the Arno River (Central Italy). Eng Geol 61(4):221–242
11. Fakhraldin MK (2013) Properties Measurements and Applications of some Geogrids in Sand, Ph.D. Thesis, Al-Nahrain University

Determining the Stress-Settlement Distribution of a Gravity Dam Foundation Considering Different Water Levels Using Finite Element Method

Hazim Alkhafaji, Meysam Imani, and Ahmad Fahimifar

Abstract This paper investigates determining the stress-settlement distribution considering different water levels using the finite element method. A full 3D model of the Shafaroud dam which is under construction in the north of Iran was created. The seepage forces were considered in the numerical models. The effect of different magnitudes of the gradient ratio $i(\gamma_w/\gamma_{sub})$ was investigated in the numerical models. The obtained results of the numerical solutions show improvement with a more precise value of about 11–13% compared with other studies using analytical solution methods. Using 3D numerical modeling for this type of problem can be a useful guideline for designing foundations when complicated geometries and seepage forces are present.

Keywords Dam foundation · Numerical method · PLAXIS 3D · Seepage · Rock mass

1 Introduction

To investigate the relationship between settlement and footing stress, it was important to determine the ultimate bearing capacity from each of the footing load tests in a consistent manner. In the absence of a well-defined plunging failure that can be used to identify the ultimate capacity, several methods may be used to interpret either the "allowable" bearing capacity or the "ultimate" bearing capacity of the foundations from footing load tests. In most geotechnical practice, the allowable bearing capacity

H. Alkhafaji (✉) · A. Fahimifar
Amirkabir University of Technology, 424 Hafez Avenue, Tehran, Iran
e-mail: hazim.alkhafaji@aut.ac.ir

A. Fahimifar
e-mail: fahim@aut.ac.ir

M. Imani
Amirkabir University of Technology, Garmsar Campus, Garmsar, Iran
e-mail: imani@aut.ac.ir

M. O. Karkush and D. Choudhury (eds), *Modern Applications of Geotechnical Engineering and Construction*, Lecture Notes in Civil Engineering 112,
https://doi.org/10.1007/978-981-15-9399-4_19

is obtained by first determining the ultimate bearing capacity and then reducing this value by applying an appropriate factor of safety. This approach does not consider the deformation of the footing, and settlement is then estimated by a separate (decoupled) calculation. If the estimated settlement is unacceptably high, the "allowable" value is further reduced until the estimated settlement is within tolerable limits set by the project. When results from an actual load test are available, either from some smaller plate load or prototype-scale tests or from a full-scale proof load test, the ultimate bearing capacity may be estimated directly from the footing performance.

The authors [1] has introduced the stress-settlement curve methods to obtain the ultimate bearing capacity value from the test data, especially when there may be several tests available. The numerical models are used where the continuum is divided into a finite number of elements; failure in one point only results in a redistribution of the stresses to the neighboring elements. Later [2, 3] used PLAXIS 3D to investigate bearing capacity using stress-settlement curve methods and showed that the obtained results using PLAXIS 3D are more precise compared with the results obtained from different methods. However, a few studies like [4] obtained the impact of seepage force on the ultimate bearing capacity of rock mass foundations, using analytical solutions, upper bound method.

The aim of the study is mainly focused on determining the stress-settlement distribution considering different water levels and seepage force for large-scale and complicated geometry problems. A dam mass foundation is considered to be a good example for this case. A gravity dam distributes the forces and stresses placed upon it downward and downstream. The mass of the gravity dam and its foundation key trench prevent it from moving under the force of water pressure. Therefore, in the present paper, a full 3D model of the Shafaroud dam which is under construction in the north of Iran was created using PLAXIS 3D [5]. The seepage forces through the rock foundation were considered in the numerical models, and the stress settlement was obtained for different seepage forces in the rock foundation of the dam.

2 Shafaroud Dam

Dam Geometry. Figure 1a shows the 128 m high, 70 m wide (**B**), and 130 m long roller compacted concrete (RCC) dam, named Shafaroud, which is under construction in Gilan Province, north of Iran. The dam site is shown in Fig. 1b. The full 3D model of this dam and its rock foundation and abutments were constructed in the current paper.

Material Properties. As shown in Fig. 2, the dam body is constructed on a rock mass with layers of two kinds of materials: a mixture of micro conglomerate and tuff sandstone with siltstone interlayers. The properties of these layers are presented in Table 1.

Rock Mass Material Models. The 2002 edition of the Hoek–Brown model [6] has been implemented in PLAXIS 3D to simulate the behavior of rock-type materials. Shear failure and tension failure are described by a nonlinear strength curve as the

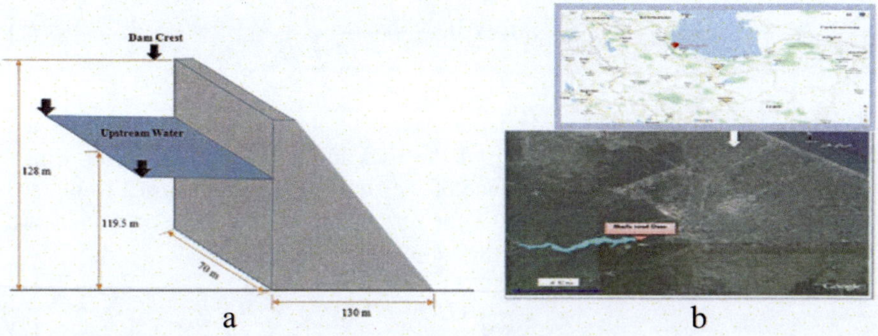

Fig. 1 **a** The Shafaroud dam geometry, **b** The Shafaroud dam location (37° 32´ N and 49° 08´ E) in Iran

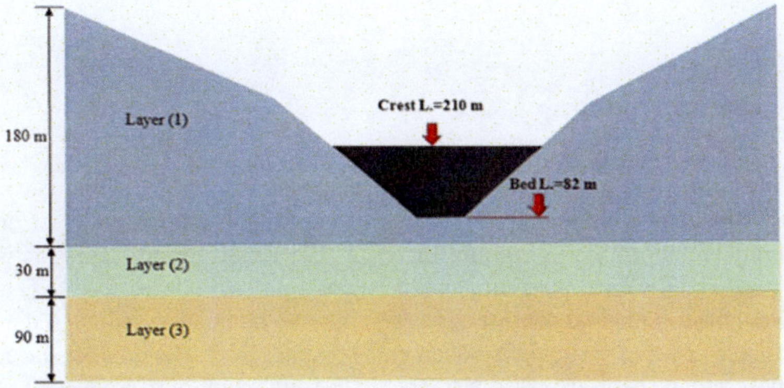

Fig. 2 The rock mass layers surrounding the Shafaroud dam

Table 1 Properties of the rock mass foundation of the Shafaroud dam

Layer	γ_{sat} (kN/m^3)	σ_{ci} (MPa)	m_i	GSI	D	Poisson's ratio
1	25.6	72.9	21	33	0	0.29
2	25.56	53.9	17	30	0	0.31

following relationship between the major and minor effective principal stresses at failure:

$$\sigma_1' = \sigma_3' \, \sigma_{ci} \left(m_b \frac{\sigma_3'}{\sigma_{ci}} + S \right)^a \tag{1}$$

where σ_{ci} is the uniaxial compressive strength of the intact rock and m_b is given by

$$m_b = m_i \exp\left(\frac{GSI - 100}{28 - 14D}\right) \quad (2)$$

in which, m_i is the value of m for intact rock and can be obtained from the experiments, GSI is the geological strength index of the rock mass, and D is a factor which depends upon the degree of disturbance of the rock mass, which varies from zero for undisturbed in situ rock masses to 1 for very disturbed rock masses. s and a are constants for the rock mass given by the following relationships:

$$s = \exp\left(\frac{GSI - 100}{9 - 3D}\right) \quad (3)$$

$$a = \frac{1}{2} + \frac{1}{6}\left(e\frac{-GSI}{15} - e\frac{-20}{3}\right) \quad (4)$$

3 Numerical Modeling

Boundary Condition. In this study, numerical analysis of the stress-settlement curve of the rock foundation subjected to the load of the Shafaroud dam was performed using a finite element-based program, PLAXIS 3D. Two different cases including dry rock mass and the rock mass subjected to seepage were modeled. The base of the models was constrained in three directions and the side boundaries were fixed in two directions in a way that vertical displacements were allowed. The contact surface of the dam base and the bedrock was considered to be rough. For this purpose, the nodes under the dam foundation were fixed in the horizontal direction. Figure 3 shows the 3D numerical model of the dam body and the surrounding rock mass.

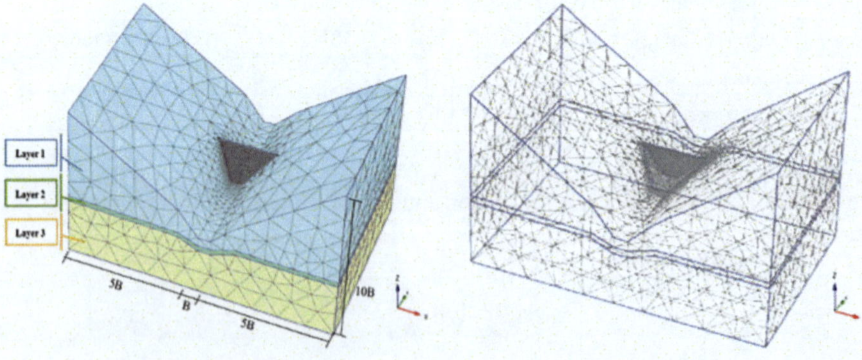

Fig. 3 3D Numerical model of the Shafaroud dam layers and problem geometry

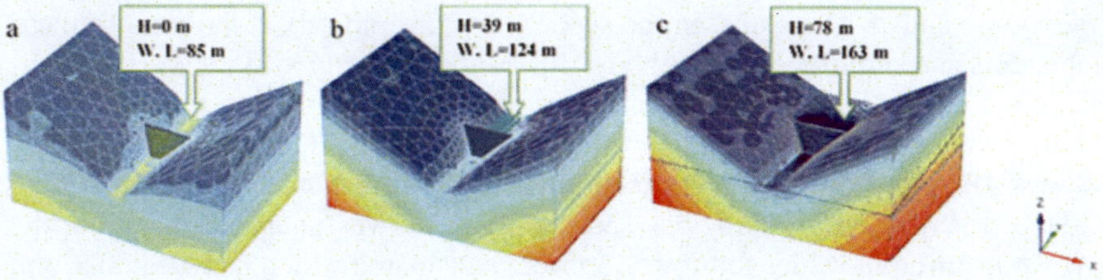

Fig. 4 PLAXIS 3D of model and simulate the dam geometry considering effect of hydraulic force, **a** case of $i(\gamma_w/\gamma_{sub}) = 0$, **b** case of $i(\gamma_w/\gamma_{sub}) = 0.3$, **c** case of $i(\gamma_w/\gamma_{sub}) = 0.6$

Flow Boundary Condition. The different values of gradient ratio verse factor of safety were studied by [7], and the worst case for factors of safety against collapse failure was introduced, as a function of four different values of the gradient ratio namely, 0, 0.3, and 0.6 for dams in a parametric study. The results showed that the factor of safety of collapse failure decreased when the gradient ratio increased. These three different values of the gradient ratio were considered in the present study.

Figure 4 shows the boundary condition for the groundwater flow for the Shafaroud dam with different upstream water levels. Three different values for the gradient ratio $i(\gamma_w/\gamma_{sub})$ were considered which include 0, 0.3, and 0.6 in which, i = H/L is the hydraulic gradient, H is the height of water behind the dam, L is the waterway path length beneath the dam foundation base, and γ_w and γ_{sub} refer to the unit weights of water and submerged rock mass, respectively.

4 Stress-Settlement Curve

Used Method. The stress-settlement curve is usually used for obtaining the bearing capacity. This curve can be obtained by applying an incremental stress to the foundation material and recording the corresponding settlement. This procedure can be performed in the field by plate load test or more simply, by using a numerical software like the latter was applied in the current research.

In the case of a simple shallow foundation subjected to a central vertical load, the incremental load can easily be applied uniformly to the bedding material. But in a foundation of a gravity dam which is subjected to the nonuniform stress of its body weight, upstream water, and multidirectional seepage, the stress distribution exerted on the foundation material is so complicated and its magnitude in different parts of the base foundation is difficult to be anticipated. Therefore, in the current study, the numerical analysis of bearing capacity comprises two steps. In the first step, several full 3D models were constructed considering different values of $i(\gamma_w/\gamma_{sub})$. Then, the stress distribution and magnitude in different parts of the contact surface exactly beneath the dam body were obtained from the numerical analysis. It is obvious that the stress beneath the dam is not uniform and it comprises different components in the x-, y-, and z- directions. In the second step, keeping fixed the ratio of these components and omitting the dam body, different stress values in the x-, y-, and z-directions were applied to the contact surface of the dam, and its base foundation and the corresponding settlement were obtained for drawing the load-settlement curve.

Stress Distribution for the Shafaroud Dam Foundation. For performing the first step described in the previous section, the case was analyzed including the dam. Three different values for $i(\gamma_w/\gamma_{sub})$ were considered which include 0, 0.3, and 0.6. The Cartesian coordinate system and the positive stress directions based on the PLAXIS 3D default are shown in Fig. 5. Based on this hypothesis, the contour of stress distribution in the rock foundation beneath the Shafaroud dam was obtained for all considered cases. As an example, the stress contours for the case of $i(\gamma_w/\gamma_{sub}) = 0.6$ is shown in Fig. 6.

Since the stress intensity is changeable in the base of the dam, it is important to divide the base foundation into some parts in a way that the variation of the stress components in each part can be ignorable. For the sake of simplicity, this partitioning was performed considering the lowest possible variation in the normal stress component, σ'_{zz} and the surface tractions σ'_{zx} and σ'_{zy}, simultaneously. As an example, Fig. 7 shows the partitioning of the base foundation for the case of $i(\gamma_w/\gamma_{sub}) = 0.6$. As shown, the base foundation was divided into six parts; in each of them, the stress component intensity is almost constant. This partitioning was performed for all considered cases and the stress components are presented in Table 2.

The data presented in Table 2 is only a guideline for applying the incremental stress components to the dam foundation to obtain the stress-settlement curve. For doing so, keeping fixed the ratio among the stress components of Table 2, incremental stress values for the three directions, x, y, and z, were applied to each part of the foundation and the corresponding stress-settlement curve was obtained. For example, Fig. 8 shows the incremental stress components exerted on the dam foundation for obtaining the stress-settlement curve.

The Stress-Settlement Curve for the Shafaroud Dam Foundation. Based on the discussions of the previous section, one can conclude that six stress-settlement curves were obtained for each $i(\gamma_w/\gamma_{sub})$ ratio, respectively, for the domain of the dam foundation. There are currently four published methods for defining failure

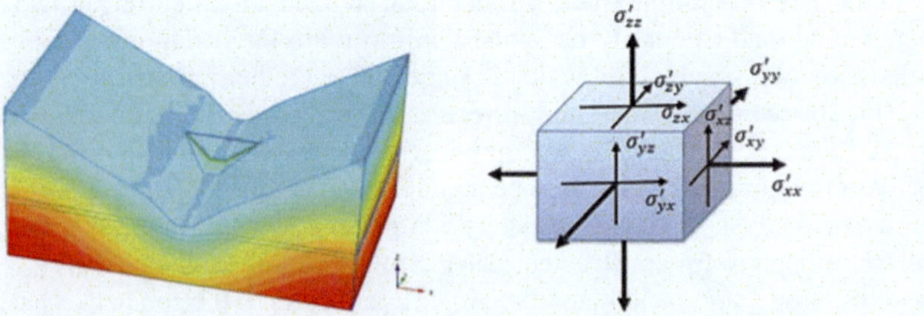

Fig. 5 The Cartesian coordinate system and the positive stress directions based on the PLAXIS 3D default

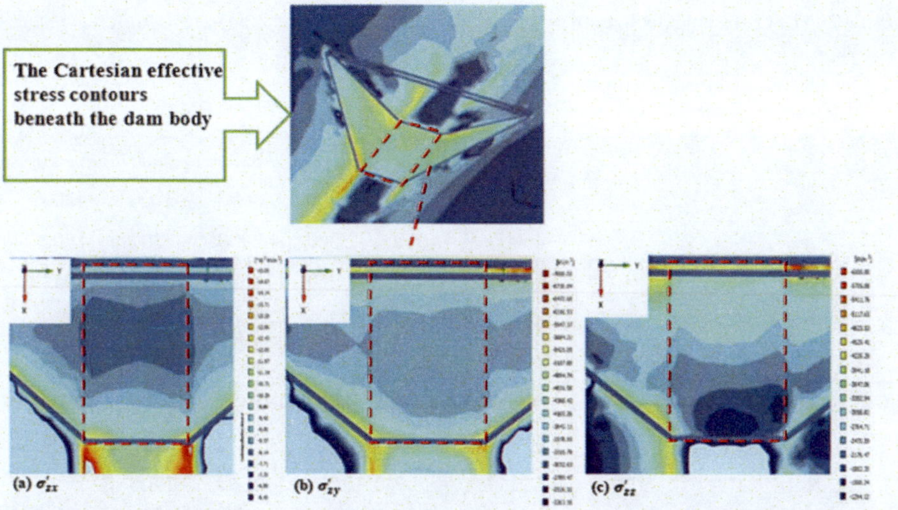

Fig. 6 The Cartesian effective stress contours beneath the dam body in the x–y plane, assuming $i(\gamma_w/\gamma_{sub}) = 0.6$

Fig. 7 The dam foundation base divided into different parts of the same σ'_{zx}, considering $i(\gamma_w/\gamma_{sub}) = 0.6$.

of a shallow foundation, all resulting in different values of q_{ult}. Methods have been introduced in [3] which included tangent intersection, Log–Log, hyperbolic, and 0.1B methods. All these methods use the stress-settlement curve, but each of them may give a different value of bearing capacity. In the present study, the method which resulted in the lowest bearing capacity was chosen. For clarifying purposes, the application of the tangent intersection method is presented in Fig. 9 for the base foundation, subjected to different $i(\gamma_w/\gamma_{sub})$ ratios. For considering this partitioning was performed and the ultimate bearing capacity components are presented in Tables 3, and the minimum values were chosen in each part.

Table 2 The Cartesian effective stress for each part of the dam foundation base for different values of $i(\gamma_w/\gamma_{sub})$

$i\frac{\gamma_w}{\gamma_{sub}}$	Cartesian effective stress (kN/m^2)	Part					
		1	2	3	4	5	6
0	σ'_{zx}	−10,500	−8500	−7500	−8500	−9500	−11,000
	σ'_{zy}	−5700	−4200	−3600	−3600	−3700	−4200
	σ'_{zz}	−4600	−4000	−3200	−2900	−2300	−1800
0.3	σ'_{zx}	−9500	−8000	−7250	−8000	−8750	−10,000
	σ'_{zy}	−4900	−4100	−3500	−3500	−3700	−4150
	σ'_{zz}	−4200	−3600	−2750	−2250	−2000	−1600
0.6	σ'_{zx}	−9000	−8500	−8000	−8500	−8700	−9500
	σ'_{zy}	−4500	−3500	−2750	−2750	−3500	−4000
	σ'_{zz}	−3800	−3000	−2500	−2500	−1700	−1400

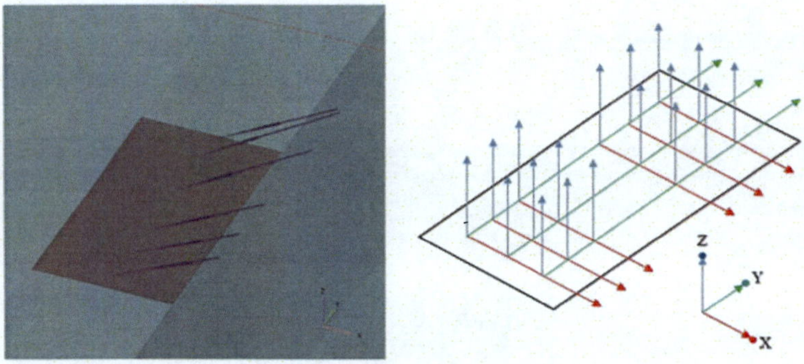

Fig. 8 Applying the incremental stress components to the dam foundation base in the case of $i(\gamma_w/\gamma_{sub}) = 0.6$

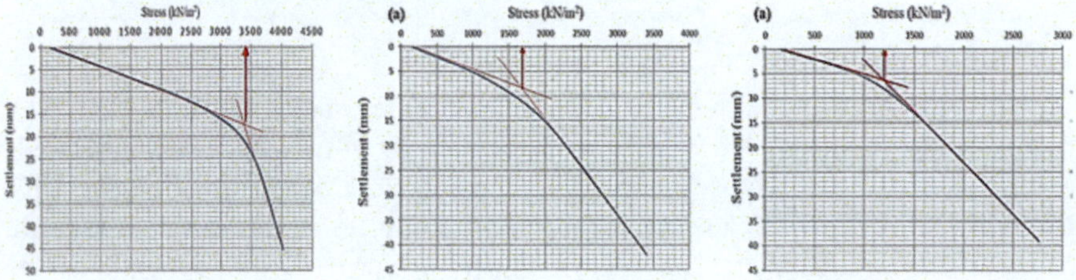

$i(\gamma_w/\gamma_{sub}) = 0.0$ $i(\gamma_w/\gamma_{sub}) = 0.3$ $i(\gamma_w/\gamma_{sub}) = 0.6$

Fig. 9 Application the Tangent Intersection method for obtaining the ultimate bearing capacity of the Shafaroud rock mass foundation for part (3)

Table 3 The ultimate bearing capacity for each part of the dam foundation base for different values of $i(\gamma_w/\gamma_{sub})$

$i\frac{\gamma_w}{\gamma_{sub}}$	Ultimate bearing capacity	Part					
		1	2	3	4	5	6
0	q_u(kN/m^2)	3950	3600	3400	3500	3750	4050
0.3	q_u(kN/m^2)	2050	1850	1700	1750	1900	2150
0.6	q_u(kN/m^2)	1650	1400	1200	1300	1500	1750

5 Discussion and Verification of the Models

Comparison with Experimental Bearing Capacity Determination. The project consultant of the Shafaroud dam decided to compute the bearing capacity of the underlying rock mass using several experimental classical methods to reach the best value for the design. The consultant calculated the bearing capacity using the methods available for dry rock masses and introduced it equal to 6 MPa. Moreover, in the present study, the bearing capacity for the dry case was calculated using the numerical analysis method obtained in the current study, i.e. the PLAXIS 3D and introduced it equal to 3.40 MPa.

Comparison with Analytical Solutions of Bearing Capacity Determination. In the domain of the authors' knowledge, the only published method that deals with the effect of seepage on the bearing capacity of rock masses was proposed by [4]. Using the upper bound method of limit analysis, and considering the failure mechanism shown in Fig. 10, they developed Eq. (5) for determining the ultimate bearing capacity of rock masses subjected to seepage, named (M2):

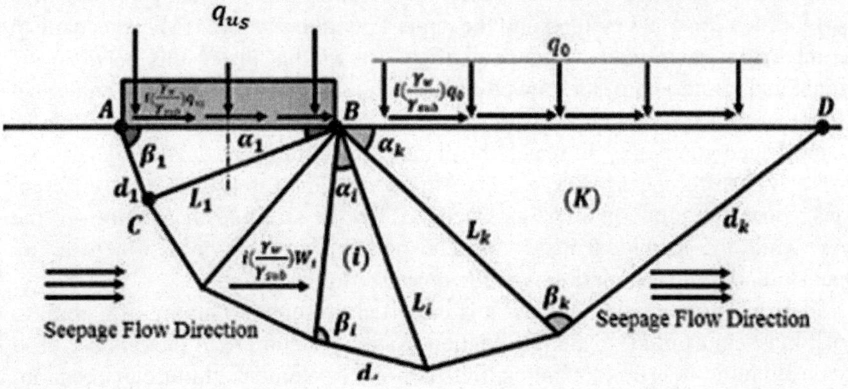

Fig. 10 The failure mechanism proposed by [4] for determining the bearing capacity of a rock mass subjected to seepage forces

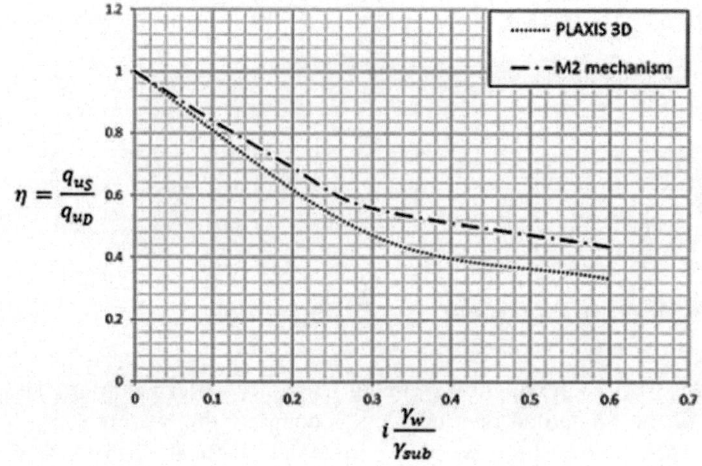

Fig. 11 Comparison between the upper bound method and PLAXIS 3D results for the Shafaroud dam foundation

$$q_{uS} = s^{0.5}\sigma_i N_\sigma^{S2} + q_0 N_q^{S2} + \frac{\gamma B_0}{2} N_\gamma^{S2} \quad (5)$$

where q_0 is the surcharge load, γ is the density of rock mass, B_0 is the foundation width, and N_σ^S, N_q^S and N_γ^S are the bearing capacity factors in the presence of water seepage. That can be obtained using the formulation developed by [4].

Figure 11 presents the comparison between the upper bound method and PLAXIS 3D results for the Shafaroud dam foundation. The magnitude of the ultimate bearing for with/without seepage cases was used to calculate the efficiency factors $\eta = q_{uS}/q_{uD}$ for the numerical method and the upper bound method for (M2) mechanism and comparisons were made. This comparison shows that the results between the analytical and numerical parts showed a numerical solution less than the analytical solution, that the upper bound 2D solution for Eq. 5 can use for large-scale problems and complicated geometries in which the plane strain conditions existed like gravity dams. Not high differences between the two methods, which are based on the different theories. Since the numerical analysis depends on the size and distribution of the meshes, while the analytical results depend on the shape of the assumed failure mechanism under the foundation, and the optimization technique.

For elastoplastic material models, it is important to separate failure of a point in the rock mass and failure of the foundation system. If numerical models are used where the continuum is divided into a finite number of elements, failure in one point only results in a redistribution of the stresses to the neighboring elements. A failure of the foundation system does not have to be the result. Besides, if analytical models are used, the system is considered as one component, and stress levels which reach failure would lead to a system failure.

6 Conclusion

The results of numerical analysis of the Shafaroud dam foundation subjected to seepage forces showed the following:

- The obtained values of the bearing capacity for the dry case was acceptable compared to the suggestion from the consultant of the Shafaroud dam project.
- Due to the existence of seepage in the dam foundation which leads to a large hydraulic gradient, there is a difference between the bearing capacity value for the dry case obtained from the consultant of the project and the real bearing capacity in the case of seepage. The difference between these two cases is about 34–38%.
- The numerical solutions are more precise than analytical solutions (11–13%).
- The results are well applicable with the Shafaroud dam (case study), in other words, the method suggested in this paper is suitable for large-scale and complicated geometry problems in which the plane strain conditions existed like gravity dams.

References

1. Lutenegger AJ, Adams MT (1998) Bearing capacity of footings on compacted sand, Proceedings of the 4th International Conference on Case Histories in Geotechnical Engineering 1216–1224
2. Mansouri M, Imani M, Fahimifar A (2019) Computers and geotechnics ultimate bearing capacity of rock masses under square and rectangular footings. Comput Geotech 111:1–9
3. Shekhawat P, Viladkar MN (2014) Behaviour of Shallow Foundations on Jointed Rock Mass, IIT Roorkee
4. AlKhafaji H, Imani M, Fahimifar A (2019) Ultimate bearing capacity of rock mass foundations subjected to seepage forces using modified Hoek–Brown criterion, Rock Mech Rock Eng
5. PLAXIS b.v. PLAXIS 3D, Reference Manual (2012)
6. Hoek E, Carranza-Torres C, Corkum B (2002) Hoek-Brown failure criterion-2002 edition. Proceedings of NARMS-Tac 1(1):267–273
7. Hansen D, Roshanfekr A (2012) Assessment of potential for seepage-induced unraveling failure of flow-through rockfill dams. Int J Geomech 12(5):560–573

Linear and Nonlinear 3-D Finite Element Analysis for Mat Foundations

Mahmood R. Mahmood, Mohammad M. Abbas, and Mohammad M. Mahmood

Abstract Linear and nonlinear three-dimensional finite element analyses have been used to perform a numerical analysis to investigate the effect of applied load on the raft foundation. ANSYS (version 18.20) program is adopted for this paper. An 8-node brick element in ANAYS is used to simulate the concrete of the foundation and the soil beneath the foundation which are denoted by "solid 65 for concrete and solid 45 for soil", and the stress–strain curve is used to simulate nonlinearity for both soil and concrete. It was found that there is a clear difference between linear and nonlinear behavior for both soil and concrete; the actual behavior is the nonlinear behavior, and therefore it is better to consider the stress–strain curves instead of a constant modulus of elastic and Poisson's ratio to simulate the actual behavior for both soil and concrete. Also, there is a clear difference in results for displacement in all directions (the slope decreases with nonlinearity); this due to the cumulative displacement for nonlinearity and reduced strength because of the decrease in modulus of elasticity. On the other hand, the shear stress in all planes increases with nonlinearity also; this due to the ultimately applied load with nonlinearity, and then the shear force that developed is greater than that of the linear material. Finally, the elastic shear strains decrease with nonlinearity due to the limitation stress in concrete for nonlinearly and the plastic shear strains developed only for nonlinear material which is necessary to determine the failure stress for materials.

Keywords Finite element · Linear · Nonlinear · Raft foundation · Stress–strain · Modulus of elasticity

---

M. R. Mahmood (✉)
Civil Engineering Department, University of Technology, Baghdad, Iraq
e-mail: mahmoudal_qaissy@yahoo.com

M. M. Abbas
Roads and Transportation Engineering Department, University of Al-Mustansiriya, Baghdad, Iraq
e-mail: mohammadmakki2003@gmail.com

M. M. Mahmood
Civil Engineering Department, Philadelphia University, Amman, Hashemite Kingdom of Jordan
e-mail: mohammadmakki2003@gmail.com

M. O. Karkush and D. Choudhury (eds), *Modern Applications of Geotechnical Engineering and Construction*, Lecture Notes in Civil Engineering 112,
https://doi.org/10.1007/978-981-15-9399-4_20

1 Introduction

The existing raft foundation under static load needs a procedure to assess the settlements, linear and nonlinear behavior of the two different materials (concrete and soil) under static load, especially when increasing the steps of the loads, and how the modified work will affect both the strength and the durability of both materials. Unlike the assessment of concrete, the problem of assessing the remaining performance of soil strength is increased by the fact that it is virtually impossible to visually inspect them for damage. So, it is necessary to introduce reliable assessment procedures for soil that simulate not only the actual behavior of the soil in the composite material but also the forces that are applied directly to the raft foundation from the superstructure. Rashed [1] developed a formulation used in analyzing building raft foundations. The soil or the elastic foundation is simulated as continuous media (according to Winkler assumption). The boundary element method is used to model the raft plate, whereas the soil is modeled by constant domain cells or elements. In formulation, both the domain and the boundary of the raft plate are discretized. The associate soil domain integral is replaced by equivalent boundary integrals along each cell contour. The matrix formulation is carried out and explained in detail. The main advantage of the formulation is the ability to analyze rafts on nonhomogeneous soils.

Limkar et al. [2] used the finite element method to analyze the interface of soil mass and raft. Simulation of soil is done by using springs below the node points to discretize the soil flexibility. SaFe 2000 application software was used for finite element analysis. They approved the effectiveness of the finite element approach by studying the moments along the x and y directions, and soil pressure beneath the raft. The study was carried out on three types of soil; loose sand, medium sand, and hard clayey soil. Their study shows that finite element formulation minimizes the error to a significant extent in the results due to correction in the assumption of linear pressure distribution. They also reveal that the finite element approach is found effective in the case of loose and medium sandy soil as the error is magnified due to the assumption of linear pressure distribution in the conventional approach.

Hemada et al. [3] analyze the problem of a piled raft foundation system, essentially a three-dimensional problem. Two simplified techniques for reducing the computational demands are associated with such large three-dimensional problems so, that it can be handled with personal computers in reasonable CPU time and without violating the solution accuracy are proposed. The first proposed technique is based on the use of three-dimensional interface elements for modeling of the field soil layer. The second proposed technique is a simplified approach for treating lateral loads due to wind and earthquakes on symmetrical piled rafts. The accuracy and efficiency of the proposed simplified approaches are verified by the analysis of three-dimensional test problems and back analysis of a piled raft foundation. Sall et al. [4] studied the influence of soil mechanical properties on the displacements of the mat foundation. They have introduced the soil–structure interaction that is simulated by two parameters, the modulus of subgrade vertical reaction (k_v) and the modulus of subgrade horizontal reaction (k_h). These two parameters are dependent on the geometrical and mechanical characteristics of the piled raft system. It appears from this study that the modulus of vertical subgrade reaction is not an intrinsic characteristic but depends on the parameters of the soil and the concrete (Es νs, Eb, and νb) and the dimensions of the plate (so dependent on the superstructure). It is clear from this analysis that the foundation soil parameters are more influential than those of the plate.

Jawad [5] studied the effect of increasing the value of the coupled moment and the rigidity of raft footing on the lateral deflection by using a 3-D finite element using the Abaqus program. The results revealed that the increase of coupled moment value leads to an increase in lateral deflection and rotational angle (α∘). Gill and Kumar [6] used three-dimensional nonlinear finite element methods to analyze the piled raft foundation. The measurement of axial load distribution in a pile in the field is very difficult and costly; the nonlinear finite element analysis solves this problem. The three-dimensional nonlinear finite element analyses predict the actual behavior of axial load distribution in piles of different length to diameter ratio at various spacing to diameter ratios. In the study, a numerical solution for raft foundation resting on medium-stiff clayey soil with linear and nonlinear solution for both materials, subjected to static loading, is presented and closed-form solutions are obtained for the problem involving boundary conditions. For the raft footing of a building of 21 columns, the column loads and spacing are shown in Fig. 1; the raft is rested on medium-stiff clay. The finite element computer program (ANSYS version II) is used to analyze the problem.

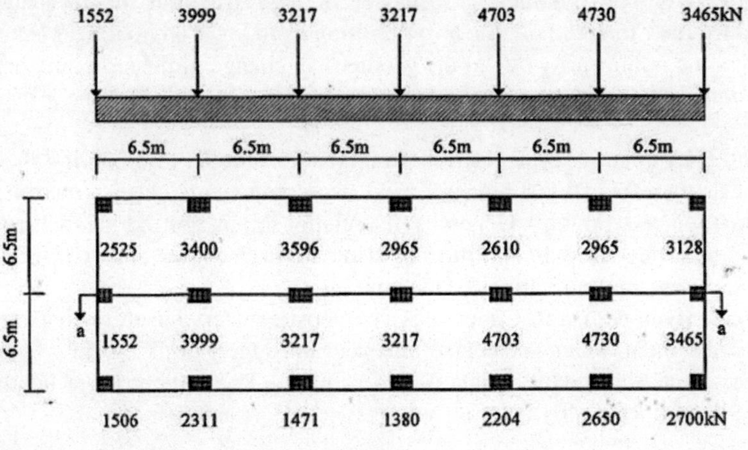

Fig. 1 The basic problem adopted in the study

2 Software Used

ANSYS (version 18.20) was used for analysis throughout this work. ANSYS (version 18.20) program can be applied to a wide variety of engineering applications. It provides the possibility to simulate the superstructure together with its foundation and foundation soil along all its profile required to be considered. The program includes many routines; the main purpose is to find a solution to an engineering problem by the finite element method.

3 Types of Analysis

Two case studies are performed in this study: Linear behavior for soil and concrete of the raft foundation and Nonlinear behavior for soil and concrete of the raft foundation.

Linearity for the soil and concrete. In some problems, the soil and concrete may behave as linear materials for an approximate procedure for analysis and design. The linearity of materials is considered (i.e. modulus of elasticity and Poisson's ratio are constant), so that the strain is not a function of stress, which means the materials do not approach the plastic one but remain in the elastic zone. The materials representing the linearity as data to the ANSYS program can be divided into foundation materials and soil materials; the properties of materials are shown in Table 1.

Nonlinearity. The nonlinearity behavior of any structural member may be attributed to either material nonlinearity resulting from material nonlinear behavior, or to geometric nonlinearity due to the geometrical change of the structure, or their combination. The theory presented here is extended to include nonlinear material behavior.

Nonlinearity of soil. Atypical stress–strain relationship for a soil during shearing is nonlinear; therefore, the strain will have two components elastic (recoverable) and plastic (non-recoverable). Generally, the relationship is approximately linear for most soils then beyond that, significant plastic strain becomes apparent. Figure 2 clarifies the stress–strain relation for medium clay.

Nonlinearity of concrete. Concrete is a composite material itself, which consists especially of cement, water, and coarse soil (sand and aggregate). Cement is hydrated to produce a binding medium, called cement paste. The cement paste bonds the aggregate to produce highly heterogeneous stone-like concrete material.

Table 1 Properties of the material used

Material	Modulus of elasticity (E) kPa	Unit weight (γ) kN/m^3	Poisson's ratio (υ)	Cohesion (c) kN/m^2
Concrete	2,500,000	24	0.15	–
medium clay	40,000	18	0.4999	40

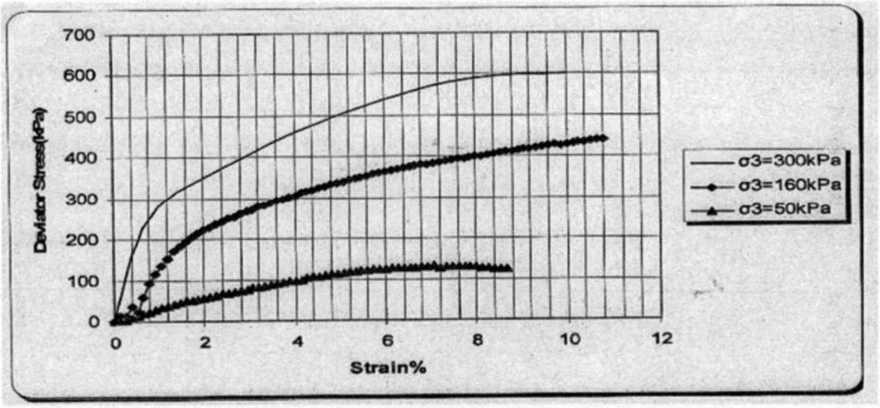

Fig. 2 Stress–strain relationship for medium clay

Capillary pores occur within the cement paste due to the incomplete hydration, which is filled with air or water. These pores can be a source of weakness in the structure of the concrete. Other sources of weakness due to repeated load are caused by microscopic cracks and these cracks grow and cause failure due to static load. The formation of microscopic cracks at higher loads belongs to the nonlinear behavior of concrete and also causes volume expansion near failure. When static load is applied, the ratio between the stress and strains is approximately linear at first, and the concrete behaves as an elastic material with virtually a full recovery of displacement when the load is removed. Eventually, the curve is no longer linear and the concrete behaves more and more as a plastic material. If the load was removed during the plastic range, the recovery would no longer be complete and a permanent deformation would remain. The ultimate strain for most structural concrete tends to be approximately 0.0035.

One of the most widely used models for the stress–strain relationship of concrete in compression is that proposed in (BS 8110), as shown in Fig. 3, where the variation of the curved portion of the stress–strain relationship is given by

$$\sigma = 5500\sqrt{f_{cu}}\,\varepsilon - 11.3 \times 10^6\,\varepsilon^2 \tag{1}$$

with an initial modulus of elasticity equal to

$$\mathrm{E_t} = 5500\sqrt{\mathrm{f_{cu}}} \tag{2}$$

Under tensile stress, the strength of concrete is approximately 10–15% of the compressive strength, or

$$\mathrm{f_{ct}} = 0.7\sqrt{\mathrm{f_c}} \tag{3}$$

Fig. 3 Stress–strain curve for concrete of (fcu in MPa)

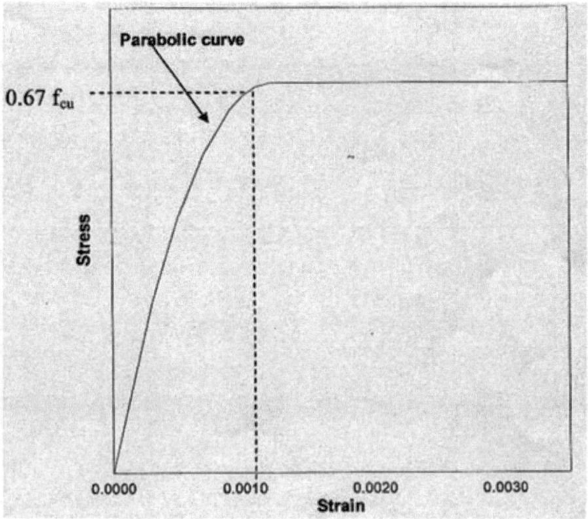

However, if it is accounted for the stress–strain curve in tension may be idealized as a straight line up to rupture strength and the modulus of elasticity can be assumed similar to that in compression. Concrete does not obey Hooke's law when subjected to static loading and therefore the stress–strain relation can be written in the form:

$$\sigma = \varepsilon E_c \tag{4}$$

4 Effect of the Applied Load (Static) on Displacements

Figure 4 shows the comparison between linear and nonlinear behavior for the relation incremental load ratio with horizontal displacement. It can be seen that the slope decreases with nonlinearity.

The difference in results between linear and nonlinear behavior is due to the reduced strength in soil under incremental load which occur from the residual curve of stress–strain relationship and due to the slip (horizontal movement) between the foundation and the top of the soil; this slip leads to a cumulative displacement in the horizontal direction during nonlinear behavior. The ratio between the result solution for maximum displacement between nonlinear and linear results is approximately 4.83. Figure 5 shows the comparison between linear and nonlinear behavior for the relation incremental load ratio with vertical displacement (settlement). It can be seen that the slope decreases with nonlinearity. The difference in results between linear and nonlinear behavior is due to the reduced strength in the soil under incremental load, which occurs from the residual curve of stress–strain relationship, and due to the vertical movement of the foundation and the soil; this movement leads to a cumulative

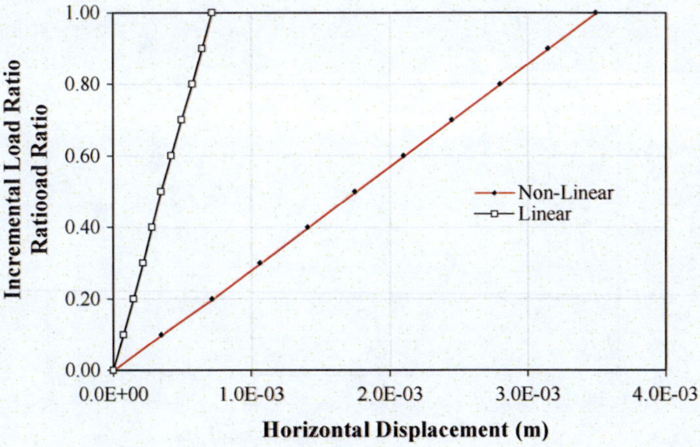

Fig. 4 Comparison between linear and nonlinear behavior for incremental load ratio with horizontal displacement (m)

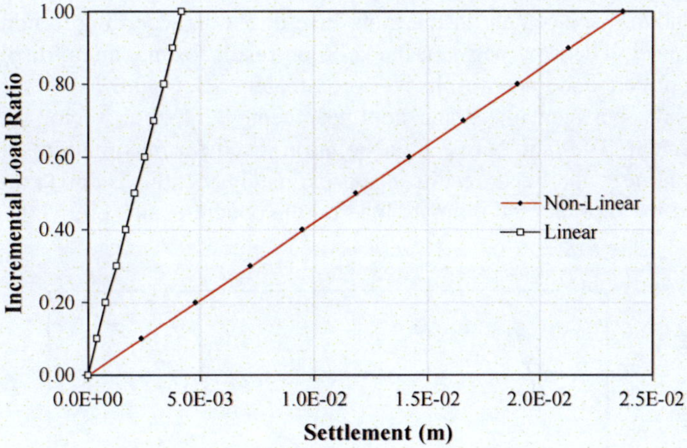

Fig. 5 Comparison between linear and nonlinear behavior for incremental load ratio with settlement

displacement in the vertical direction during nonlinear behavior. The ratio between the result solutions for maximum displacement between them is approximately 5.57.

Figure 6 shows the comparison between linear and nonlinear behavior for the relation between incremental load ratios with transverse displacement. It can be seen that the slope decreases for nonlinearity.

The difference in results between linear and nonlinear behavior is also due to the reduced strength in the soil under incremental load which occurs from the residual curve of stressstrain relationship, and due to the transverse movement (due to the applied load) between the foundation and the top of the soil; this movement leads

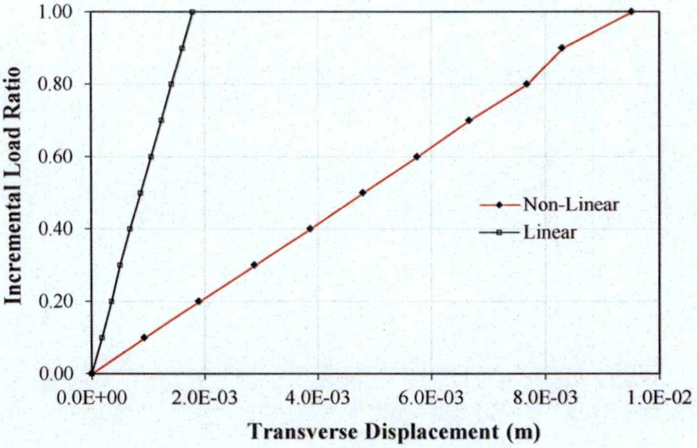

Fig. 6 Comparison between linear and nonlinear behavior for incremental load ratio with transverse displacement

to a cumulative displacement in the transverse direction according to nonlinearity behavior aspect. The ratio between the solution result for maximum displacement between nonlinear and linear results is approximately 5.3. Figure 7 shows the difference in results between linear and nonlinear behavior for the vector summation of displacement. The ratio between the solution result for maximum displacement between nonlinear and linear results is approximately 5.4; this is due to the nonlinearity effect by applying the ultimate load on the concrete and soil, while in linear

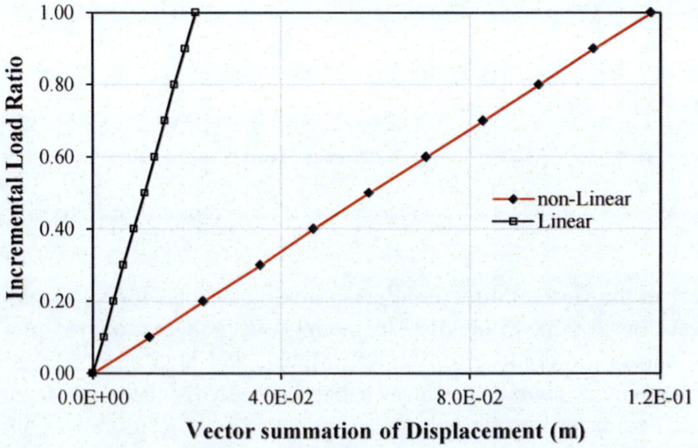

Fig. 7 Comparison between linear and nonlinear behavior for incremental load ratio with the vector summation of displacement (m)

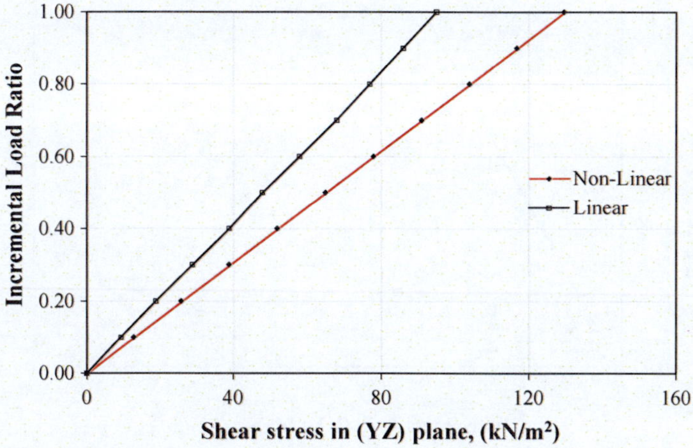

Fig. 8 Comparison between linear and nonlinear behavior for incremental load ratio with shear stress (kN/m^2) in the YZ plane

ones the applied load is equal to 40–50% from the ultimate load, therefore the strength is reduced with nonlinearity.

Figure 8 shows the comparison between linear and nonlinear behavior for the relation between incremental load ratios with shear stress in the YZ plane. It can be seen that the slope decreases for nonlinearity.

The difference in results between linear and nonlinear behavior is also due to the reduced strength in the soil under incremental load for nonlinear behavior which occurs from the additional load applied until reaching the failure, and due to the shear force (as a result of applied load). The ratio between the solution result for maximum shear stress between nonlinear and linear results is approximately 1.4. Figure 9 shows the comparison between linear and nonlinear behavior for the relation between incremental load ratios with shear stress in the XY plane. It can be seen that the slope decreases with nonlinearity due to the reduced strength for nonlinear behavior. The ratio between the solution result for maximum shear stress between nonlinear and linear results is approximately 2.7.

Figure 10 shows the comparison between linear and nonlinear behavior for the relation between incremental load ratios with shear stress in the XZ plane. It can be seen that the slope also decreases for nonlinearity. The difference in results between linear and nonlinear behavior is also due to the reduced strength in the soil under incremental load for nonlinear behavior which occurs from the additional load applied until reaching the failure, and due to the shear force (as a result of applied load). The ratio between the solution result for maximum shear stress between nonlinear and linear results is approximately 1.2.

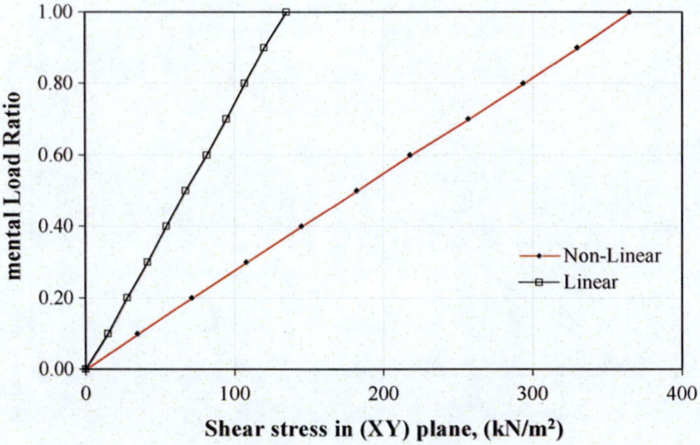

Fig. 9 Comparison between linear and nonlinear behavior for incremental load ratio with shear stress in the XY plane

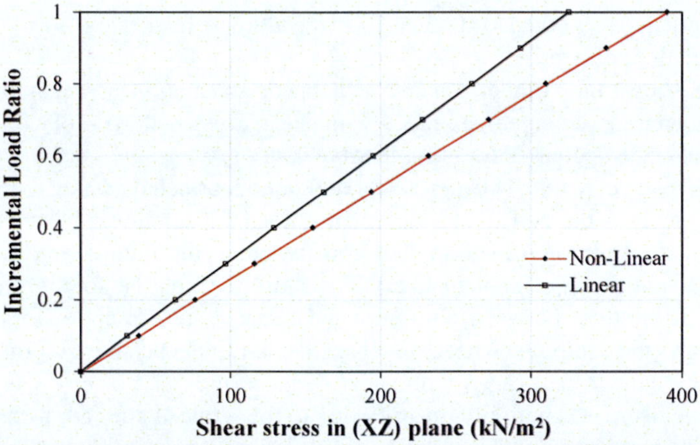

Fig. 10 Comparison between linear and nonlinear behavior for incremental load ratio with shear stress in the XZ plane relationship

5 Effect of the Applied Load on Shear Strains

Figure 11 shows the comparison between linear and nonlinear behavior for the relation between the incremental load ratio with elastic shear strain in the YZ plane. It can be seen that the slope decreases with linearity. The ratio between the solution result for the maximum elastic shear strain between linear and nonlinear results is approximately 3.26.

The difference in results for elastic shear strain between linear and nonlinear behavior is due to the limitation in the stress nonlinearity of concrete instead of

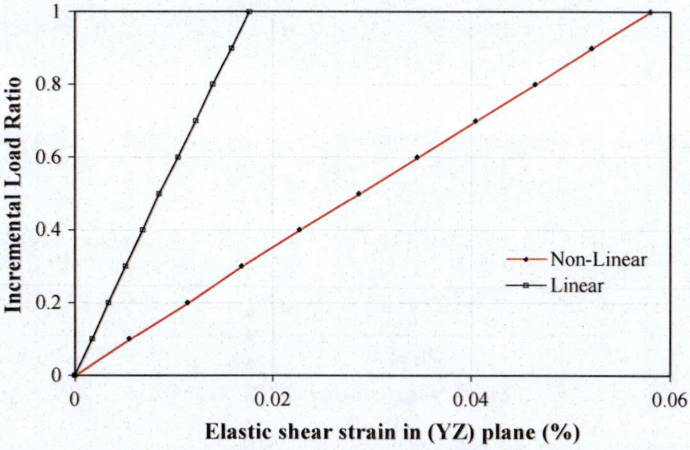

Fig. 11 Comparison between linear and nonlinear behavior for incremental load ratio with elastic shear strain (%) in the YZ plane

soil. Figure 12 shows the comparison between linear and nonlinear behavior for the relation between incremental load ratios with elastic shear strain in the XY plane. It can be seen that the slope also decreases with linearity. The difference in results between linear and nonlinear behavior is due to the limitation in the stress nonlinearity of concrete instead of soil. The ratio between the solution result for the maximum shear strain between linear and nonlinear results is approximately 3.2. Figure 13 shows the comparison between linear and nonlinear behavior for the

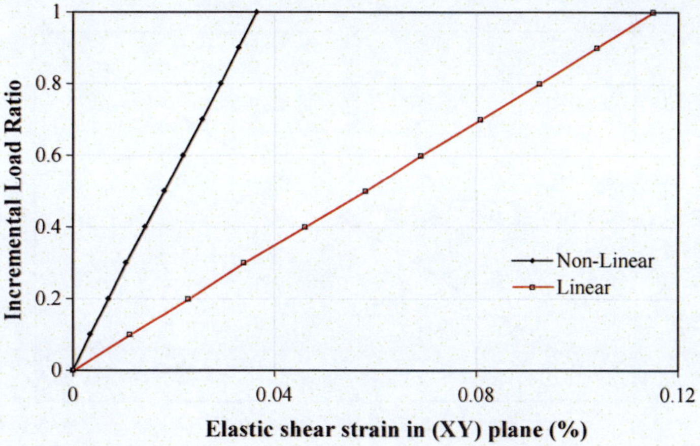

Fig. 12 Comparison between linear and nonlinear behavior for incremental load ratio with elastic shear strain (%) in the XY plane

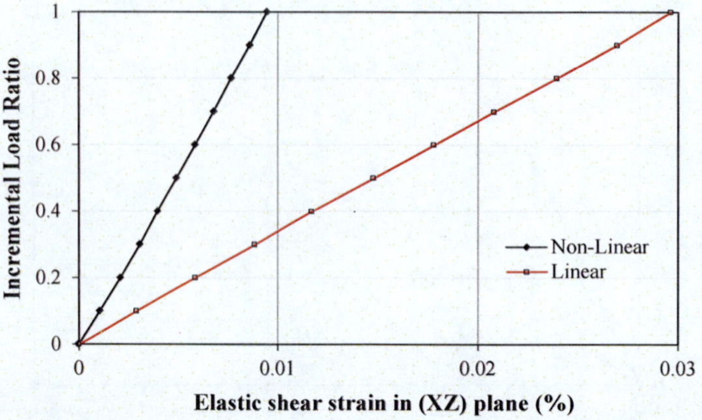

Fig. 13 Comparison between linear and nonlinear behavior for incremental load ratio with elastic shear strain (%) in the XZ plane

relation between incremental load ratios with elastic shear strain in the XZ plane. It can be seen that the slope also decreases with linearity.

The difference in results between linear and nonlinear behavior is due to the limitation in the stress nonlinearity of concrete instead of soil. The ratio between the solution result for the maximum shear strain between linear and nonlinear results is approximately 3.2. Figure 14 clarifies the effect of applied load on the plastic shear strain in the YZ plane; it can be seen that the plastic shear strain increases when the load ratio is increased. Plastic zone means that the materials reach failure which means concrete reaches 0.67fcu from its ultimate stress and the soil reaches

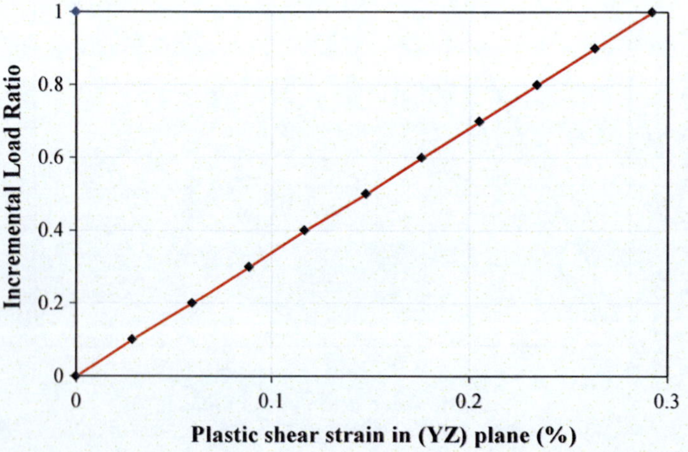

Fig. 14 Relationship between the incremental load ratio and plastic shear strain (%) in the YZ plane for nonlinear behavior

its ultimate strength; this stage is necessary to determine the failure stress for both concrete and soil. Figure 15 clarifies the effect of the applied load on the plastic shear strain in the XY plane. Also, the plastic shear strain increases when the load ratio is increased. The plastic shear strains occur due to applied load at planes parallel to load (i.e. YZ and XY). Figure 16 represents the effect of the applied load on the plastic shear strain in the XZ plane. The plastic shear strain in the XZ plane is due to the loss of material interaction as a result of the applied load.

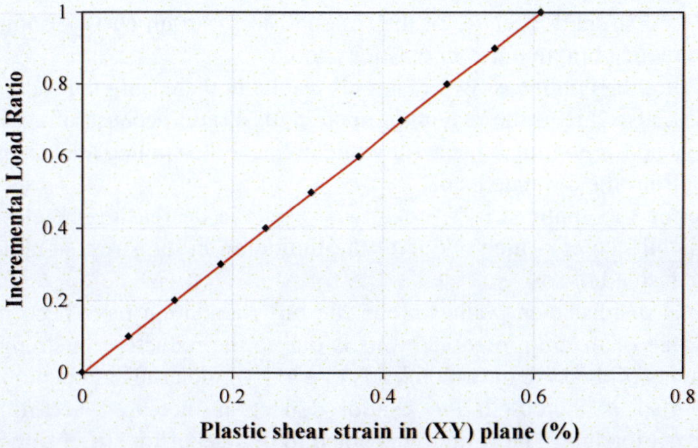

Fig. 15 Relationship between the incremental load ratio and plastic shear strain (%) in the XY plane for nonlinear behavior

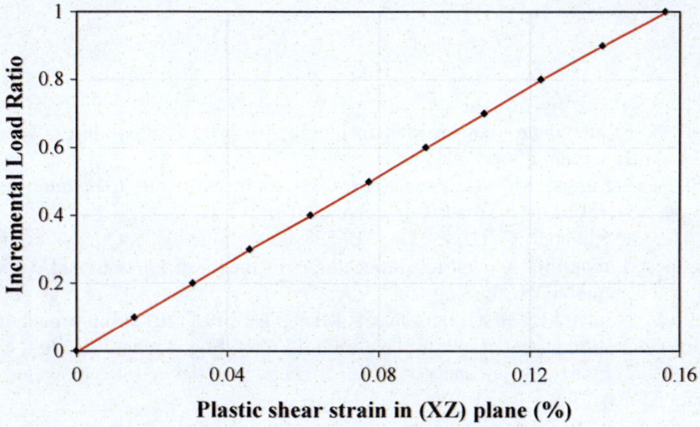

Fig. 16 Relationship between the incremental load ratio and plastic shear strain (%) in the XZ plane for nonlinear behavior

6 Conclusions

The conclusions which can be drawn from this study are summarized below:

- Considering the nonlinearity of materials is an important point in the foundation design to understand the behavior of materials, and neglecting the nonlinearity may result in a clear difference in the results.
- Considering the nonlinearity of materials simulates the actual condition results in increasing the displacements in all directions due to the cumulative displacement and reduced strength because of the residual stress–strain curve of nonlinearity and also reduced in modulus of elasticity too.
- Considering the nonlinearity of materials results in increasing the shear stress in all planes, also due to the reduction in strength of soil because of applying the ultimate load for nonlinear behavior while in linear the applied load is only about 40–50% from the ultimate load.
- Considering the nonlinearity of materials results in decreasing the elastic shear strains in all planes, which is due to the limitation in the stress nonlinearity of concrete instead of soil.
- Horizontal displacement values along the raft foundation have been increased as the value of loading increased; this is due to the reduction in strength of all components of the system, which leads to a loss of materials interaction.
- As the load is increased, the central deflection increased accordingly for incremental loading. The variation is noted to be steeper for stiffer sections.
- The plastic shear strain in each plane is considered with nonlinear behavior only, and it is necessary to decide the failure stress.

References

1. Rashed YF (2005) A boundary/domain element method for analysis of building raft foundations. Eng Anal Boundary Elem 29(9):859–877
2. Limkar SD, Kalyanshetti MG, Halkude SA (2005) Analysis of raft foundation using finite element approach. Int J Latest Trends Eng Technol 8(3):014–028
3. Hemada AA, Abd El Fattah TT (2008) Three dimensional finite element analysis of large piled rafts: a proposed simplified modelling, Proceedings of the sixth international conference on engineering computational Technology
4. Sall OA, Ba M, Ndiaye M, Fall M, Berthaud Y, Mbaye I, Fall M (2015) Finite element computation of the behavioral model of mat foundation. Int J Civil Eng. Technol (IJCIET, 6(5):10–22
5. Jawad FW (2018) Finite element analysis of raft foundation under coupled moment. Journal of Engineering 24(10):73–87
6. Gill S, Kumar R (2018) Finite element analysis for axial load distribution in pile in a piled raft foundation. Int J Civil Eng Technol (IJCIET) 9(6):309–314

Observation of Foundations of Historical Buildings: The Cultural Heritage of St. Petersburg

R. Mangushev

Abstract The engineering and geological conditions of the central part of St. Petersburg are considered. The analysis of the main characteristics of the foundations of historical buildings (eighteenth to nineteenth century) is made based on field investigations. The data of dependence between the size of foundations, the number of floors, and the condition of soil are given. The pressure at the bottom of the foundation and the bearing capacity of the soil of the base are analyzed. The examples of the foundations of some historical buildings in St. Petersburg are given. This paper also provides monitoring results of the behavior of historical buildings, especially settlements of foundations. These characteristics of foundations that are important for the understanding of the interaction of historical old buildings with soft soil ground of St. Petersburg are described.

Keywords Foundation · Historical · Building · Heritage · Geotechnical

1 Introduction

The origin of the St. Petersburg City is connected with the events of the North War (1700–1721) between Russia and Sweden. Peter and Paul fortress erected on one of the islands in Neva delta for the protection of the won territory became the first building in the future City. Starting with the eighteenth century, all the buildings in the City were constructed from stone in accordance with project specifications. These specifications were developed with the participation of Peter I—the founder of the City. The soil of the territory of the future City was characterized by thick layers of soft water-saturated silty soil settled at the bottom of the glacial Baltic lake and covered by alluvial soil formed by silty and fine sand from Neva River.

R. Mangushev (✉)
Geotechnical Department, Saint-Petersburg State Architectural and Civil Engineering University, Saint-Petersburg, Russia
e-mail: ramangushev@yandex.ru

M. O. Karkush and D. Choudhury (eds), *Modern Applications of Geotechnical Engineering and Construction*, Lecture Notes in Civil Engineering 112,
https://doi.org/10.1007/978-981-15-9399-4_21

2 Types of the Foundation

According to these historical specifications it was recommended to use wood sills when making rubble foundations in soft soil below the level of underground waters. The pine, fur tree, and larch trunks were used as wood sills. When the foundations of the old (eighteenth century) buildings were opened, it was revealed that the trunks located lower the level of the groundwater were in relatively good condition. In the areas with low level of ground waters rubble plates were done in opened trenches. The main types of foundations used till the 30 years of the twentieth century were rubble foundations formed from limestone, sand, or sometimes granite and diabase fragments, of 80 cm diameter. The stones were carefully selected and laid on lime solution fitting each other.

Starting with the twentieth century the cement solution was used. In eighteenth to nineteenth century in the central part of St. Petersburg where the soft soil prevails, wooden piles 10 m long were used.

3 Settlements of the St. Isaacs Cathedral

St. Isaacs Cathedral may be taken as an example. The period of construction was 1818–1858 by architect A. Montferrand. It is one of the largest domed structures in the world of 1015 m high, having an area of 4000 m^2, and it can hold 12,000 people. When erecting the foundation, 6 m piles were put under the rubble plate of 7.5 m thickness. The plate was laid at a depth of 5 m. Under the pylons, the length of oak piles was 8.4 m. Totally, 24,000 wooden piles with an average diameter of 26 cm were used (Fig. 1).

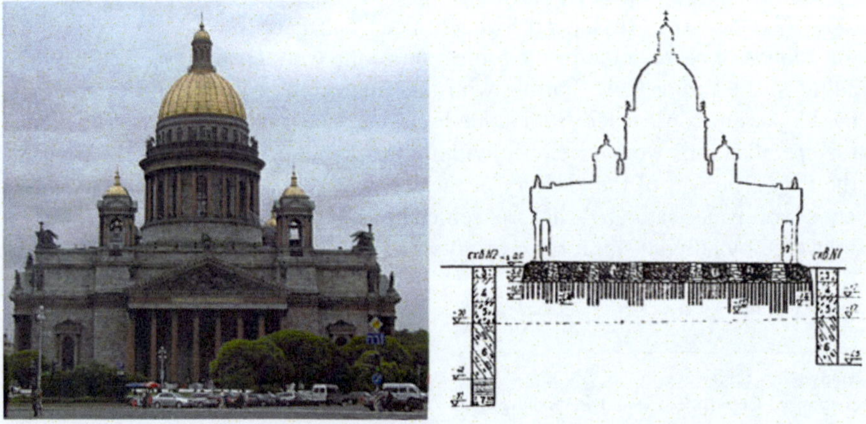

Fig. 1 General view of the Isaacs Cathedral and schematic cross-section of the foundation

Measured in the middle of 1980s of the twentieth century the difference in the surface marks near the building and 50 m from the building was 85 cm, and the slope of the plate toward the south-west was registered. The maximal difference between the surface marks at a distance of 78 m ($L = 78$ m) inside the Cathedral was 30 cm (Figs. 2 and 3).

It means that the average settlement of the Cathedral for more than 150 years was less than 100 cm. But thanks to the rigidity of the structure of the Cathedral, there were no serious deformations of the upper structures and there was no need in strengthening the foundation and bearing structures. The reason for the large foundation settlement was big vertical deformation of soft water-saturated layers laid above till sediments.

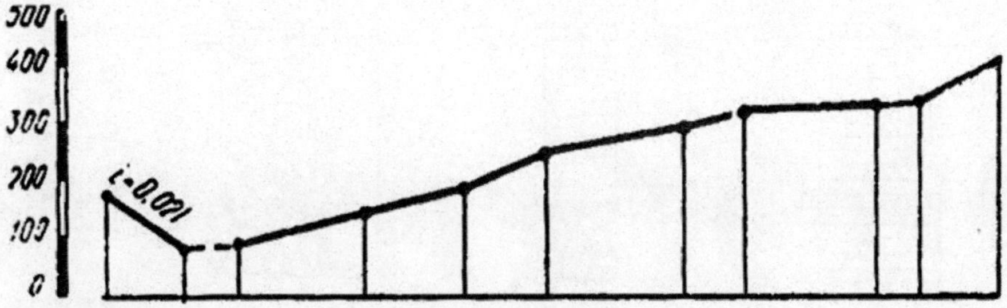

Fig. 2 The floor profile inside the Cathedral in the direction "A-A"

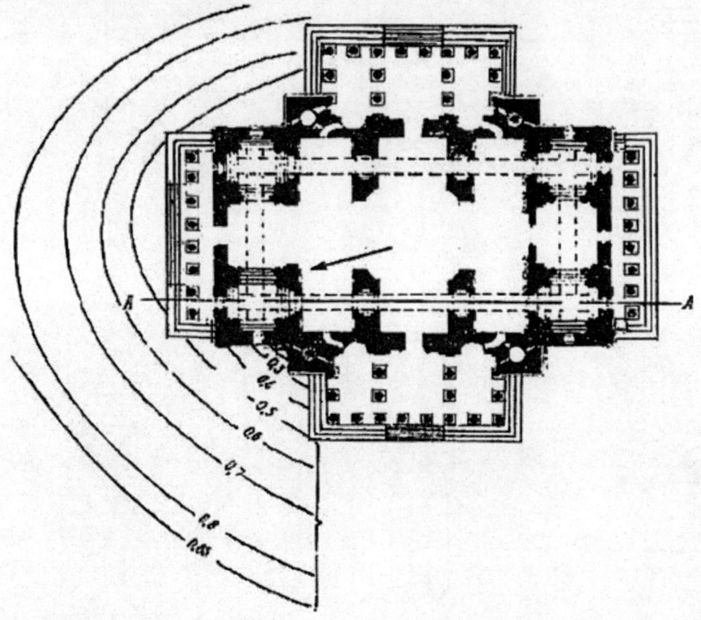

Fig. 3 The plan of the Cathedral on the floor level. The arrow shows the slope direction of the pile raft

The thickness of soft soil with deformation modulus less than 10 MPa reached more than 45 m in the construction area of the Cathedral [1, 2, 3].

In our days the reconstruction and repair of the old buildings and also compact construction of new buildings in old parts of the City made it necessary to examine and monitor the foundations of old buildings. The average depth of the foundations in most of the examined buildings was 1.5–3.4 m and the depth of the artificial layer in some regions of St. Petersburg reach 3 m.

The cross-section of the typical belt and plate foundations of the civil and apartment buildings constructed before the twentieth century in St. Petersburg is shown in Fig. 4.

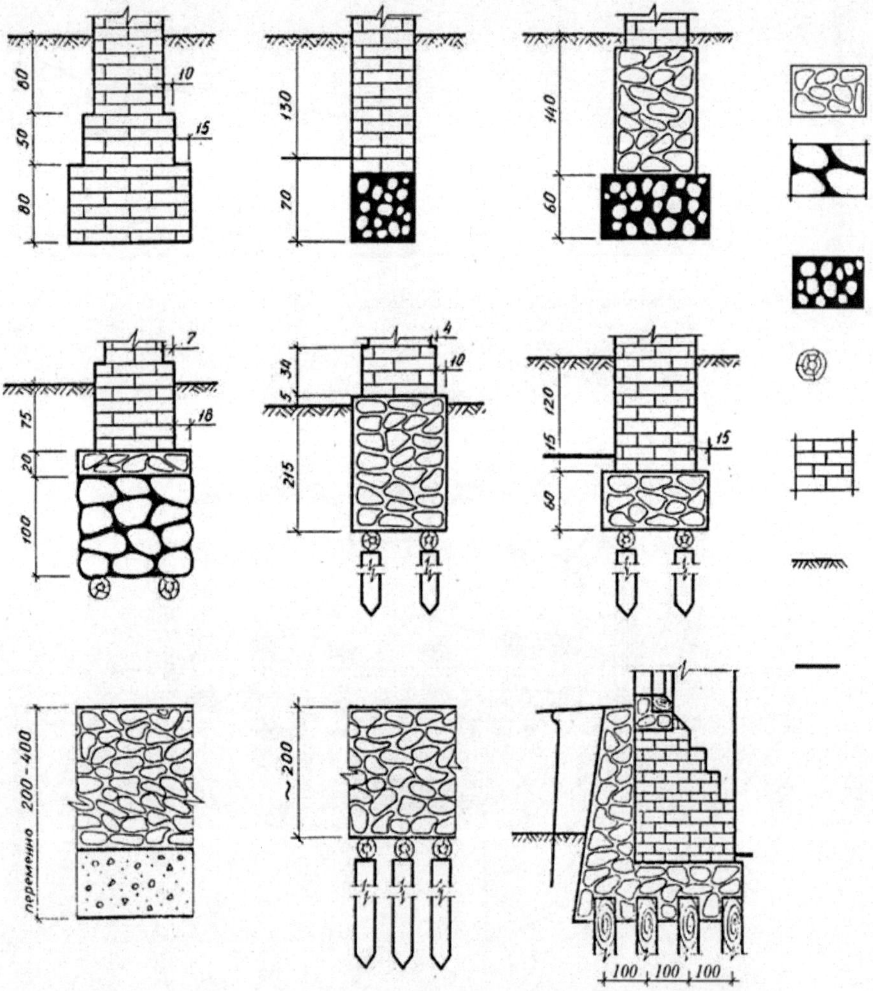

Fig. 4 Types of foundations of old buildings in St. Petersburg

4 Analysis of Results of Foundations Inspections

The generalized experience and the analysis of the inspection of 64 buildings constructed in the period of eighteenth to beginning of the twentieth century revealed some regularity, which characterizes the buildings referring to this period and was constructed in the central part of the City. The buildings usually have 2–6 floors. The foundation of 44% of the buildings was laid on a silty soil, 56% on a sand of different course. The foundations of 82% are formed from rubble, granite, or slime stones, and only 18% from burned bricks. Wood sills were found under the foundations of 22 buildings (34%), and wooden piles under the foundation of 9 buildings (14%). The photo of opened wood sills and wooden piles under five-floor building is shown in Fig. 5.

The analysis allowed us to estimate the main characteristics of the examined foundations depending on the number of floors in the building. Figure 6 shows the dependence between the width of the foundation bottom made on sand and silty-sand base and the number of floors. Buildings with wood sills under the foundation bottom are marked by a separate line.

The width of the foundation bottom in the inspected buildings "**b**" varied from 0.5 to 2.8 m for foundations on sand base and from 0.8 to 2.3 m for silty soil. The most common width of the foundation bottom is $b = 1–1.5$ m (36%—sand base and 54%—silty soil) (Fig. 7).

They followed by width of the foundation $b = 1.5–2.0$ m, 20, and 25% correspondingly. The wood sills were found under the 2–6 floor buildings. Besides, there was a case when the width of the foundation bottom was 4 m, and it was rubble/concrete foundation under which the wooden piles were found. Lobanov-Rostovsky House

Fig. 5 Photo of wood sills and wooden piles under the foundation of five-floor building

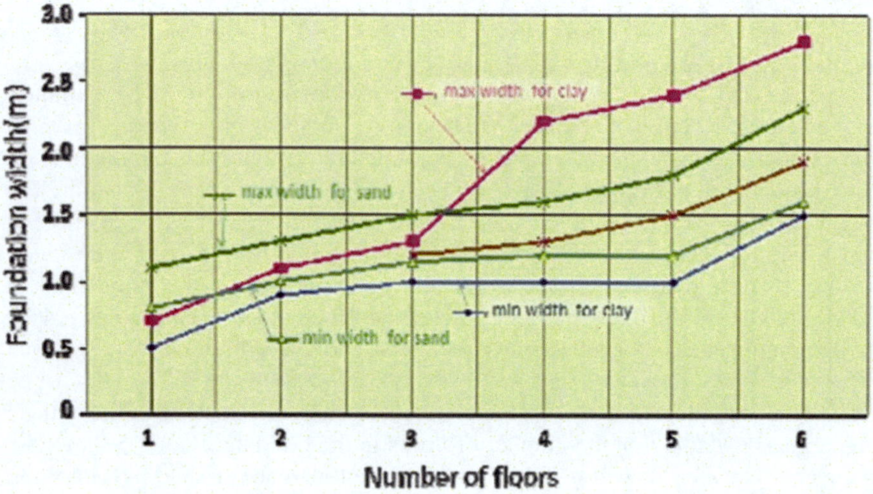

Fig. 6 Dependence between the width of the foundation bottom and the number of floors

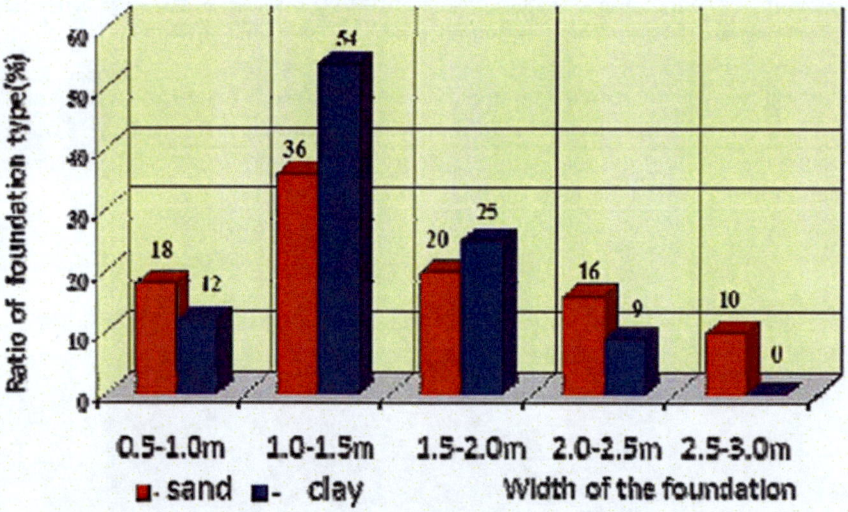

Fig. 7 Subdivision of the width of foundation bottom in percentage

(constructed in 1817–1820) within two centuries of the building has not received any deformation [4, 5]. The dependence between the depth of the foundation "**d**" and the number of the floors in the buildings is shown in the diagram in Fig. 8. The depth of the foundation varies from $d = 0.3$ to 3.0 m for sand base and from 0.7 to 2.5 m for silty soil.

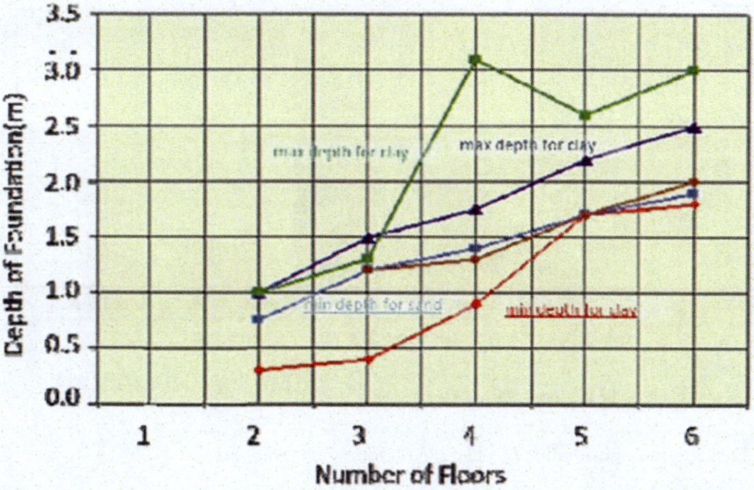

Fig. 8 The dependence between the depth of the foundation bottom and the number of floors

As a rule, the wood sills were found when the depth of the foundation bottom was 1.2–1.8 m. The diagram of the subdivision of the foundation depth of examined buildings shows that the most common depth "*d*" is equal to 1.5–2.0 m. This depth corresponds to 44% of examined buildings on sand base and 53% on silty soil (Fig. 9).

When you compare the action of the actual average pressure "**p**" at the bottom of the foundation of the inspected buildings and the pressure specified in the modern construction documentation, the value "**R**" in particular, it is possible to find the following (Fig. 10). The value "**p**" for 62% of the buildings constructed on sand

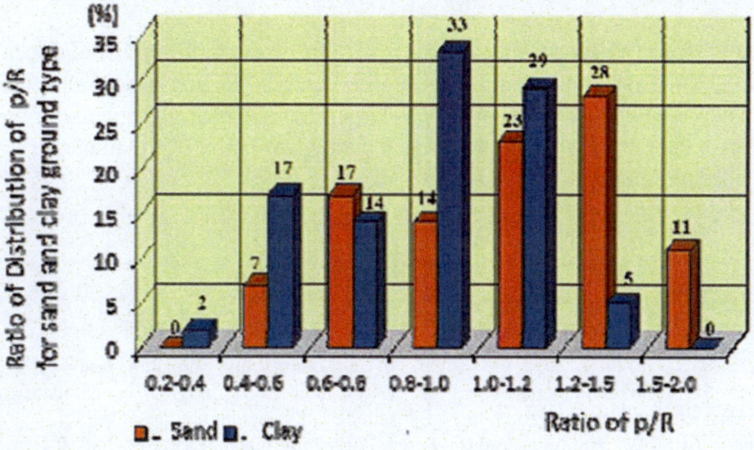

Fig. 9 Subdivision of the depth of the foundation bottom in percentage

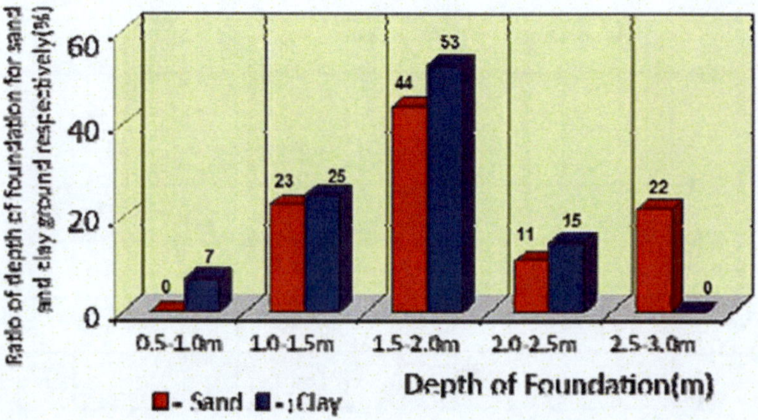

Fig. 10 Distribution of p/R value of examined buildings in percentage

base in the central part of St. Petersburg exceeds the value "**R**" permitted by modern Russian norms. It was registered in 34% of cases of bases formed by silty soil.

These results do not correspond to the similar values for Moscow and Moscow region. According to his estimation, the value **p/R** > 1 was registered in 8.3% of examined buildings, and 85.7% corresponds to the value **p/R** < 0.8, which means that the bearing capacity was not sufficiently used. During the examination of the old buildings of St. Petersburg, the **p/R** < 0.8 value was found in 24% of buildings on sand base and 33% on silty soil base.

5 Conclusion

The results of the investigation confirm that most of the buildings of eighteenth to beginning of the twentieth centuries were constructed in such a way that the real pressure at the base of the foundation was larger than the bearing capacity of the soil base. This is one of the reasons for cracks and deformation of the upper structures of many old buildings.

Reference

1. Mangushev R, Osokin A (2018) Geotechnics of Saint-Petersburg. Publishing house ASV, p 377
2. Sotnikov SN (1986) Settlements of the Isaakievsky Cathedral in Leningrad//Fundamentostroenie in a moustache-lovijah weak soils: Interhigh school. LICI, Leningrad, pp 7–18
3. Dalmatov BI (2002) The bases and the bases. Publishing house ASV, SPbGASU, Uch-to for High Schools, p 387

4. Sotnikov SN (2003) About a palace of Rostov-Lobanova, SPbGASU
5. Konovalov PA (2004) The bases and the bases of reconstructed buildings, 4th edn. VNIINTPI, Moscow, p 317

Numerical Modeling of Honeycombed Geocell Reinforced Soil

Mushriq F. K. AL-Shamaa, Ammar A. Sheikha, Mahdi O. Karkush, Mohammed S. Jabbar, and Ayad A. Al-Rumaithi

Abstract Shallow foundations have been commonly used to transfer load to soil layer within the permissible limits of settlement based on the bearing capacity of the soil. For most practical cases, the shape of the shallow foundation is of slight significance. Also, friction resistance forces in the first layers of soils are negligible due to non-sufficient surrounding surface area and compaction conditions. However, the bearing capacity of a shallow foundation can be increased by several techniques. Geocell is one of the geosynthetic tool applied mainly to reinforce soil. This study presents a numerical approach of honeycombed geocell steel panels reinforcing the sandy soil under shallow foundation, and several parameters are investigated such as the size and depth of honeycombed steel panels. The numerical results showed that honeycombed geocell reinforcement can increase the bearing capacity of soil by 65% and decrease the displacement of shallow foundation by 45%. This improvement of soil behavior under load resulted from the confinement of soil under foundation and increases the friction between soil and walls of geocells.

Keywords Geocell · Honeycomb · Shallow foundation · Sandy soil · Numerical analysis

---

M. F. K. AL-Shamaa · A. A. Sheikha (✉) · M. O. Karkush · M. S. Jabbar · A. A. Al-Rumaithi
Civil Engineering Department, University of Baghdad, Baghdad, Iraq
e-mail: aaqassim@coeng.uobaghdad.edu.iq

M. F. K. AL-Shamaa
e-mail: mushriqf@coeng.uobaghdad.edu.iq

M. O. Karkush
e-mail: mahdi_karkush@coeng.uobaghdad.edu.iq

M. S. Jabbar
e-mail: mohammed.s.jabbar@gmail.com

A. A. Al-Rumaithi
e-mail: ayad.a@coeng.uobaghdad.edu.iq

M. O. Karkush and D. Choudhury (eds), *Modern Applications of Geotechnical Engineering and Construction*, Lecture Notes in Civil Engineering 112,
https://doi.org/10.1007/978-981-15-9399-4_22

1 Introduction

There are many different practical applications of geocells as construction materials in civil engineering. They are often incorporated into geotechnical constructions, such as retaining walls, landfills, and roads, to provide strength and reinforcement, or they can be used for erosion control, rockfall nets, and gabion cages to prevent the construction from collapse. Also, geocell can be used to improve the bearing capacity of the soil and make a reduction in the settlement. Besides, geocell enriches the stiffness and shear strength of the base course layer, which will help to distribute the load uniformly onto the soft layers. Geocells are usually manufactured from polymeric composites. For transport purposes, many of geocell items have a foldable three-dimensional shape; almost it is honeycomb formed after extension as shown in Fig. 1a. However, in some cases, it can be supplied as rigid panels as shown in Fig. 1b.

In the 1970s for the first time, the US Army Corps of Engineers improved geocell to reinforce cohesionless soil quickly [1]. Subsequently, many studies are carried out to investigate the behavior of geocell-reinforced soil. These studies entail the topics of the kinds of soil, the geometric shape, the material components of geocell, and the loading states to identify the influence of the different key parameters in assessment of the performance and mechanism of soil reinforcement. In the last few decades, limited attempts have been investigated to model geocell reinforcement simulation based on the three-dimensional analysis, and the work in this arena generally concerns itself with more complications due to the difficulties in modeling the real configuration. Yang et al. in 2010 presented idealization of 3D honeycomb shape of the single-cell geocell; in this study, the soil was simulated through stress-dependency and nonlinearity inside cells. The numerical predictions compared well with the test data [3]. In 2105, Hegde and Sitharam analyzed geocell-reinforced sand beds using a three-dimensional numerical approach based on the finite difference method. They obtained that the load was laterally distributed in the depth of shallow foundation reinforced with geocell, also the numerical approach was successfully

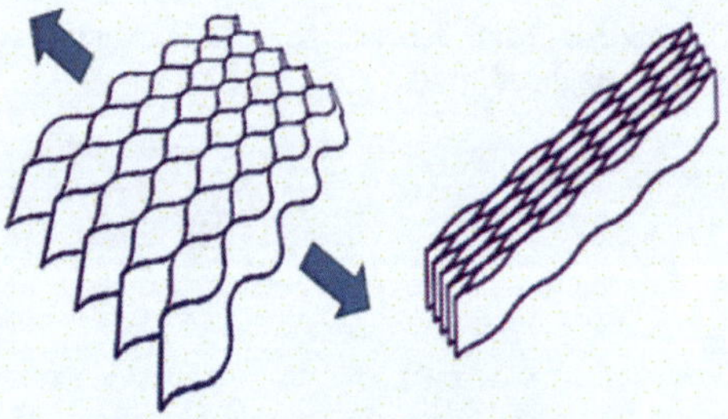

(a) Folded-Expand panel

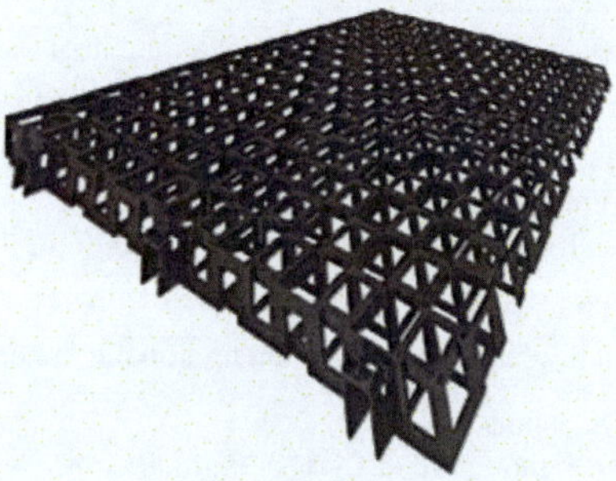

(b) Rigid panel

Fig. 1 Typical geometry of geocell panel

verified with the experimental results, as well as the performance of the reinforced bed was affected by the size and surface roughness of geocell [4].

In 2016, Hedge et al. [5] used LAXIS3D and FLAC3D software to investigate the complex geotechnical cases. They reviewed the benefit and weakness of using these programs during the investigation of geocell. The simulation of geocells reinforcing soil has been made using this software and linked with the experimental work. The results showed that the bearing capacity of the reinforced sand got increasing by four times with a 70% reduction in the settlement. They found that FLAC3D is more appropriate than LAXIS3D to simulate geocells because it includes robust structural elements.

In 2019, Keshmiri et al. [6] used the Abaqus program to analyze the behavior of circle and square foundation placed on a sand reinforced with geocell. The 3D model was evaluated and compared well with the results of experimental studies. It was found that the bearing capacity has been increased by 65% with a 15% decrease in settlement for geocell-reinforcing soil under circle foundation. Also, there was a better influence of geocell (up to 12% increase in the bearing capacity) in square foundations when compared with circle foundations. The proposed method has been examined under an old structure with unequal settlement and compared with the reinforced concrete piles. The settlement was paused for six months after improvement in soil. There are several studies that investigated the behavior of soil reinforced by geocell using experimental modeling or numerical analysis, such as references [7–9].

2 Numerical Methods and Element Modeling

Numerical modeling has the advantages over experimental modeling, that the parameters may be varied easily and details of stress and deformations throughout the system may be studied. This is particularly valuable for investigating the mechanisms acting at the unreinforced and reinforced interface between the soil and geocell, which is extremely difficult to do in a model test. Thus, a preliminary numerical approach would be required for each case as the first step to provide a realistic configuration. This work is an attempt to present a theoretical study of the performance of shallow foundation on sandy soil reinforced with the honeycomb geocell steel panel.

A numerical approach was carried out using a commercial program, *plane strain*, and *axial symmetry* (PLAXIS). PLAXIS is a finite-element software used for solving geotechnical engineering problems [10]. In the present work, numerical analysis was used to model the load on unreinforced and honeycomb-reinforced sand soil. To study the mechanism of reinforcement in greater detail, it is instructive to inspect a thorough parametric investigation at the soil, reinforcement, and interface elements. It is first necessary to establish the most appropriate numerical approach for these elements.

Table 1 Properties of sand soil

Parameter	Value
Elastic modulus (E), kPa	9000
Poisson's ratio (ν)	0.3
Dry unit weight, kN/m^3	15.6
Angle of internal friction (φ), degree	28
Drainage type	Drained

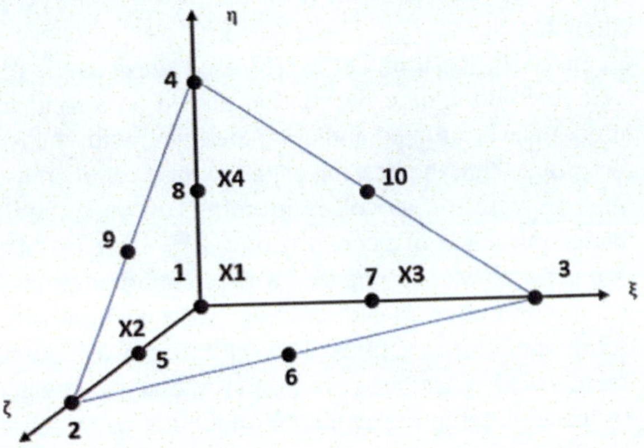

Fig. 2 Three-dimensional soil elements (10-node tetrahedrons)

2.1 *Sand*

Compared to clay, the sandy soil is more suitable to be reinforced by honeycombed geocell due to the friction forces imposed by the frictional interface elements through the top to bottom of the geocell steel plate. Honeycombed geocell steel panels could be combined action of soil reinforcement and driven structure. The soil properties used in this study were a medium pure sand as shown in Table 1.

The sandy soil is modeled by a mesh of finite elements, both two- and three-dimensional (2D and 3D) elements are available in PLAXIS library. However, the behavior of sand in this research can be adequately investigated by tetrahedral elements with 10 nodes, as shown in Fig. 2.

2.2 *Geocell*

The honeycombed geocell used in this study consists of multi-thin plates as regular hexenal shape. Plates are elements used to model slender structures in the top layer of reinforced soil with a considerable flexural rigidity. The plates are modeled by six nodes of triangular plate element with six degrees of freedom per each node:

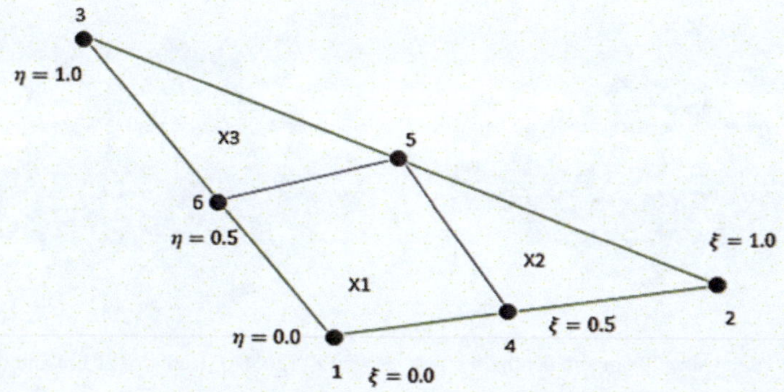

Fig. 3 Six nodes plate triangle element

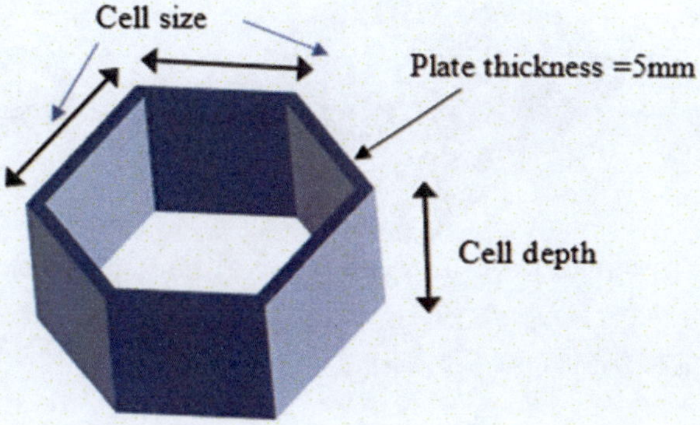

Fig. 4 Typical dimension of individual geocell

three displacement degrees of freedom (u_x, u_y, and u_z) and three rotating degrees of freedom (φ_x, φ_y and φ_z). Figure 3 shows the plate element configuration.

Three different cell sizes ($B = 100, 200, 400$ mm), with embedment depth ratios ($D/B = 0.5$ to 2) are used, and the thickness of the geocell plate is 5 mm, as shown in Fig. 4.

Several cells per each size are combined to form a section of geocell panel, as shown in Fig. 5. In total, for the current study, five numerical simulations have been carried out: four of them reinforced soil and one unreinforced soil.

2.3 *Interface Element*

To investigate the mechanisms of interaction between the reinforcement and soil, which incorporates a geosynthetic membrane, the stress acting at the interfaces of the different materials needs to be calculated accurately. This requires an accurate modeling of the soil–structure interaction. The interface element comprises pairs of nodes, corresponding with soil element or plate nodes.

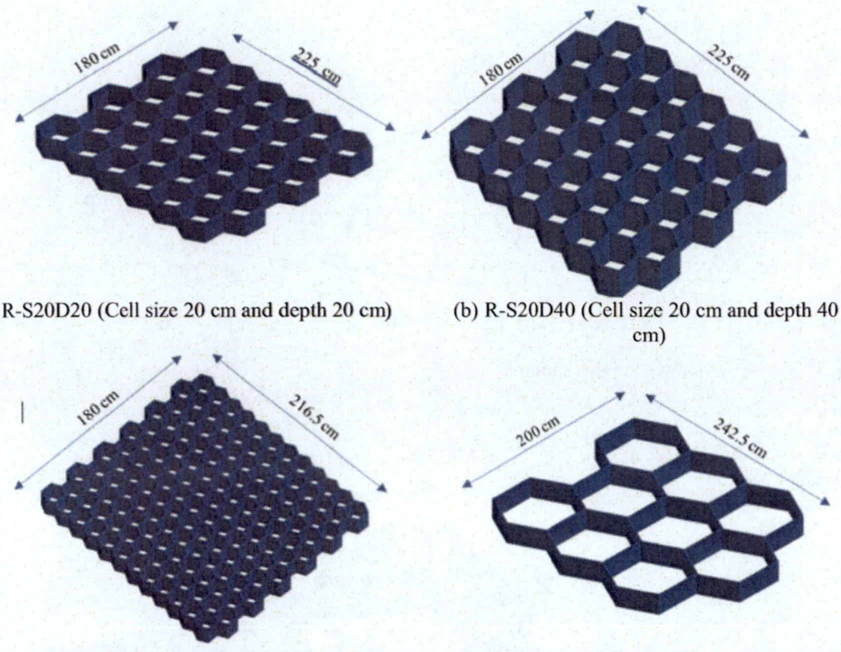

(a) R-S20D20 (Cell size 20 cm and depth 20 cm)

(b) R-S20D40 (Cell size 20 cm and depth 40 cm)

(c) R-S10D20 (Cell size 10 cm and depth 20 cm)

(d) R-S40D20 (Cell size 40 cm and depth 20 cm)

Fig. 5 Details of geocell panels

3 Numerical Mesh and Boundary Conditions

The numerical mesh model was created in PLAXIS3D, as shown in Fig. 6. To obtain an accurate comparison through numerical simulation, the full model dimensions of the soil represented in this study were used: 10 m in depth, 10 m in length, and 10 m width. The simulation of this model produced by the finite-element analysis can be considered reliable and precise. Displacement control was applied at the center of the top surface of the model (within the blue area in Fig. 6), so that a constant final vertical footing displacement of 50 mm is considered, and the corresponding failure load was measured for each of studied cases. A velocity boundary for the vertical direction was used to simulate a uniformly distributed load per unit area of a rigid foundation.

Displacement in all directions was restrained at the bottom boundary of the model, and the four side boundaries were fixed only in the lateral direction (X and Y axes), such that the displacement was permitted to occur along the vertical direction (Z-axis).

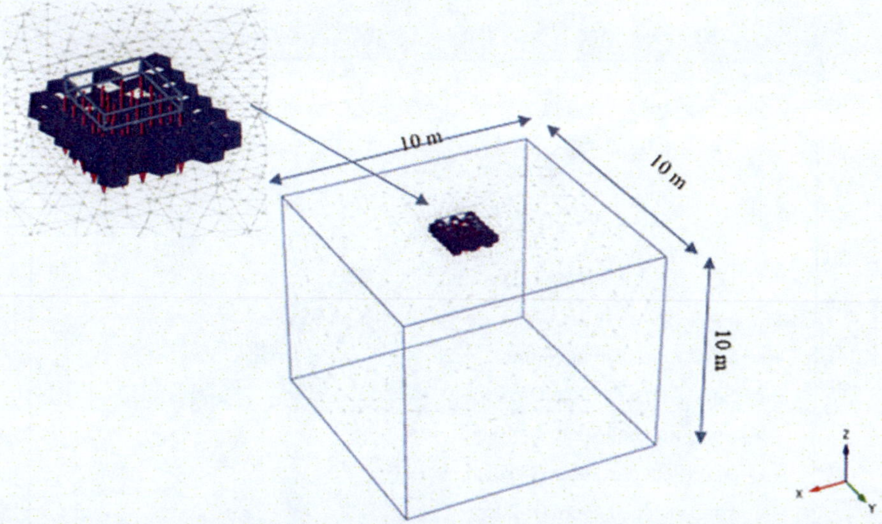

Fig. 6 Typical numerical mesh for the static loading R-S20D20 (cell size 20 cm, cell depth 20 cm)

4 Numerical Results and Discussion

The load–displacement responses for each of the four geocell panels reinforcement and the unreinforced case are illustrated in Fig. 7, where displacement is the value of vertical displacement beneath the foundation. It is clear that the reinforcement has a minor influence on the behavior of the shallow foundation during the first stage of loading, but it has significant influence in the later stages of loading, where the footing displacement increases and plastic soil deformations start to occur. On the other side, a reduction in the magnitude of the displacements at the same loading increments for the reinforced cases is noticed, compared with the unreinforced case, and the full set of summary of results is presented in Table 2.

Based on parametric result, depth and size of geocell can be slightly effective, for example, increasing the depth from 20 to 40 cm changes the failure load from 299 to 309 kN (it is 5.5% higher) and decreasing the size from 40 to 20 cm changes the failure load from 291 to 299 kN (it is 4% below). Also, it can be noticed that the continued reduction of cell size could produce the opposite effect. For example, decreasing the size from 20 to 10 cm changes the failure load from 299 to 265 kN (it is 19% below). This result concludes that increasing the stiffness of the reinforcement is not proportional to the corresponding failure load. However, the criteria for the selection of depth and size of the cell would be done by more factors, such as the performance and economic aspects. Figures 8 and 9 show the spread of vertical displacements in three dimensions (z-direction) for reinforced and unreinforced models. It can be noticed that under the same rigid foundation (displacement of 5 cm) on top of the

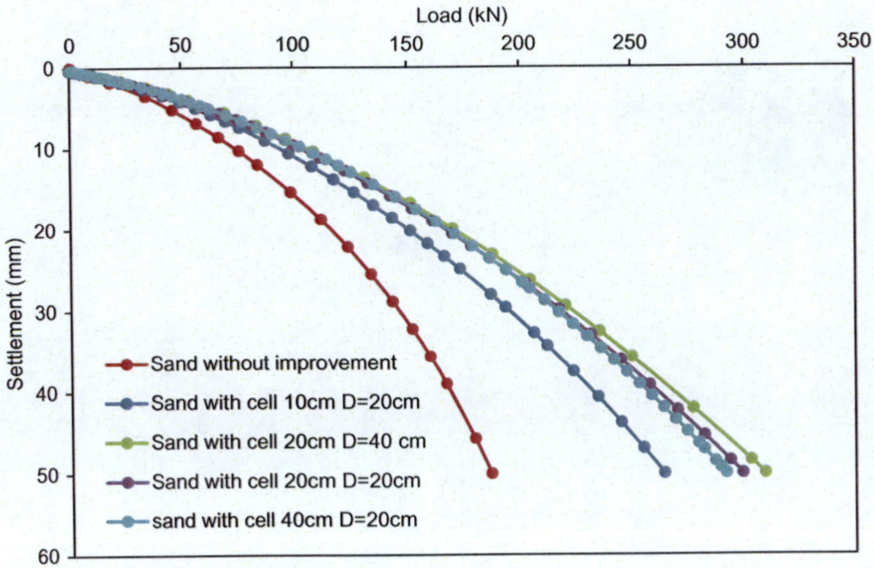

Fig. 7 Load–displacement for reinforced and unreinforced soil

Table 2 Load and displacement comparison for reinforced and unreinforced soil

Case	Load at 5 cm settlement (kN)	% Increasing[b]	Displacement at 150 kN (mm)	% Reduction[c]
UR[a] (Ref)	187	–	31.4	–
R-S20D20	299	59.9	17.0	45.9
R-S10D20	264.5	41.4	19.8	36.9
R-S20D40	309.3	65.4	16.7	46.9
R-S40D20	291.9	56.1	16.8	46.5

[a]Unreinforced, [b]% = (R-UR)/UR, [c]% = (UR-R)/UR

model, the reinforced sand is spread more horizontally on ineffective area than unreinforced sand. To present more details in all cases, a cross-section of displacements distribution under load through the model are shown in Fig. 10, compared to the results, and it can be observed that spread of displacement in reinforced soil has a similar distribution zone with little differences in the boundaries of in effective area (Fig. 10).

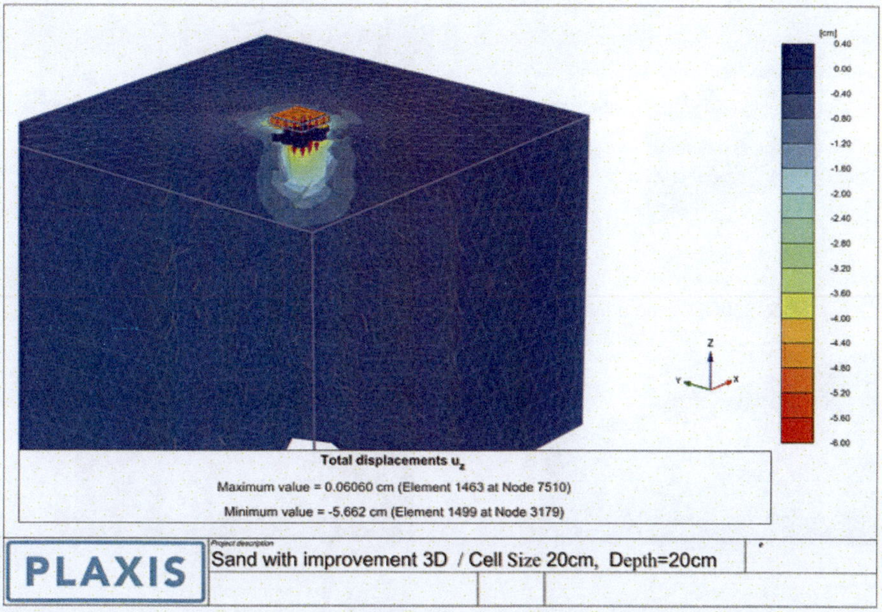

Fig. 8 Vertical displacement in *Z* direction (cell size = 20, depth 20 cm)

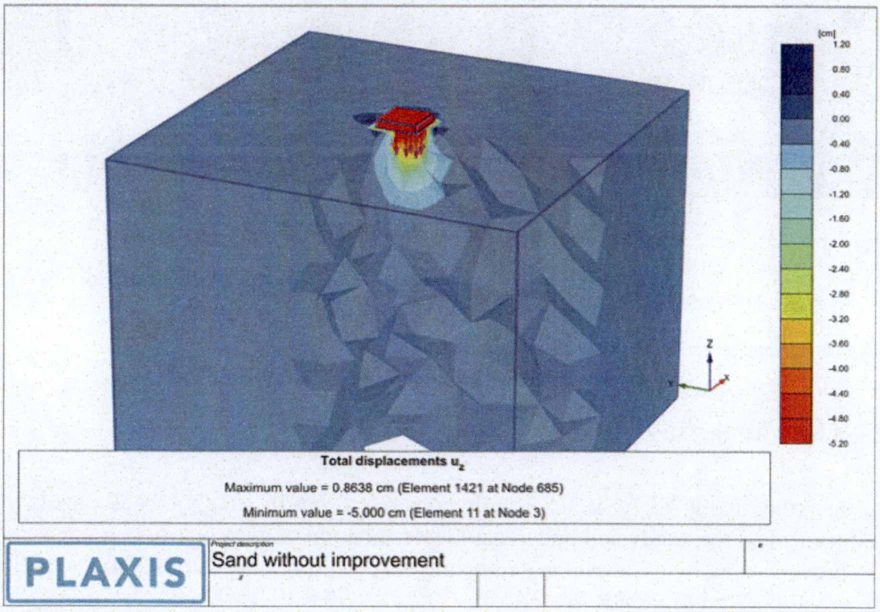

Fig. 9 Vertical displacement in *Z*-direction (without reinforcement)

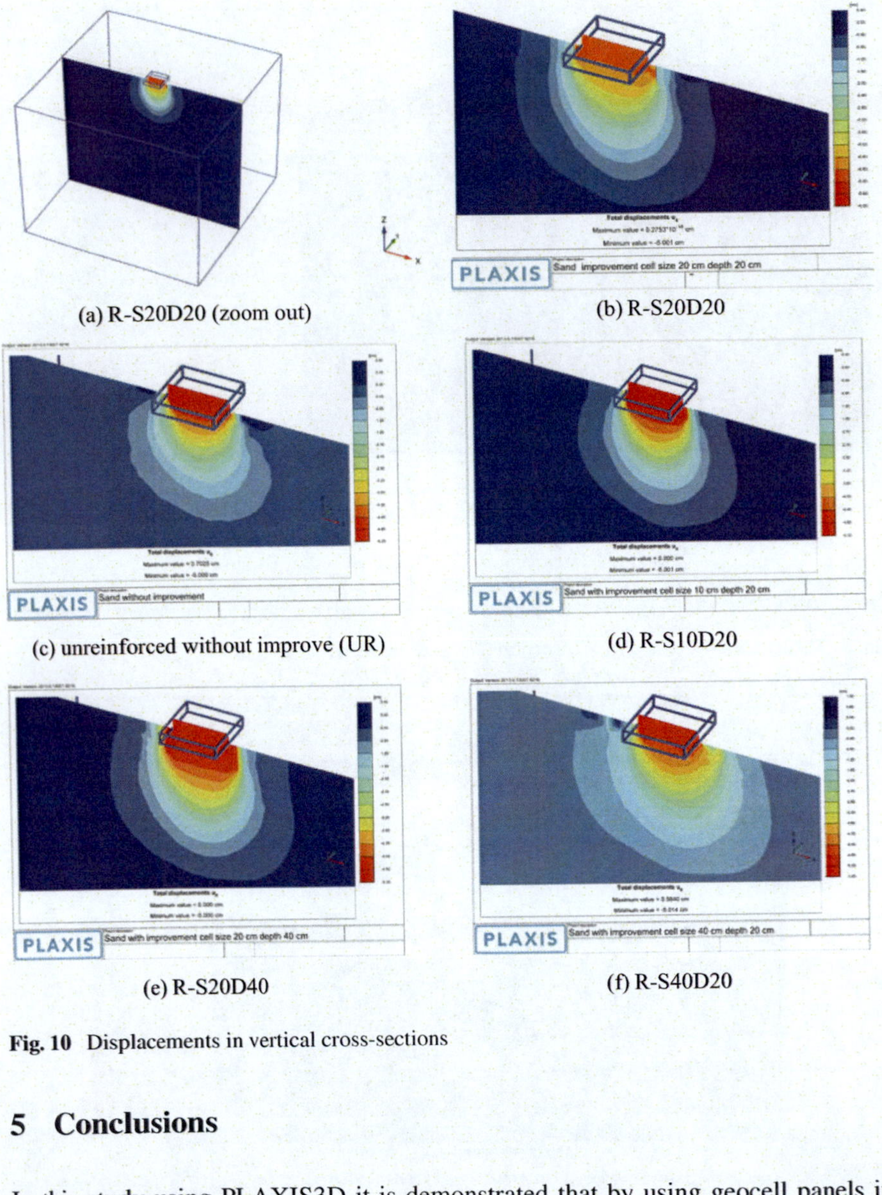

(a) R-S20D20 (zoom out)

(b) R-S20D20

(c) unreinforced without improve (UR)

(d) R-S10D20

(e) R-S20D40

(f) R-S40D20

Fig. 10 Displacements in vertical cross-sections

5 Conclusions

In this study using PLAXIS3D it is demonstrated that by using geocell panels in reinforcement the sandy soil under shallow foundation work properly on increasing the bearing capacity of the soil and reducing the corresponding settlement. The main conclusions from this study are:

- The numerical analysis by PLAXIS3D can model some main characteristics in the performance of the geocell-reinforced soil, such as size, depth, and thickness of the wall of geocell.

- The bearing capacity of the soil significantly improved with the presence of geocell, where the increase in bearing capacity resulted from the confinement of soil and due to the friction between the soil and the walls of geocell. Also, the stresses were concentrated at the surfaces of geocell reinforcement.
- An important finding from the parametric study is that the depth of geocell has a secondary role while the size of geocell could be more effective on the behavior of reinforced soil.

References

1. Yang X (2010) Numerical analyses of geocell-reinforced granular soils under static and repeated loads. Ph.D. thesis, University of Kansas
2. Information on https://www.geoplastglobal.com/
3. Yang X, Han J, Parsons RL, Leshchinsky D (2010) Three-dimensional numerical modeling of single geocell reinforced sand. Front Archit Civ Eng China 4(2):233–240
4. Hegde A, Sitharam TG (2015) 3-Dimensional numerical modelling of geocell reinforced sand beds. Geotext Geomembr 43(2):171–181
5. Hedge A, Shivdev S, Sitharam TG (2016) Numerical simulation of geocell reinforced foundation beds: a comparative study using PLAXIS-3D and FLAC3D. In: Indian geotechnical conference, IGC
6. Keshmiri N, Ghareh S, Kazemian S, Hosseinian AR (2019) 3D numerical analysis of loading geometry on soil behavior reinforced with geocell element. J Test Eval 47(3):1645–1657
7. Ting PAN, Changjun YIN, Peng XIAO (2019) Settlement analysis of geocell reinforced soils using asymptotic homogenization method. Soil Eng Found 33(1):54
8. Mamatha KH, Dinesh SV (2019) Performance evaluation of geocell-reinforced pavements. Int J Geotech Eng 13(3):277–286
9. Song F, Tian Y (2019) Three-dimensional numerical modelling of geocell reinforced soils and its practical application. Geomech Eng 17(1):1–9
10. Brinkgreve RB, Engin E, Swolfs WM (2013) PLAXIS 3D manual. PLAXIS, Delft

Production of Waste Rubber-Made Geogrid Reinforcement for Strengthening Weak Soils

Mohammed Y. Fattah, Mustafa A. Yousif, and Anwar L. Mohammed Rasheed

Abstract The research aims at the production of local materials from recycled rubber by subjecting these materials to high levels of temperature and pressure. The researchers believe that the standard geogrid used in the reinforcing of the soil under the foundations of hardness (stiffness) and resistance to the withdrawal (tensile strength) is much higher than the levels of stress (stress) and emotion (strain) which exposed. So, the trend has been toward the geogrid manufactured from materials of less hardness and the strength of polymers should perform the same purpose that standard geogrid plays. Hence, the thinking was to take advantage of waste rubber material in the manufacture of alternative geogrid and conduct the tensile test of manufactured geogrid as per the American Society for Testing and Materials specifications (ASTM-D6637 A) to compare between the produced and the standard geogrid. This requires manufacturing of the molds taking the shape of geosynthetics (geogrids) used in soil reinforcement. After that, the product has to be tested to ensure its stiffness and tensile strength. Several mixing proportions of new and old recycled rubber has been tried for the production of geogrids (20% new: 80% recycled, 40% new: 60% recycled, 60% new: 40% recycled, 80% new: 20% recycled) for two types of geogrids: geogrids with rectangular openings and geogrids with honeycombed openings. The product was subjected to a tensile test as per the American Society for Testing and Materials specifications (ASTM-D6337A). To make use of this product in the applications of foundation engineering, it was used in several plate loading tests to investigate the improvement in bearing capacity because of using layers of geogrids beneath foundations. It was found that both types of geogrids, rectangular and honeycombed, maintained an increase in the bearing capacity of the foundation compared with unreinforced soil and for all proportions (20, 40, 60, and 80%). It

M. Y. Fattah (✉)
Civil Engineering Department, University of Technology, Baghdad, Iraq
e-mail: myf_1968@yahoo.com

M. A. Yousif
College of Engineering, Al-Mustansiriyah University, Baghdad, Iraq

A. L. M. Rasheed
Building Research Center, Ministry of Reconstruction and Housing, Baghdad, Iraq

M. O. Karkush and D. Choudhury (eds), *Modern Applications of Geotechnical Engineering and Construction*, Lecture Notes in Civil Engineering 112,
https://doi.org/10.1007/978-981-15-9399-4_23

was concluded that the case of insertion of geogrid of the same size of footing at a depth B beneath the footing exhibits the best improvement in bearing capacity.

Keywords Plate load test · Geogrid · Sand · Layer · Improvement · Settlement

1 Introduction

The various types of geosynthetics are: geotextiles, geogrid, geonets, geofoam, geomembranes, geocomposites, and so on. The polymeric source of the products makes them adequate for application in the soil where high levels of durability are needed. Geosynthetics perform five major tasks, like separation, reinforcement, drainage, filtration, and moisture retention barrier. One of the categories of geosynthetics, in particular geogrids are applied to improve the soil engineering properties.

Geogrids are considered a small but rapidly developing type in the geosynthetics field. Rather being a woven, nonwoven, or knit textile (or textile-like) fabric, geogrids represent plastics produced into a very open, grid-like shape; that is, they contain apertures larger than 6 mm to permit interlocking with surrounding soil, rock, earth, and other materials. Usually, they are pulled in one or two directions to improve their physical properties [1]. To apply reinforcement in the soil, there are three categories. The first is through a physical method that is conducted by vibration, freeze and thaw, or thermo-electrical. The second is through the mechanical method by the use of fiber materials from the geosynthetic group (geotextile, geogrid, geonet, geocomposite, and geocell). The third is by chemical method that usually make use of used materials, polymeric resins, and enzymes. Soil reinforcement is a very ancient and influential technique.

Binquet and Lee [2, 3] investigated the mechanisms of using reinforced ground slabs to improve the durability of granular soils. They designed strip patterns on sandy foundations reinforced with broad strips cut out of aluminum foil. An analytical method was presented to estimate the increased endurance based on the tests. Fragaszy and Lawton [4] also used aluminum reinforcing bars and model strip foundations to study the effects of sand density and length of reinforcing bars on durability. Many authors also studied the foundations of the tape but reinforced it with various materials such as steel bars (Verma and Char [5]), steel grids (Dawson and Lee [6], Abdel-Baki et al. [7]), geotextiles (Das [8]), and geogrids (Khing et al. [9], Ismail and Raymond [10]). Other researchers adopted the circular (Kazerani and Jamnejad [11]), square (Guido and Christou [12], Adams and Collin [13]), or strip footing (Omar et al. [14], Fattah et al. [15]). A set of results of laboratory-scale model footing tests conducted to determine the cyclic load resistance of sand beds reinforced with horizontal sheets of geogrid was presented by Saran et al. [16]. The test results indicated that the total settlement decreases and the bearing capacity increases, with the increase in size of reinforcements and number of layers. With the inclusion of reinforcing sheets, the coefficient of elastic uniform compression decreases slightly;

but this decreased value is valid up to the increased bearing capacity of the reinforced sand bed. There is significant improvement in the damping capacity upon reinforcing the sand bed, as indicated by the comparison of the strain energies under the pressure–settlement curves obtained from cyclic plate load tests.

A research on bearing capacity and compression properties of cohesive soil, strengthened by geogrid and carrying a square foundation loads was carried out by Ghiassian and Jahannia [17]. The missed suitable frictional resistance between the clayey soil and reinforcement bars was treated by adding a thin layer of sand (lens) surrounding the geogrid mesh. So, the tension forces produced in the reinforcement were transmitted to the underlying clay layer throughout the particles of sand and thus geogrid reinforcement was improved. Tests were carried out on two groups of samples: one group consisting of 1 × 1 × 1 m dimension and the foundation dimensions were 190 × 190 mm (series-I), and the other group of 150 × 150 × 150 mm dimensions and the foundation dimensions were 37 × 37 mm (series-II). The systems of load applications for the samples were stress-controlled and strain-controlled, respectively. All samples were saturated and entirely loaded under an undrained situation. The results proved the effectiveness of the sand layers on increasing the bearing capacity and settlement properties of the model foundation. In series-I, the maximum improvement in the bearing capacity caused by the existence of the sand layer was 17%; while in series-II, the value of improvement was 24%. The percentage decrease in settlement was 30 and 46%, respectively.

Al-Sanidi and Ali [18] attempted to provide details of the investigation into the geographic performance of the soil. For this purpose, tests of loading the isolation model on soil with and without multiple layers of geogrid network were carried out at different depths below the footing. The load-leveling properties were observed for each soil geogrid composition. The effect of different selected parameters on pregnancy settlement behavior was studied and critically evaluated for their practical importance. The paper discussed the mechanisms of this system using a large-scale model for case studies in which soil is supported by the geogrid network. This site contains very soft/hard loamy soil, and the plate loading test has been extensively tested according to the ASTM-D1194 procedure. Also, the paper presented a successful application in the use of geogrid reinforcement. Field observations demonstrated that the geogrid supported system generates very soft/soft soil reinforcement and reduces differential leveling. The geogrid supported system is more economical and attractive and shows superior performance compared to most other land improvement techniques and is optimal for rapid construction and/or strict and differential aggregation of the structure and/or newly developed thick packaging. The objective of the present study is to investigate the effect of geogrid layers in improving the bearing capacity of foundations and decreasing the load transferred to sub-soil layers. The research aims at the production of local materials from recycled rubber by subjecting these materials to high levels of temperature and pressure. The researchers believe that the standard geogrid used in the reinforcing of the soil under the foundations of hardness (stiffness) and resistance to the withdrawal (tensile strength) is much higher than the levels of stress (stress) and emotion (strain) which exposed.

2 Experimental Work

2.1 Stages of Production of Rubber Gratings

1. Manufacturing of the metal mold
The metal mold was manufactured in a special industrial workshop of iron steel. The metal mold was 1.2 m long and 0.6 m wide and thickness 0.06 m. The engraving required for research was in the form of hexagonal (hive), and each engraving was of 0.6 (0.6) m, that is, the metal mold produced two engravings at the same time and the metal mold consisted of a thick part in which the engravings were made. Figures 1 and 2 show the stages of metal mold manufacturing. The rubber used in packaging is a waste of rubber tires, and it is natural that the rubber material originally used in the tire industry is subject to laboratory tests and conform to the required

Fig. 1 Metal mold for honeycombed mesh

Fig. 2 Metal mold for mesh with rectangular openings

Table 1 Physical properties of the used rubber

Property	Value	Unit	Specification
Specific gravity	1.16	–	ASTM D792
Density	1.16	g/cm^3	ASTM D1895-03
Ultimate tensile strength	9	MPa	ASTM D412-02
Elongation	150	%	ASTM D412-02
Hardness	64	–	ASTM D2240-05
Absorption	1.12	%	ASTM D540

international standards. Physical tests were conducted on the rubber material used in the production of the grids and the results are shown in Table 1.

2. Customizing the rubber materia

The rubber used in the research consists of mixing the amount of recycled rubber used and the original rubber at different rates and are as follows:

1. 80% new rubber and 20% recycled rubber.
2. 60% new rubber and 40% recycled rubber.
3. 40% new rubber and 60% recycled rubber.
4. 20% new rubber and 80% recycled rubber.
5. 80% new rubber and 20% recycled rubber.

The recycled rubber takes the form of a substance closest to the powder, which reacts to the recycled rubber and softens until it turns into a powder-like substance, while the original rubber takes the form of gelatin. When mixing the two materials, they are placed in the mixing device. This device consists of two revolves that rotate around their axes in opposite directions, the rubber material between them, and using the pressure with the rotation constantly to get rubberized materials together to produce a new rubber material with homogeneous properties. It is not easy to separate or distinguish between the rubber and re-original after kneading to take one form homogeneous similar to rubber rug. Figures 3, 4 and 5 show the stages of molding rubber in the metal mold.

3. The method of rubber engraving

The rubber mixture is placed on the metal mold and pressure with the heat is applied. The temperature generated by the electric heater connected to the mold ranges from 80 to 60 °C and with an air pressure of 225–250 lb/in^2 (12–14 air pressure). In the presence of heat, the mold is left with rubber inside until all the spaces are filled and the material is formed and the rubber form is formed by the required texture. The durability of the rubber model in the mold is between 4 and 24 h to ensure the level of rubber model. The dimensions of the mesh openings for the rectangular openings are (3.3 × 2.3 cm) inside, and for the honeycomb networks are 2.6 cm for the big diameter and 2.34 cm for the small diameter. After extracting the rubber form from the metal mold, it is cut as per the demand and then used in the base models.

Fig. 3 The raw rubber used in manufacturing geogrids

Fig. 4 Placing of rubber in the metallic mold

Fig. 5 The raw rubber placed in the metallic mold before applying temperature

Figures 6, 7, 8 and 9 show the pieces of rubber that have been poured and are ready for use.

Evaluation of the Use of New Nets in Strengthening Weak Soils. In this section, the possibility of using locally manufactured grids to test soft clays under the foundations will be tested. The grids will be used under the foundations at a depth representing the width of the foundation (B) and the length of time as far as the dimensions of the base square. The plate loading test is a semi-direct method for estimating the permissible carrier pressure of the soil to induce a certain amount of leveling. Circular panels of varying sizes, from 25 to 46 cm and thickness of about 2.5 cm are used for testing. The load is applied to the plate with a 50-ton hydraulic socket. The crane load is reacted by a truck. Plate leveling shall be measured by two measures of 0.01 mm sensitive disk, set at a distance of 180°. The dial gauges are installed on standalone support that remains unobstructed during testing. The test is performed according to ASTM-D1196-93 specifications [19].

Fig. 6 Removing the geogrid mesh after the engraving process

Fig. 7 The manufactured mesh of honeycombed type

Fig. 8 The manufactured geogrid mesh of rectangular openings

Fig. 9 A piece of manufactured geogrid mesh of rectangular openings ready for use below the foundation

2.2 *Testing Procedure*

Plate loading tests provide a direct measure of compression and sometimes soil carrying capacity that cannot be easily sampled. The technique adopted in this investigation is described to perform a plate loading test by following ASTM-D1196-93 standards [19]. The procedure followed in this work is as follows:

(1) Test pit area (1 × 1 m) was selected and the natural soil was excavated to a depth of 1.0 m. This depth was chosen based on Boussinesq stress distribution theory where stress below a footing dissipates to zero effectively at a depth of about 3B below the footing.
(2) Plate bearing of 300 mm diameter was carefully centered under the jack assembly. Other plates of lesser diameter were also placed at the center of the main plate to increase rigidity.

(3) Three dial gauges were attached to the reference beam and fixed over the plate in an appropriate location to indicate the average settlement results.
(4) When the equipment has been properly arranged, pressure was applied to the soil in cumulative equal increments of approximately 90 kPa.
(5) The applied load was held for 10–15 min and the dial gauges readings are recorded at an interval of 5 min. The same process was done to all loading increments reaching maximum load.
(6) The load was released at the same rate above and the readings of dial gauges are recorded; then reloading was done as above procedure.
(7) The above loading and unloading were repeated in the repetitive process.

The tests were carried out in a site within Al-Jadiriyah complex in Baghdad city.

3 Results and Discussion

Figure 10 shows a comparison between the pressure–settlement curves of the plate constructed on the natural ground either in its natural water content (approximately dry) or wetted with water. Figures 11, 12, 13, 14 and 15 present the pressure–settlement curves of the footing (plate) constructed on the soil reinforced with different types of geogrid. In Fig. 11, it is evident that the case of insertion of geogrid of the same size of footing at a depth B beneath the footing exhibits the best improvement in bearing capacity. Extension of the geogrid outside the footing (to a distance of B/2) from each side led to opposite behavior. Also, the manufactured geogrid made of rubber showed similar behavior to the original geogrid (Tensar) produced using plastic material. Figures 12 and 13 indicate that placement of the geogrid reinforcement at a depth of B/2 produces lower improvement in bearing capacity than the case of placement at a depth of B.

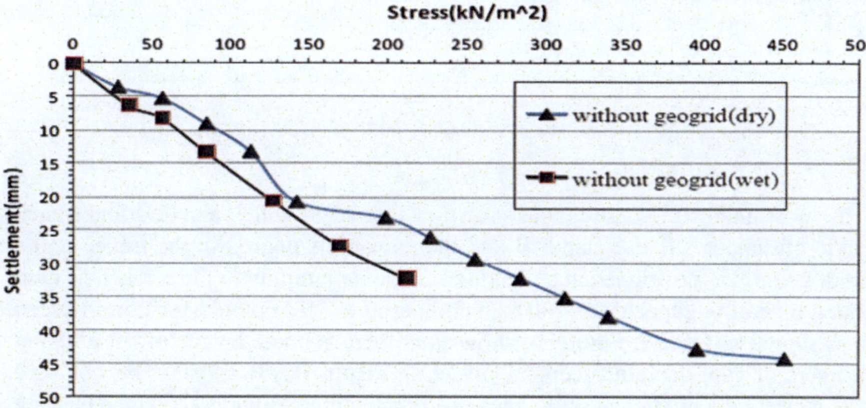

Fig. 10 Comparisons between pressure–settlement curves of plate on natural dry and wet soil

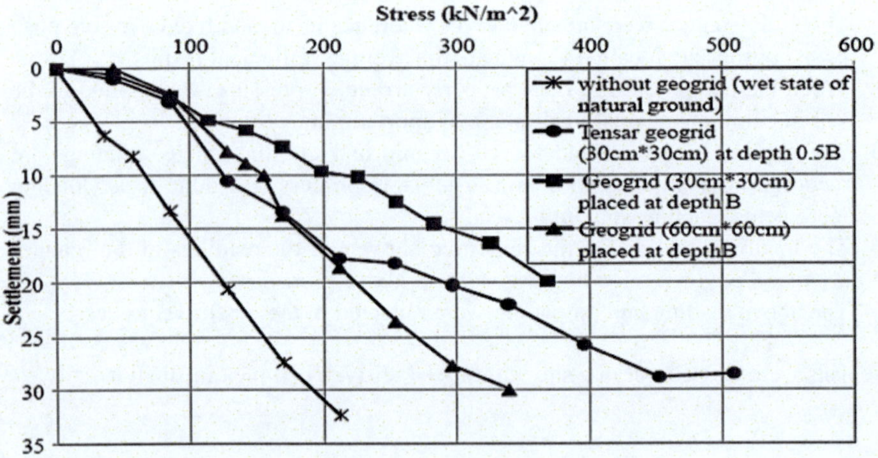

Fig. 11 Comparisons between natural soil and with different types of geogrids

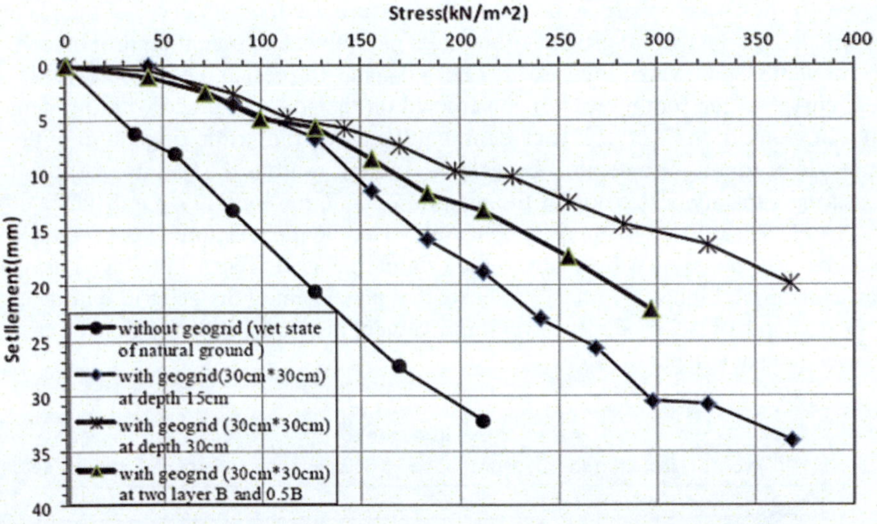

Fig. 12 Comparisons between natural soil with geogrid (30 × 30 cm) at different depths

The most interesting case is that shown in Fig. 13 in which a cylindrical trench is made of diameter B to a depth B and then a sand is poured in the trench and a layer of geogrid is positioned at a depth B with some tamping to the sand. This case revealed a bearing capacity of 240 kPa compared to 20 kPa only for unreinforced soil, as shown in Table 2. Figure 14 shows that there is no beneficial effect of using two layers of geogrid reinforcement placed at depths 0.5 B and B. The obtained results showed the efficiency of the increase in the laying of the grids on the bearing of the foundation in addition to reducing the amount of vertical displacement at the

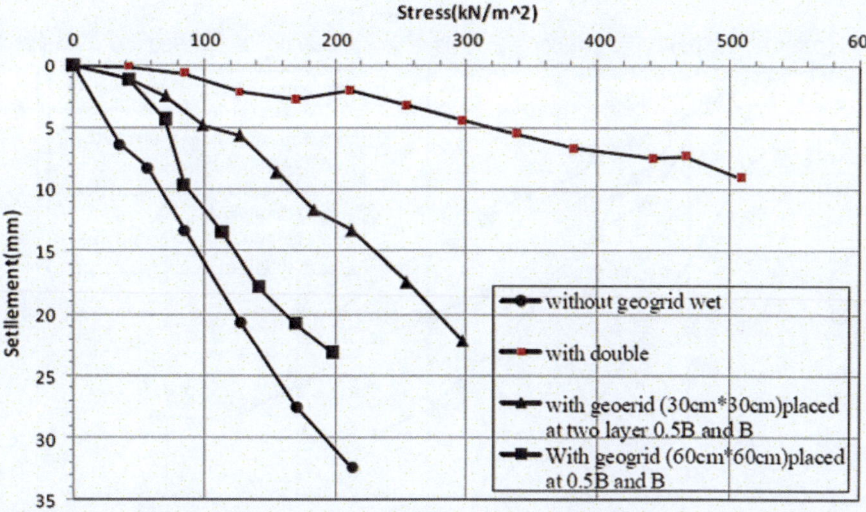

Fig. 13 Comparisons between pressure–settlement curves of plate on natural wet soil and soil reinforced with two layers of geogrid

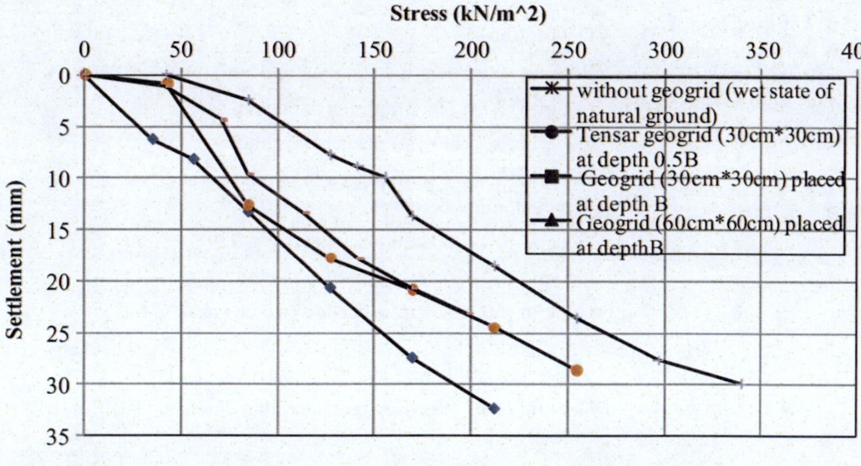

Fig. 14 Comparisons between pressure–settlement curves of plate on soil reinforced with geogrids of dimensions (60 × 60 cm) placed at different depths

different loading stages and all the ratios and both types of rectangle or hive. The reason may be due to increasing the area of overlap between the soil and the nets and thus increasing the friction between the soil and the nets. It was also observed that the effect was clearer as the proportion of new rubber increased to the user.

Geogrid works in two ways: reinforcement and separation which are techniques for improving poor soils with a geographic network, to increase the hardness and

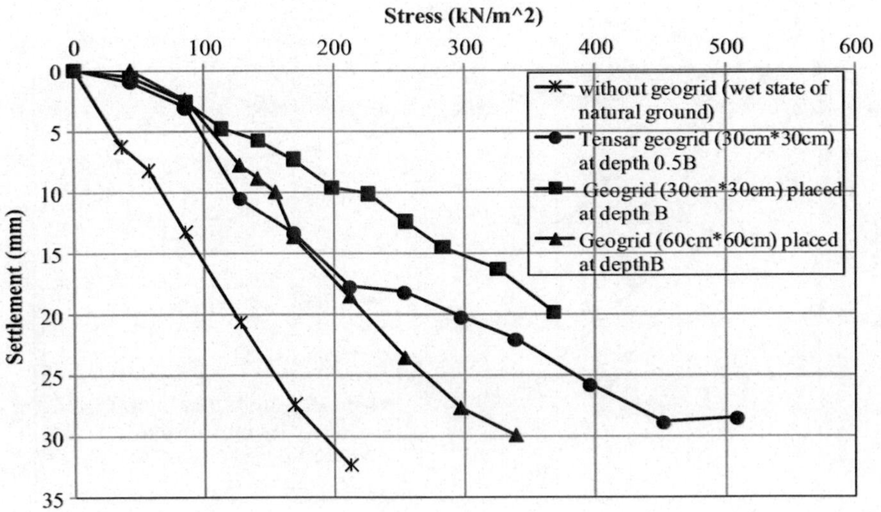

Fig. 15 Comparisons between pressure–settlement curves of plate on soil reinforced with different types of geogrid

Table 2 Ultimate bearing capacity of foundations (q_{ult}) reinforcement with different sizes of geogrid and different layers

Description	q_{ult} (kN/m^2)
Without geogrid (wet state of natural soil)	20
Geogrid (30 × 30 cm) placed at depth 0.5B	80
Geogrid (30 × 30 cm) placed at depth B	90
Geogrid (30 × 30 cm) placed in two layers at depths B/2 and B	80
Geogrid (60 × 60 cm) placed at depth 0.5B	50
Geogrid (60 × 60 cm) placed at depth B	92
Geogrid (60 × 60 cm) placed in two layers at depths 0.5B and B	62
Tensar geogrid placed at a depth 0.5B	78
Geogrid (30 × 30 cm) placed at within a trench of sand	240

bearing capacity of the soil through frictional interaction between the soil and the materials of the geogrid. Soils reinforced by the geogrid network are stronger and more solid and give more strength than the equivalent soil without strengthening the geogrid. Geogrids provide optimized aggregate meshing in road infrastructure stabilization through sub-base tuning enhancement applications. The geological armament presented between the primary track and the sub-grade soil carries the shear stress caused by the vehicle's loads.

The effect of extending the grids outside the base area was investigated when reinforcing. The reinforcing dimensions were increased by half the width of the base in all directions. The results obtained showed the efficiency of the increase in the laying of the grids to withstand the base, in addition to reducing the amount of vertical displacement at different loading stages, either the rectangle or the hive. The reason may be due to increased area of overlap between the soil and the grids and thus increased friction between the soil and the grids. It was also observed that the effect was clearer as the proportion of new rubber increased to the user. Geogrid mesh supplies better entanglement with soil particles, ensuring adequate mooring during loading. The improvement in the carrying capacity can be attributed to the improvement of the dispersion of the load through a reinforced sub-base to the sub-layer. This in turn leads to less stress in obtaining transport to the sub-layer, resulting in less stress in the suburbs.

Due to the better particle-geogrid interlock, the vertical stresses in sand decrease when the geogrid is placed in one layer in comparison to unreinforced sand. Low vertical pressures in sand in the case of sand reinforced by a geogrid network also mean a subsequent decrease in basement pressures. This indicates the role of reinforcement in dispelling applied vertical pressures, to an acceptable level in the lower soil, which is a particularly important observation in the case of structures to be built on soft soils. In essence, the geographic reinforcement of the soil will divert part of the applied vertical pressure that will be transferred to the lower soil, which increases the pressure confined to the soil and thus enhances the stability of the structure.

4 Conclusions

- There is a decrease in the measured settlement with the inclusion of geogrid layers.
- Both types of geogrids; rectangular and honeycombed maintained an increase in the bearing capacity of the foundation compared with unreinforced soil and for all proportions (20, 40, 60, and 80%).
- The case of insertion of geogrid of the same size of footing at a depth B beneath the footing exhibits the best improvement in bearing capacity. Extension of the geogrid outside the footing (to a distance of B/2) from each side led to opposite behavior.
- The manufactured geogrid made of rubber showed similar behavior to the original geogrid (Tensar) produced using plastic material.
- The effect of extending the grids outside the base area by half the width of the base and all directions showed the efficiency of the increase in the laying of the grids to withstand the base, in addition to reducing the amount of vertical displacement at different loading stages, either the rectangle or the hive.
- When geogrid reinforcement was placed in a trench of sand, the ultimate bearing capacity was increased by about 300% compared with conventional reinforcement layers.

References

1. Ramteke NB, Saxena A, Arora TR, Review A (2014) Effect of geo-grid reinforcement on soil. Int J Core Eng Manage (IJCEM) 1(4):35–47
2. Binquet J, Lee LL (1975a) Bearing capacity tests on reinforced earth slabs. J Geotech Eng Div, ASCE 101(12):1241–1255
3. Binquet J, Lee KL (1975b) Bearing capacity analysis of reinforced earth slabs. J Geotech Eng Div, ASCE 101(12):1257–1276
4. Fragaszy RJ, Lawton EC (1984) Bearing capacity of reinforced sand subgrades. J Geotech Eng, ASCE 110(10):1500–1507
5. Verma BP, Char ANR (1984) Bearing capacity tests on reinforced sand subgrades. J Geotech Eng, ASCE 112(7):701–706
6. Dawson A, Lee R (1988) Full scale foundation trials on grid reinforced clay. In: Holtz RD (ed) Geosynthetics for soil improvement, pp 127–147
7. Abdel-Baki S, Raymond GP, Johnson P (1993) Improvement of the Bearing capacity of footings by a single layer of reinforcement. In: Geosynthetics 93 conference, vol 2, Vancouver, Canada, pp 407–416
8. Das BM (1988) Shallow foundation on sand underlain by soft clay with geotextile interface. In: Holtz RD (ed) Geosynthetics for soil improvement, pp 112–126
9. Khing KH, Das BM, Puri VK, Cook EE, Yen SC (1993) The bearing capacity of a strip foundation on geogrid reinforced sand. Geotext Geomembr 12:351–361
10. Ismail I, Raymond GP (1995) Geosynthetic reinforcement of granular layered soils, vol 1. In: Geosynthetics '95, Nashville, TN, USA, IFAI, St. Paul, MN, USA, pp 317–330
11. Kazerani B, Jamnejad GH (1987) Polymer grid cell reinforcement in construction of pavement structures, vol 1. In: Geosynthetics '87 Conference, New Orleans, USA, pp 58–68
12. Guido VA, Christou SN (1988) Bearing capacity and settlement characteristics of geoweb-reinforced earth slabs. In: Das BM (ed) Special topics in foundations, pp 21–36
13. Adams MT, Collin JG (1997) Large model spread footing load tests on geosynthetic reinforced soil foundations. J Geotech Geoenviron Eng, ASCE 123(1):66–72
14. Omar MT, Das BM, Yen SC, Puri VK, Cook EE (1993) Ultimate bearing capacity of rectangular foundation on geogrid-reinforced sand. Geotech Test J 16(2):246–252
15. Fattah MY, Al-Baghdadi W, Omar M, Shanableh A (2010) Analysis of strip footings resting on reinforced granular trench by the finite element method. Int J Geotech Eng 4(4):471–482
16. Saran S, Lavania BVK, Sharma RK (1995) Cycle plate load tests on reinforced sand. In: Proceedings of the third international conference on recent advances in geotechnical earthquake engineering and soil dynamics, vol Ill, April 2–7,1995, St. Louis, Missouri, pp 1123–1126
17. Ghiassian H, Jahannia M (2004) Influence of encapsulated geogrid-sand system on bearing capacity and settlement characteristics of reinforced clay. Int J Civ Eng 2(1):45–53
18. Al-Sinaidi AR, Ali AH (2006) Improvement in bearing capacity of soil by geogrid—an experimental approach. In: IAEG2006 Paper number 240, The Geological Society of London
19. ASTM D1196–93 (Reapproved 2004) Standard test method for nonrepetitive static plate load tests of soils and flexible pavement components, for use in evaluation and design of airport and highway pavements. Annual Book of ASTM Standards: Volume 04.08, 2004, Soil and Rock

The Influence of Nailing on the Seismic Response of a Superstructure with Underground Stories

HamidReza Bolouri Bazaz, Ali Akhtarpour, and Abbas Karamodin

Abstract Soil nails are mainly designed for excavations for a limited period and are not designed to withstand excavation forces permanently, specifically seismic forces. Despite this issue, when the excavation and retaining walls are completed the permanent effects of nails remain in the soil. This study investigates the effects of soil nailing in soft soils on the seismic response of a superstructure with four underground stories. The results show that soil nailing has a significant effect on the response of the superstructure which should be considered in future soil–structure interaction studies especially when soft soils are encountered.

Keywords Superstructure · Soft soil · Nailing · Seismic response · Soil–structure interaction

1 Introduction

Soil–structure interaction studies have been improved considerably over the last few decades. Due to the rapid development of computational calculations, it is now possible to investigate sophisticated soil and structure systems simultaneously [1]. For tall structures resting on soft soil, the effects of considering soil–structure interactions are significant, and due to limited studies in this field, there is no certain estimation for the response of the aforementioned systems [2]. Buildings with underground stories are known to have better seismic responses than those without it. This is mainly caused by two reasons: (1) the structure is generally resting on a denser

H. B. Bazaz
Faculty of Civil, Water and Environmental Engineering, Shahid Beheshti University, Tehran, Iran
e-mail: h_bolouri@sbu.ac.ir

A. Akhtarpour (✉) · A. Karamodin
Civil Engineering Department, Faculty of Engineering, Ferdowsi University of Mashhad, Mashhad, Iran
e-mail: akhtarpour@um.ac.ir

A. Karamodin
e-mail: a-karam@um.ac.ir

M. O. Karkush and D. Choudhury (eds), *Modern Applications of Geotechnical Engineering and Construction*, Lecture Notes in Civil Engineering 112,
https://doi.org/10.1007/978-981-15-9399-4_24

and stiffer soil and (2) the integrity of the structure and the retaining wall commonly reduces the seismic loads to the structure. The effect of reduction of seismic loads can be related to the change of seismic base's location in the equivalent-linear method.

Ganainy and Naggar studied the impact of underground stories for 5, 10, and 15 story buildings. The SSI effects for soft soil deposits led to a 10–25% increase in the base shear and moment demand of buildings, yet as the number of underground stories increased, the effects were decreased compared to fixed base conditions. This fact can be associated with the rigidity of the underground slabs and basement walls delivering a rigid box, hence fixing the structure [3]. Moreover, Saad et al. conducted similar results for a 10-story building with three basements embedded in soft soils [4].

Soil nailing is widely used to reinforce geotechnical structures in problems such as slopes and excavation [5, 6]; for example, numerous slopes have been stabilized in Hong Kong since the 1970s with nails [7]. Although soil nails are commonly designed for static situations, their reinforcement increases the bearing capacity of soil; therefore their seismic effects are undeniable. In this paper, the seismic effects of soil nailing have been investigated by studying a 20-story benchmark building with four underground stories. The building is resting on soft soil and the effect of nailing is analyzed by the seismic structural response in both the situations. A two-dimensional finite-difference software, namely FLAC2D, has been adopted in this research which can model the soil and structure simultaneously to study their seismic interaction effects on each other.

2 Soil Characteristics

To investigate the effects of nailing on soft soil, the soil profile of a megaproject in Bangkok, namely Sukhumvit MRT station, has been selected (shown in Fig. 1). The details of the project can be found in previous studies (i.e. [8, 9, 10, 11]).

The constitutive model selected for this study is a combination of two models: (1) Mohr–Coulomb model and (2) capped-hardening soil model (CYSoil model). The capped-hardening soil model is employed for modeling soft soil behavior. Although this model is much more complicated than the Mohr–Coulomb model, its parameters can be determined by common soil mechanics tests. The calibration process of this model for drained and undrained parameters of Bangkok soft clay has been performed by Bolouri Bazaz et al. with the use of a series of triaxial and odometer tests [12]. Tables 1 and 2 show the parameters obtained for this study based on previous researches [8, 9, 12].

The maximum dynamic shear modulus of soil also known as the small-strain shear modulus plays an important role in the seismic response of a soil–structure system. This parameter can be determined by many experimental or field tests [13]. In this study, the dynamic shear modulus of the soil profile was acquired by the determination of the shear wave velocity that resulted from a series of cross-hole and down-hole in situ tests [14, 15] ($G_{\max} = \rho {V_s}^2$, where ρ is the density and V_s is the

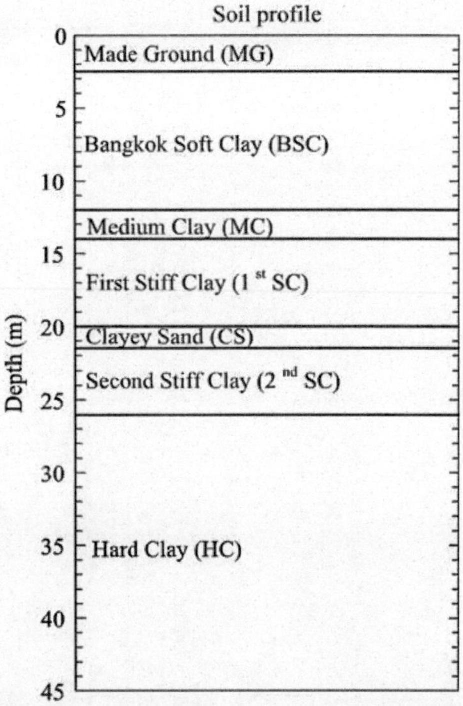

Fig. 1 Soil profile at the location of Sukhumvit Station [9]

shear wave velocity). The shear wave velocity of the soil profile is illustrated in Fig. 2 [15]. The strain-dependent modulus and damping factors of the soil are another aspect which considerably affects the seismic soil–structural response. These functions have been defined by a set of experimental tests performed by previous researches [16] and were implemented in FLAC by the equivalent-linear method and a sigmoidal model, namely sig3 [17]. The parameters obtained are: $a = 1$, $b = -0.5$ and $x_0 = -0.67$. Moreover, to overcome the deficiency of the predicted damping ratio than the experimental results in low cyclic shear strains, it is recommended to employ an additional 1% local or Rayleigh damping [17].

3 Structural Characteristics

A typical 20-story steel benchmark building has been taken into account for this study, which consists of 5 and 6 bays in the N-S and E-W direction, each 6.1 m in length [18]. The structure originally has 20 stories and two additional basements. To study the effects of underground stories, the 20-story superstructure has been considered with four basements. The specifications of the two additional basements are considered the same as the original, and this assumption has been verified following

Table 1 The drained and undrained properties of the soil profile [12]

Layer no	Soil type	Depth	Density	Elasticity modulus	Drained parameters			Undrained parameters			
					c'	**φ'**	**υ**	**c**	**φ**	**S_u**	**υ**
1	MG	0–2.5	18	8	1	25	0.2	1	25	–	0.3
2a	BSC1	2.5–7.5	16.5	10	Capped Hardening Soil Model Parameters						
2b	BSC2	7.5–12	16.5	20.5							
3	MC	12–14	17.5	27.5	10	25	0.2	–	–	55	0.495
4	1^{st}SC	14–20	19.5	40	25	26	0.2	–	–	80	0.495
5	CS	20–21.5	19	53	1	27	0.2	1	27	–	0.25
6	2^{nd}SC	21.5–26	20	72	25	26	0.2	–	–	120	0.495
7	HC	26–60	20	240	40	24	0.2	–	–	240	0.495

Table 2 The drained and undrained parameters of Bangkok soft clay for the hardening soil model (CYSoil Model) for layers 2a and 2b [12]

Parameter	Drained parameters	Undrained parameters
P_r	100 kPa	100 kPa
K_r	5.87	4.41
m	0.9975	0.9975
R	4.5	4.5
α	0.87	0.87
c_p	0	0
φ_p	23.6	29
φ_0	0	10
ψ_p	0	0
$\varepsilon_s{}^f$	0.293	0.293

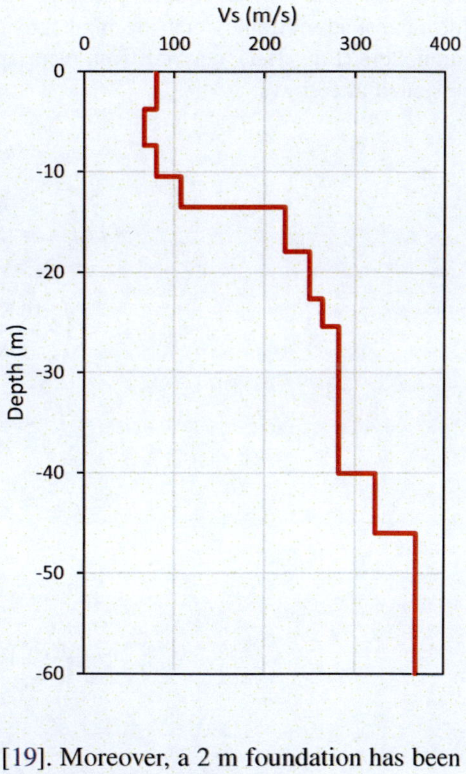

Fig. 2 Shear wave velocity of Bangkok subsoil [15]

the construction norms using SAP2000 [19]. Moreover, a 2 m foundation has been considered following common regulations using SAFE [20].

In the latest version of the finite difference software (FLAC2D) the possibility of considering an elastic-perfectly plastic model for structural elements has been provided. Hence, in this research, this model has been employed for structural elements in terms of their plastic moment as $M_p = ZF_y$, where Z is the section modulus and Fy is the yield stress [21]. The damping of the structural system was considered as Rayleigh ($C = \alpha.M + \beta.K$; where C, M and K are damping, mass and stiffness matrices and α and β are constants) [17], where α and β were, respectively, calculated as 0.1435 and 0.0076 utilizing modal analysis of the superstructure and 5% damping.

4 Soil and Structural Elements

The interface between soil and structural elements (i.e. foundation, piles, and retaining wall) can be simulated by shear and normal coupling springs [22]. The structural elements and the grid are connected using force and motion via the spring-slider system interface [17]. The characteristics of the interface system are presented as cohesive strength, frictional resistance, and stiffness in the normal and shear orientation. Also, the normal coupling springs can simulate a gap between pile and soil during ground motion by considering a tension strength [22]. The definition of these parameters (Eqs. 1–7), derived from prior studies (i.e. [17, 22, 23, 24, 25, 26]) are presented as follows:

$$cs_{ncoh} = 9\mathrm{c}_{\mathrm{soil}}.D \tag{1}$$

$$cs_{nfric} = \varphi_{\mathrm{soil}} \tag{2}$$

$$cs_{nstiff} = \left(\frac{10\left[\mathrm{K} + \frac{4}{3}\mathrm{G}\right]}{\Delta \mathrm{z}_{\mathrm{min}}}\right) D \tag{3}$$

$$cs_{nten} = \left(\frac{\mathrm{c}_{\mathrm{soil}}}{\tan(\varphi_{\mathrm{soil}})}\right).D \tag{4}$$

$$cs_{scoh} = \mathrm{c}_{\mathrm{soil}}.\mathrm{P} \tag{5}$$

$$cs_{sfric} = \varphi_{\mathrm{soil}} \tag{6}$$

$$cs_{sstiff} = \left(\frac{10\left[\mathrm{K} + \frac{4}{3}\mathrm{G}\right]}{\Delta \mathrm{z}_{\mathrm{min}}}\right) P \tag{7}$$

where c_{soil} and φ_{soil} are the cohesion and friction angle of the adjacent soil, K and G are the bulk modulus and shear modulus of the adjacent soil, P and D are the perimeter and diameter of the structural element and $\Delta \mathrm{z}_{\mathrm{min}}$ is the smallest width of the adjoining zone.

Table 3 The selected earthquakes and their specifications

Earthquake incident	Station name	Alias	Magnitude (Mw)
San Fernando (1970)	Pasadena—Old Seismo Lab	San	6.61
Morgan Hill (1984)	Gilroy—Gavilan Coll	Mor1	6.19
Morgan Hill (1984)	Gilroy Array #1	Mor2	6.19
Loma Prieta (1989)	UCSC	Lom	6.93
Northridge-01 (1994)	LA—Wonderland Ave	Nor1	6.69
Northridge-01 (1994)	Vasquez Rocks Park	Nor2	6.69
Iwate-Japan (2008)	IWT010	Iwa	6.9

5 Selected Earthquake Records

In this study, seven earthquake records have been adopted. In order to minimize the amplification effects of soil, the earthquake records were selected from stations located on rocky grounds. A summary of the specifications of the earthquakes is presented in Table 3. Given that the earthquakes are applied at the bottom of the model, the earthquake records were scaled to match the ASCE class (A) response spectra corresponding to hard rock. S_1 and S_s were selected as 1.61 and 1.15 based on a site in downtown Los Angeles where the benchmark building has been originally designed according to them.

6 Model Characteristics

In this study, two models were simulated in 2D finite-difference program (FLAC2D), one with nailing and one without it. The effects of nailing were investigated by comparing the structural response of both models in terms of base shear, peak horizontal acceleration, drift, and displacement. The geometric dimensions of the model were selected as 250 m in width and 60 m in deep [22, 27, 28] and the structural elements have been scaled in a perpendicular direction to fulfill proper 2D modeling [22, 17]. A schematic illustration of the aforementioned models is presented in Fig. 3 (a) without nailing and (b) with nailing. The characteristics of the retaining walls have been obtained regarding FHWA-NHI-14–007 [29]. The retaining wall is assumed to be 15 m deep concrete-reinforced wall with a thickness of 0.3 m for the model with nailing and 0.5 m for the model without them. It is noted that in both models the retaining wall is connected to the superstructure. For the model with nailing, the nails are considered as 12 m with an angle of 15° and the vertical and horizontal space of the nails are 1.5 and 2 m, respectively. The nails are assumed as 240 MN/m^2 ribbed steer bar with a diameter of 32 mm.

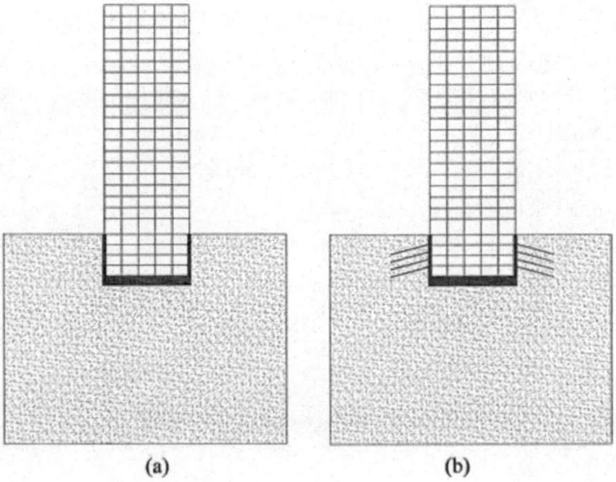

Fig. 3 Different modes of study

7 Results and Discussions

The 2D soil-structure interaction models presented in the previous section have been analyzed in the time domain, undergoing seven-scaled earthquake records by the direct method in the finite difference software FLAC2D. This section presents and discusses the results of the superstructure. The response is expressed in terms of "Max" and "Mean" values which are the maximum and average of the maximum values obtained in each earthquake, respectively. The results disclose the significant effects of nailing in soft soils for superstructures.

In the current study, base shear is the maximum lateral force of the ground level that takes place in a seismic motion. A comparison of the base shear gives a view on the shear force induced in a structure during an earthquake. The base shear of both models is illustrated in Fig. 4. It can be observed that the mean and max values of base shear have been decreased to about 9% and 7%, respectively, by assuming nailing in soft soils.

The peak horizontal acceleration (PHA) of the left-side nodes of the structure have been recorded during every seismic motion. PHA corresponds to the maximum values obtained for the horizontal accelerations of each node. The mean and max values of PHA are illustrated in Fig. 5. It should be noted that on an average basis the PVA of the superstructure with nailing has decreased to about 7% regarding the superstructure without nailing.

Displacements of structures during earthquakes are important because of the disruptive impacts of adjacent structures on each other. Besides, structural inter-story drifts throughout earthquakes are substantial to study for the case of sustainability and integrity of the structure. Herein, the drift and displacements of the left column of the superstructure have been observed to give a perspective on the seismic structural response. Figure 6 shows the mean and max values of drift and displacements. It can

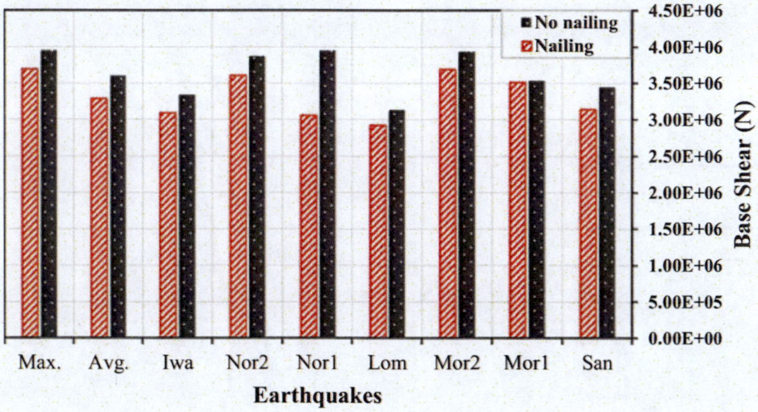

Fig. 4 The comparison of base shear in six studied models

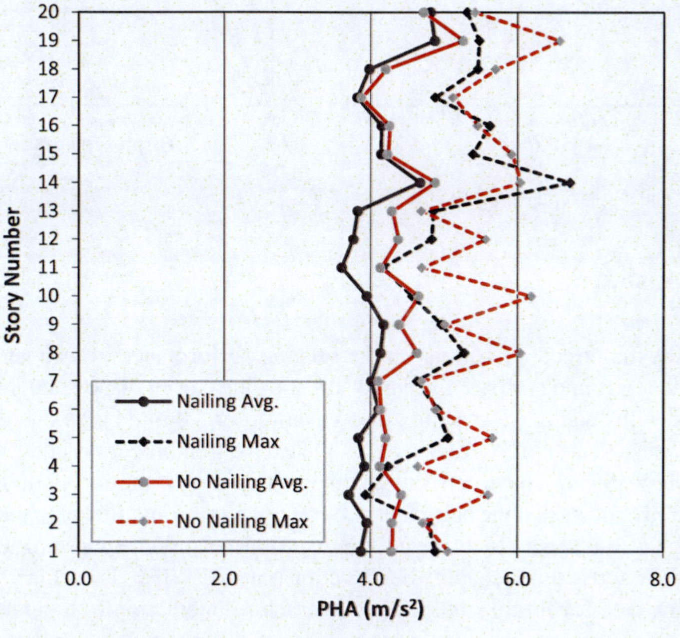

Fig. 5 The peak accelerations observed in the superstructure

be observed that the drifts and displacements of the superstructure decrease 10 and 40% on average by considering nailing.

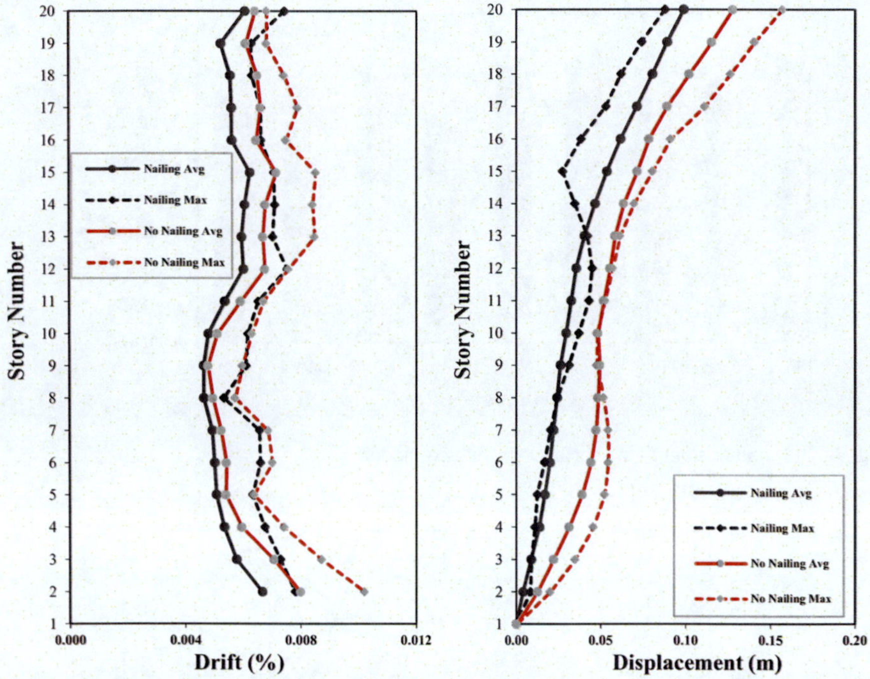

Fig. 6 Drift and displacements of the structure

8 Conclusion

In this study, the effects of nailing on the seismic performance of a superstructure with four underground stories adjacent to soft soils have been discovered. Nailing is designed to withstand the destruction of excavation for a limited period and it is not designed to resist earthquake forces, but as the excavation process is completed and the retaining walls are constructed the permanent effects of the nails remain in the soil. This study shows that the permanent effects of nails are significant to the seismic performance of the superstructure and it reduces the seismic response of it. For the situation discussed in this research, the reduction is up to 7–10%. Therefore, this issue can be considered for future studies of soil–structure interaction in superstructures, especially when they are encountered with soft soils.

References

1. Peizhen L, Hongmei R, Yueqing C, Xilin L, Heping S (2007) Shaking table testing of hard layered soil-pile-structure interaction system 34(3):307–313
2. Zhang L, Liu H (2017) Seismic response of clay-pile-raft-superstructure systems subjected to far-field ground motions. Soil Dyn Earthq Eng 101(1):209–224
3. El Ganainy H, El Naggar M (2009) Seismic performance of three-dimensional frame structures with underground stories . Soil Dyn Earthq Eng 29(9):1249–1261
4. Saad G, Saddik F, Najjar S (2012) Impact of soil-structure interaction on the seismic design of reinforced concrete buildings with underground stories. In: 15 WCCE, Lisboa
5. Xu D, Liu H, Luo W (2018) Evaluation of interface shear behavior of GFRP soil nails with a strain-transfer model and distributed fiber-optic sensors. Comput Geotech 95:180–190
6. Ehrlich M, Silva RC (2015) Behavior of a 31 m high excavation supported by anchoring and nailing in residual soil of gneiss. Eng Geol 191:48–60
7. Cartier G, Gigan J (1983) Experiments and observations on soil nailing structures. In: Proceedings from the 8th European conference on soil mechanics and foundation engineering, Helsinki
8. Surarak C (2010) Geotechnical aspects of the Bangkok MRT blue line project. Griffith University, Queensland
9. Likitlersuang S, Surarak C, Wanatowski D, Oh E, Balasubramaniam A (2013) Finite element analysis of a deep excavation: a case study from the Bangkok MRT 53(5):756–773
10. Phienwej N (2008) Ground movement in station excavations of Bangkok first MRT. In: Proceedings of the 6th international symposium on tunnelling for urban development (IS-Shanghai 2008). Shanghai, China
11. Suwansawat S, Chaiwonglek C, Horny U (2007) Design aspects of NATM and cut and cover underground stations for the Bangkok MRTA blue line extension. In: Proceedings of the 7th international symposium on tunnelling for urban development (IS-Pattaya 2007). Pattaya City, Thailand
12. Bolouri Bazaz H, Akhtarpour A, Kharaghani S (2019) A study on the efficiency of a capped hardening elasto-plastic model for soft clays. Soils Found 59(1):122–135
13. Atkinson J (2000) Non-linear soil stiffness in routine design. Geotechnique 50(5):487–508
14. Likitlersuang S, Teachavorasinskun S, Surarak C, Oh E, Balasubramaniam B (2013) Small strain stiffness and stiffness degradation curve of Bangkok clays. Soils Found 53(4):498–509
15. Likitlersuang S, Kyaw K (2010) Shear wave velocity correlations for the Bangkok subsoil. Obras Y Proyectos 7:24–30
16. Teachavorasinskun S, Thongchim P, Lukkunaprasit P (2002) Shear modulus and damping of soft Bangkok clays. Can Geotech J 39:1201–1208
17. Itasca Consulting Group (2016) Fast Lagrangian Analysis of Continua. Minneapolis, Minnesota
18. Ohtori Y, Christenson R, Spencer B, Dyke S (2004) Benchmark control problems for seismically excited nonlinear buildings. J Eng Mech 130(4):366–385
19. CSI, SAP2000 (2016) Integrated solution for structural analysis and design, version 19.0. Computers and Structures Inc., Berkley
20. CSI, SAFE (2016) Design of slabs, beams and foundations, reinforced and post-tensioned concrete. Computers and Structures Inc., Berkley
21. McCormac J, Csernak S (2012) Structural steel design. Pearson Education
22. Yeganeh N, Bolouri Bazaz J, Akhtarpour A (2015) Seismic analysis of the soil–structure interaction for a high rise building adjacent to deep excavation. Soil Dyn Earthq Eng 79(A):149–170
23. Luo W, Liu C, Han X (2007) Numerical simulation of behaviour of piled foundation influenced by tunneling. Rock Soil Mech 28:403–407
24. Naeini A, Choobbasti A, Saadati M (2013) Seismic behaviour of pile in three-layered soil (case study: Babol City Center Project). Arab J Geosci 6(11):4487–4497
25. Chen C, Martin G (2002) Soil–structure interaction for landslide stabilizing piles. Comput Geotech 29(5):363–386

26. Chen L, Poulos H (1993) Analysis of pile-soil interaction under lateral loading using infinite and finite elements. Comput Geotech 15(4):189–220
27. Tabatabaiefar HR, Samali B, Fatahi B (2010) Effects of dynamic soil-structure interaction on inelastic behaviour of mid-rise moment resisting buildings on soft soils. In: Australian Earthquake Engineering Society Conference, Perth
28. Briaud J, Lim Y (1997) Soil-nailed wall under piled bridge abutment: simulation and guidelines. J Geotech Geoenviron Eng 123(11):1043–1050
29. FHWA-NHI-14–007 (2015) Geotechnical engineering circular no. 7 soil nail walls—reference manual. Federal Highway Administration, Washington

Experimental and Numerical Analysis of Laterally Loaded Pile Subjected to Earthquake Loading

Rusul Salman Hussein and Bushra Suhil Albusoda

Abstract In this study, a dynamic soil pile interaction has been adopted to capture and address the pile response under the effect of combined lateral and axial loads during earthquake motion. A laterally loaded pile is subjected to shaking motion with a laminar shaking box installed in dry sand prepared at different initial relative densities. The pile–soil system was subjected to different excitation intensities (Kobe and Ali-Al gharbi). The pile behavior and the performance of pile–soil interaction were further discussed and compared with numerical modeling using three-dimensional finite element code (ABAQUS CAE) software. The test results indicated an increasing trend in pile (for both earthquakes) displacement as lateral loads decrease, while the acceleration increases progressively upward when input motion propagates and slightly affects vertical displacement. On the other hand, there is a good agreement between shaking table tests and finite element modeling.

Keywords Earthquake motion · Laterally loaded pile · Acceleration · ABAQUS CAE software · Numerical simulation

1 Introduction

Pile is a column element used to transfer load from superstructure through weak strata of soils to stiffer or more compact soils or rock. Piles used in retaining walls, abutments, and marine structures are subjected to transfer lateral and vertical loads [1]. The effect of lateral load on the design of a single pile is studied, and it has been found that the lateral displacement at working loads should be within the permissible limit. Many approaches including shaking table and centrifuge tests and numerical methods have been developed for the response analysis of single and group piles

R. S. Hussein (✉) · B. S. Albusoda
Department of Civil Engineering, University of Baghdad, Baghdad, Iraq
e-mail: salamrusal@gmail.com

B. S. Albusoda
e-mail: dr.bushra_albusoda@coeng.uobaghdad.edu.iq

M. O. Karkush and D. Choudhury (eds), *Modern Applications of Geotechnical Engineering and Construction*, Lecture Notes in Civil Engineering 112,
https://doi.org/10.1007/978-981-15-9399-4_25

under dynamic loading [e.g. 2, 3, and 4]. Under lateral load, the failure criterion of short pile is full rotation around the pile base in case of free pile head, while in the case of fixed head, the pile failed when fully moved and small rotation near pile base [5]. The behavior of pile foundations under combined (vertical and lateral) loads is mentioned in many literatures, describing combined lateral value, slender ratio, and pile materials under static condition [6, 7]. This study aims to investigate the behavior of laterally loaded single pile during earthquake motion under different conditions of loading for lateral and vertical loads and studying the performance of pile–soil interaction using three-dimensional finite element code (ABAQUS CAE) software.

2 Shaking Table Tests

Test Apparatus and Similitude Design Ratio. The dimension of the shaking table is 800 × 800 mm. The maximum acceleration of the shaking table was (3 *g*) and a maximum payload of 1000 kg with a total weight of shaking table equal to 300 kg. The maximum frequency of the input motion ranged from 0 to 100 Hz. A laminar shear box is used to hold the tested soil, and the laminar shear box was constructed from 12 square steel laminates. The shear soil box is shown in Fig. 1. The shaking table [8] was used to simulate different real earthquake shake history. The geometry scale ratio of the model to the prototype was (1:35) according to [9], and the scale factors for different simulated parameters for (1-g) shaking table studies are presented.

Model Geometry and Soil Properties. The soil employed for the model tests was Karbala sand (from south-west of Baghdad). The sand was air-dried and then screened on sieve no. 10 using two relative densities (30 and 70%). Figure 2 illustrates the grain-size distribution of the sand, while the physical properties of the soil are shown in Table 1.

An aluminum pile tube model was embedded into the soil layer to a depth of 400 mm with an outer diameter of 16 mm and an inner diameter of 13 mm with (L/D) equal to 25. Table 2 shows the pile properties. A static combined vertical and horizontal loading was applied under dynamic motion that has different intense, namely Ali Al-gharbi and Kobe. Figure 3 and Table 3 discussed their history.

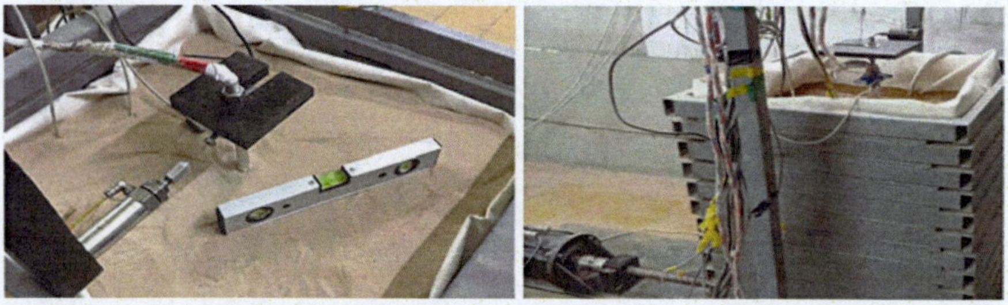

Fig. 1 Shaking table box with complete setup

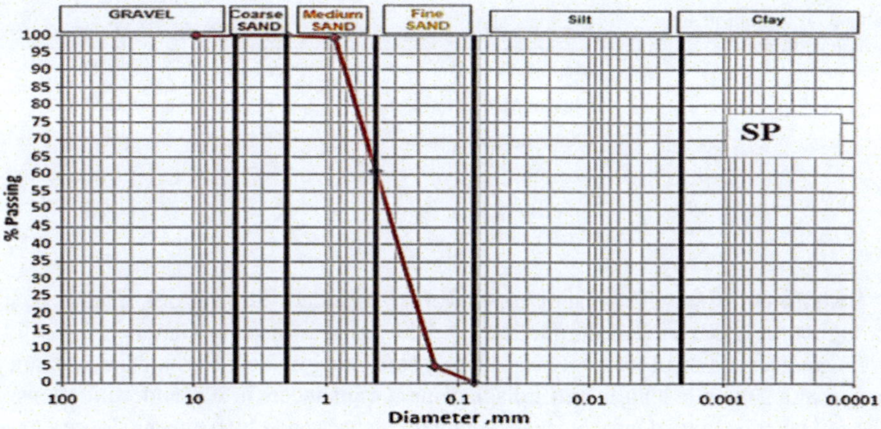

Fig. 2 Grain-size distribution of sand

Table 1 Physical properties of sandy soil used for testing

Property	Value	Standard of the test
Specific gravity, G_s	2.64	ASTM D854-2005
Maximum void ratio, e_{max}	0.86	ASTM D422-2001
Minimum void ratio, e_{min}	0.66	ASTM D4254-2000
Angle of internal friction	32° (loose), 35° (dense)	ASTM D3080-1998

Table 2 Mechanical properties of aluminum pile used

Length, mm	Poisson's ratio	Modulus of elasticity, GPa
500	0.25	67

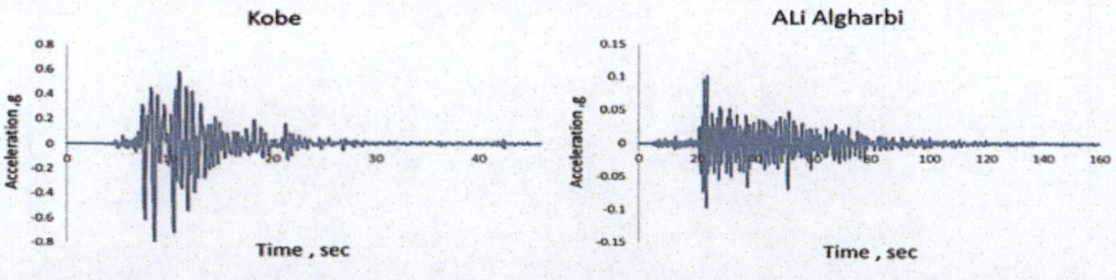

Fig. 3 Acceleration histories data

Table 3 Information about earthquakes data

Earthquake	Ali-Al gharbi	Kobe
Magnitude, (Mw)	4.9 Mw	6.9 Mw
Maximum acceleration, (g)	0.1	0.82
Shake duration, (sec)	160	48
Region	Iraq	Japan
Epicenter depth, (km)	10	17.9

Instrumentation. The laboratory works included all various parameters throughout a series of shaking table tests, with sensors (accelerometers, load cell, strain gauges, and displacement transducers) consisting of 3 accelerometers, 8 strain gauges along the pile length, and 3 displacement transducers in *x*, *y*, and *z* directions. The soil layer is divided into two layers, namely: dense sand (430) mm and loose sand (320) mm from the bottom to the top of the laminar shear box. The accelerometers 2 and 3 were located at depths of 100 and 320 mm in loose sand layer, while accelera-tion 1 was placed at the pile cap. An earthquake acceleration history was selected by using simulator software (Lab view, 2015) to simulate earthquake motion. The total pile loads (vertical and lateral) are equal to 65 and 7 N, respectively. The lateral load system was manufactured to represent the lateral load, while the mass was placed on the pile cap to represent the vertical load, and the mass perfectly constrained with a pile cap. Figure 4 exhibited the arrangement of sensors in the shaking table tests. Table 4 summarized all test codes with different input motion.

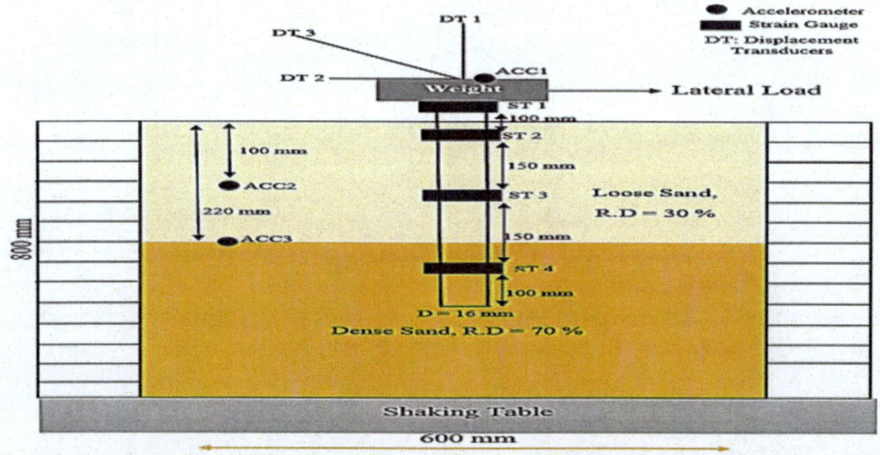

Fig. 4 Arrangement of sensors in the shaking table used in this study

Table 4 Test cases codes

Code	Description	Code	Description
K1	Kobe 50% vertical load, 50% lateral load	K3	Kobe 100% vertical load, 50% lateral load
K2	Kobe 50% vertical load, 100% lateral load	K4	Kobe 100% vertical load, 100% lateral load
A1	Ali Al-gharbi 50% vertical load, 50% lateral load	A3	Ali Al-gharbi 100% vertical load, 50% lateral load
A2	Ali Al-gharbi 50% vertical load, 100% lateral load	A4	Ali Al-gharbi 100% vertical load, 100% lateral load

3 Results and Discussion

Acceleration Response. Two different input motions were investigated (Kobe and Ali Al-gharbi) using a shaking table at three different depths, that is, pile cap, 10,

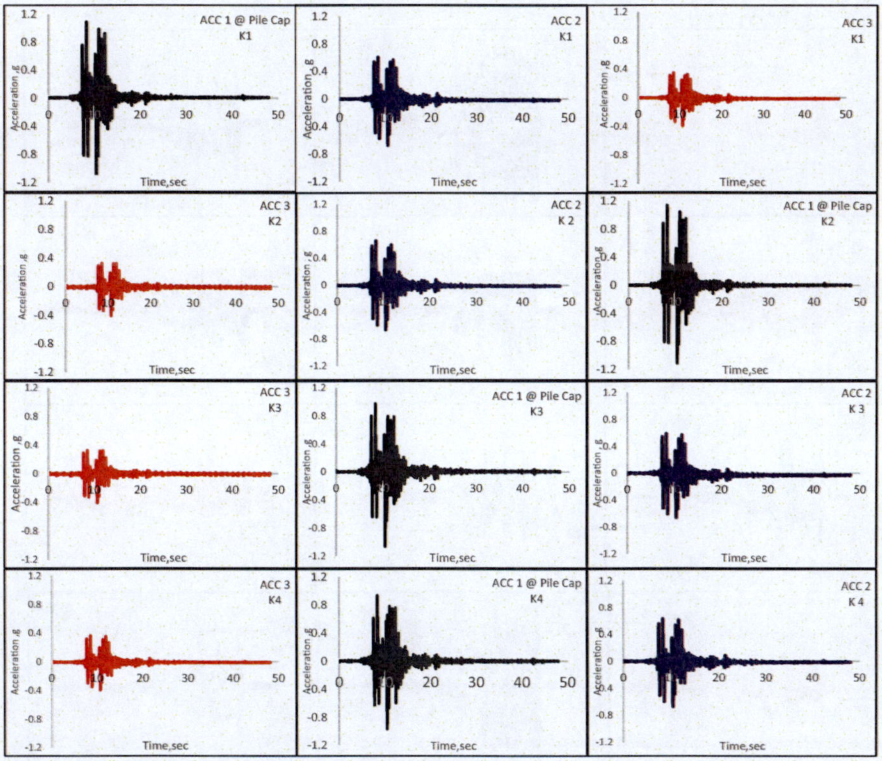

Fig. 5 Acceleration response for Kobe earthquake

and 320 mm from the surface of sandy soil. Figure 5 exhibits the variation of acceleration response of single pile under Kobe earthquake event. From the results, it is indicated that there is no clear effect observed when changing the conditions of loading, while the clear effect in Ali Al-gharbi is observed in 100% vertical load; it gives high acceleration compared with 50% vertical load. Finally, the lateral load did not affect in both earthquakes. Figure 5 and 6 exhibit the acceleration of Kobe and Ali-Al-Gharbi earthquakes, which decrease and then increase progressively when acceleration propagates upward along the shaking table, whereas the acceleration on the pile cap is greater than input acceleration of the table. Also, it should be noted that the peak acceleration tends to decrease with soil strata depth since its behavior is in nonlinear manner, and these results are in well agreement with other researches [8, 14].

Pile Lateral Response. Figures 7 and 8 show that the lateral displacement increases, as well as lateral loading decreases since the vertical load is constant. Based on the variation of lateral displacement on the pile cap in the time domain, the peak lateral displacement of the pile cap was 60and 6 mm for K1 and A1, respectively, at an allowable lateral load of 50%. It should be noted that the peak of

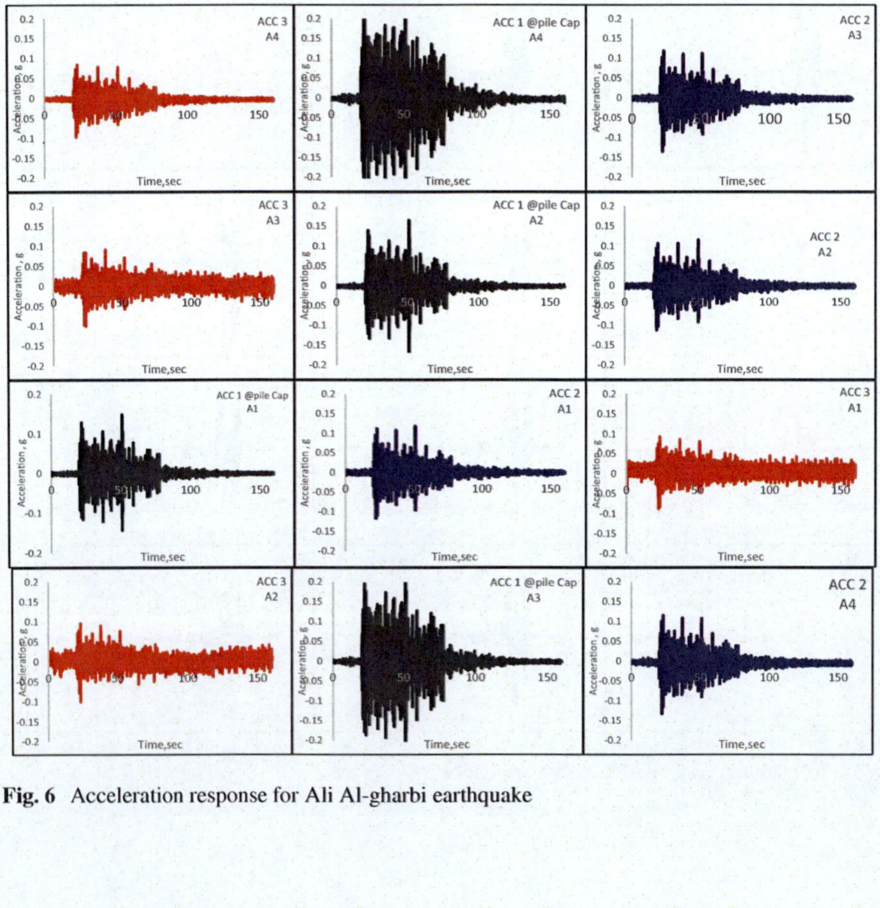

Fig. 6 Acceleration response for Ali Al-gharbi earthquake

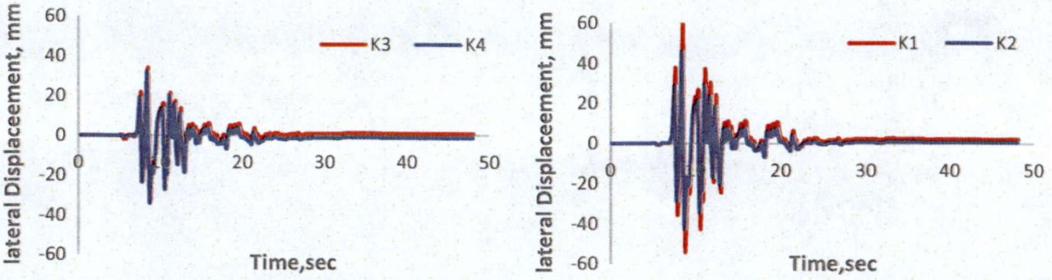

Fig. 7 Lateral displacement during Kobe earthquake

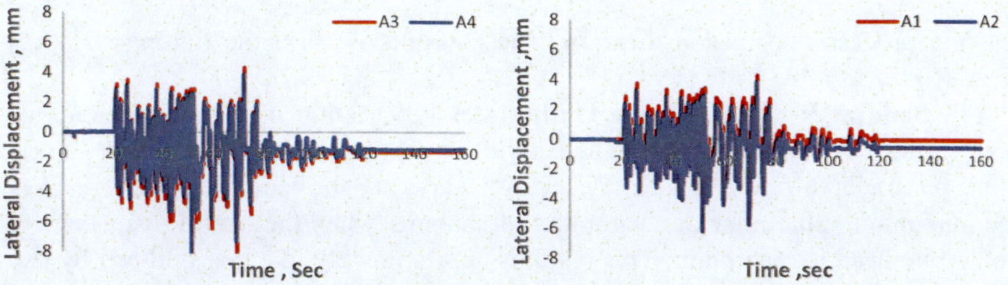

Fig. 8 Lateral displacement during Ali Al-gharbi earthquake

lateral displacement of the pile cap for Kobe earthquake greater than Ali Al-gharbi earthquake could be attributed to the intensity of the acceleration.

Pile Vertical Response. Figure 9 illustrates the maximum values of the vertical displacement on the pile cap during earthquake motion. The maximum values of the vertical displacement range between 5 and 7.5 mm during Kobe earthquake, while their range is between 0.5 and 1.5 mm during Ali Al-gharbi earthquake. From the results, it is observed that the vertical displacement in Kobe earthquake is greater than the vertical displacement during Ali Al-gharbi earthquake. This behavior may be due to an increase in the frequency of the earthquake motion. The settlement of pile is not high values at different conditions of loading compared with lateral displacement under dry soil state in both earthquakes. Soil is gradually strengthening during the

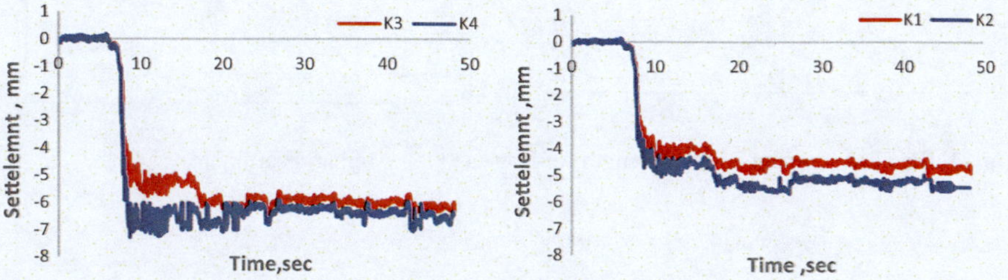

Fig. 9 Vertical displacement during Kobe earthquake

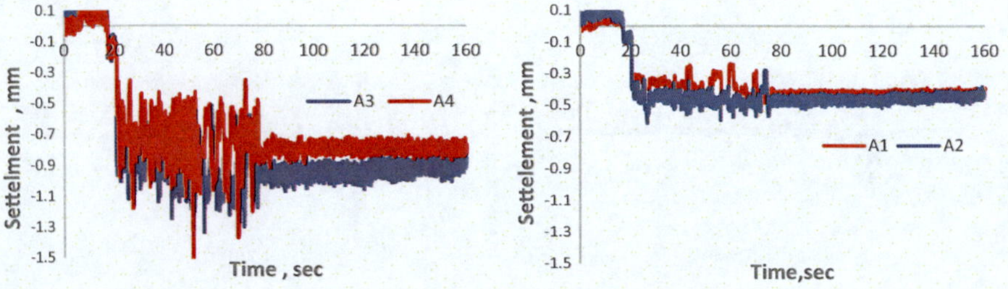

Fig. 10 Vertical displacement during Ali Al-gharbi earthquake

shaking process due to densification. These results confirm the findings by [15] (Fig. 10).

Pile Bending Response. Figure 11 illustrates the variation of bending moments for both earthquake motions through the bending strains measured by the strain gauges along pile under different depths. These results show that the bending moment reaches the maximum value after the beginning of shaking, when the lateral displacement at the pile head is maximum. This result is approximately like that findings by [8]. The bending moment in all cases peaked at 200 mm depth from the soil surface. The maximum bending moment closed in K2 reached to 5.7 Nm, while in A1 reached to 3.8 Nm.

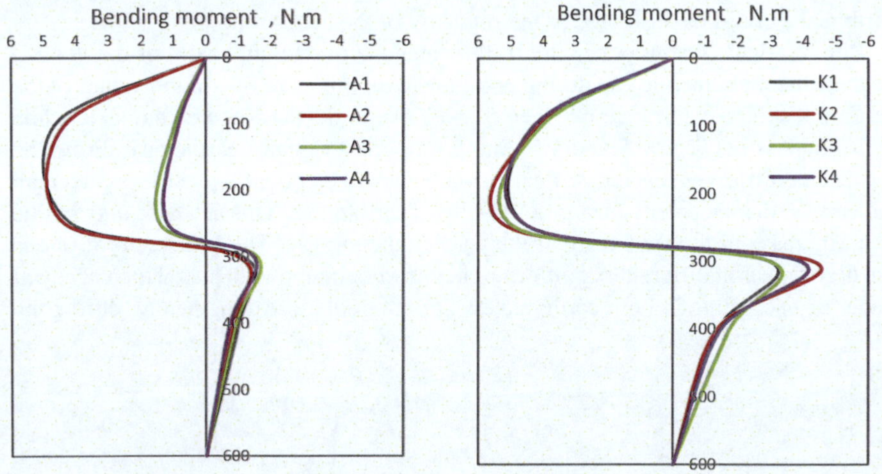

Fig. 11 Bending moment with pile length for Kobe and Ali Al-gharbi earthquakes

4 Finite Element Modeling

The finite element analysis investigated in this study was introduced as a 3D nonlinear dynamic model via ABAQUS. The soil domain was modeled using hexahedral element – C3D8R 8-node linear brick, reduced integration, the hourglass was implemented with (600 × 600 × 800) mm and used. As a reflecting wave, a quiet boundary condition of CIN3D8 infinite elements was used for reflected waves and to provide real analysis. It should be noted that soil domain with these boundaries extend to more than half of real soil domain strata [16]. The behavior of soil was modeled using Mohr–Coulomb for general failure constitutive model. A single circular pile element of 16 mm diameter embedded into the soil to a depth of 400 mm was considered and proposed to behave as linear elastic material with an 8-node brick element. On the other hand, care must be taken for soil and pile interface that laterally follows the Law of Coulomb Friction (normal and tangential behavior with penalty) [17]. Figure 12 and Table 5 summarize the main desired properties.

The analysis was separated into three stages: the first one, the gravity of pile and domain were applied to present the in situ geostatic stress, while the second stage tends to offer a static vertical and lateral loads which then propagated into the third stage by applying an input motion at the base in terms of time–acceleration for two different seismic waves. Finally, soil domain damping was calculated based on Fourier amplitude. For Kobe input motion the maximum predominant frequency is equal to 1.45262 Hz, and for this case: $\beta = 2\xi/W_0$, where W_0 is the predominant frequency of Kobe.and ξ is the material damping ratio assumed to be 5%, and hence

Fig. 12 Finite element model

Table 5 Summary of soil and pile properties used in finite element model

Properties	Soil		Pile	
Young Modulus, E (kPa)	10,000		67,000,000	
Poisson ratio, ν	0.33		0.25	
Friction Angle, φ (°)	32 loose, 35 dense	Dilation Angle, ψ (°)		2
Mohr–Coulomb Hardening				
Cohesion yield Stress, kPa	1	Absolute plastic strain		0

$\beta = 0.06884$. For this study, the numerical analysis is applied for Kobe earthquake motion using ABAQUS output test to investigate the worst effect for acceleration, lateral, and vertical displacement at the pile cap.

5 Results and Analyses

Figures 13 and 14 show the acceleration, lateral displacement, and vertical displacement at the pile cap for Kobe input motion, which are measured and computed. As can be seen in the figures, there is a good agreement between shaking table tests and finite element modeling. Finally, the lateral displacement was measured along pile depth in terms of maximum displacement during peak ground acceleration at a time of 4.23 s for both K1 and K2 scenario at pile cap equal to 60.3 and 45.8 mm, respectively. This behavior could be attributed due to small resistance of loose sand at a higher elevation of the soil domain. Figure 15 demonstrates the variation of lateral displacement along the depth of the pile under K1 and K2 earthquake motion.

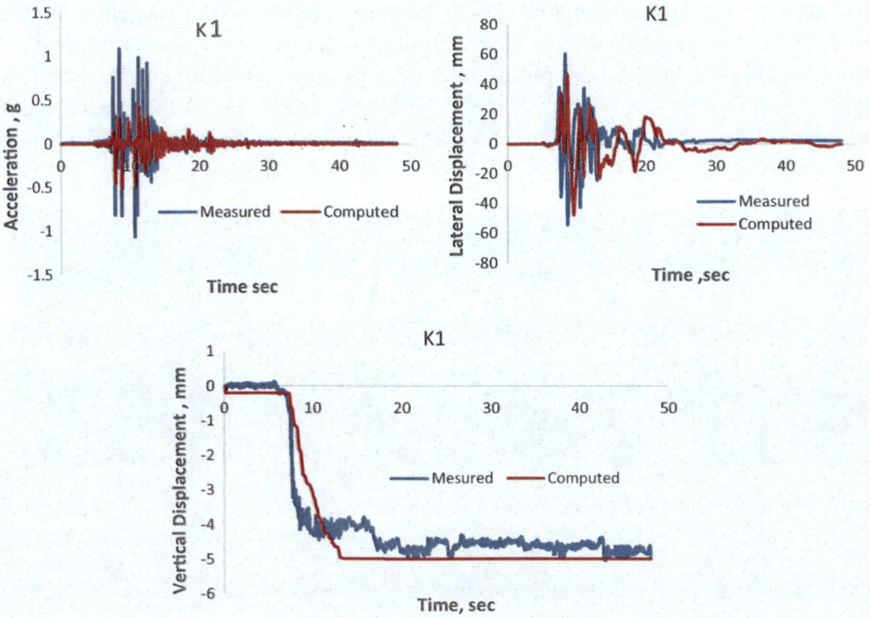

Fig. 13 Acceleration, lateral displacement, and vertical displacement at pile cap for K1

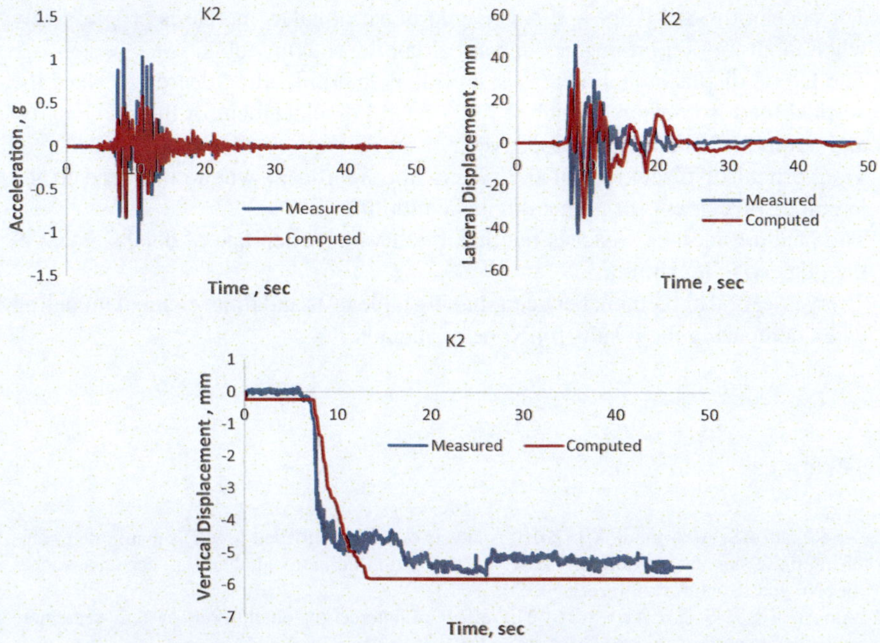

Fig. 14 Acceleration, lateral displacement, and vertical displacement at pile cap for K2

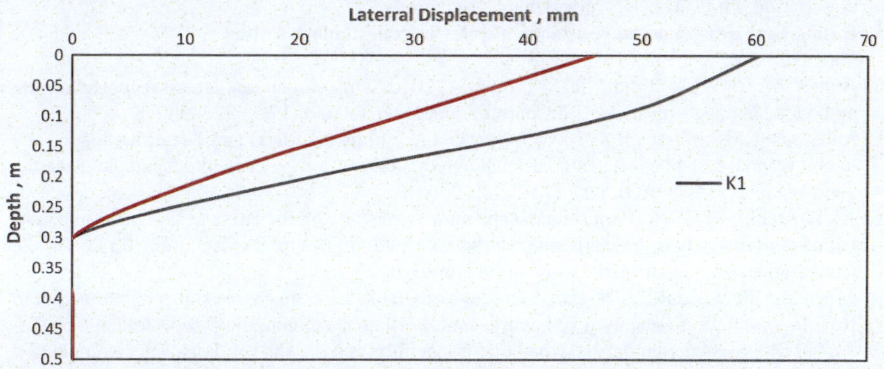

Fig. 15 Variation of lateral displacement along the depth of pile for K1 and K2

6 Conclusions

In this study, the response of laterally loaded pile subjected to different earthquake motions were investigated using shaking table tests, and simulation is done with ABAQUS CAE software. Based on the results of experimental and numerical analyses, the following findings and observations can be summarized:

- The acceleration of Kobe and Ali Al-gharbi earthquakes increases progressively when input motion propagates upward along the shaking table.
- The lateral displacement increases as well as lateral loading decreases since the vertical load is constant, which led to a lateral displacement of the pile cap: 60 and 6 mm for K1 and A1, respectively.
- The vertical displacement of the pile is not significant when compared to the lateral displacement for both earthquake motions.
- The bending moment reaches the maximum value when lateral displacement at the pile cap is maximum.
- There is a good agreement between shaking table tests and finite element modeling when comparing the results for Kobe earthquake.

References

1. Al-Rekabi AS, Al-Aboodi AH (2016) Analysis of laterally loaded (2× 2) square pile groups using finite element method. In: Proceedings of sixth international of conference on advances in civil and structural engineering-CSES
2. Albusoda BS, Salem AK (2016) The effect of interaction on pile-raft system settlement subjected to earthquake excitation. Appl Res J 2(4):205–214
3. AL-Recaby MK (2016) Dynamic response to lateral excitation of pile group model in sandy soil. Doctoral dissertation, Ph.D. Thesis, University of Technology
4. Elewi AS (2017) Response of single pile and pile groups to lateral soil movement. Doctoral dissertation, Ph.D. Thesis, University of Technology
5. Broms BB (1964) Lateral resistance of piles in cohesionless soils. J Soil Mech Found Div 90(3):123–158
6. Abbas JM, Chik ZH, Taha MR, Shafiqu QM (2010) Time-dependent lateral response of pile embedded in elasto-plastic soil. J Central South Univ Technol 17(2):372–380
7. Rajagopal K, Karthigeyan S (2008) Influence of combined vertical and lateral loading on the lateral response of piles. In: International association for computer methods and advances in geomechanics (IACMAG), pp 3272–3282
8. Al-Tameemi SM (2018) Experimental and numerical study on the effect of liquefaction potential of piles in sandy layers soil under earthquake loading. Doctoral dissertation, Ph.D. Thesis, Civil engineering department, Al-Nahrain University
9. Maymand PR, Seed R (2000) Large scale shaking table tests of seismic soil-pile interaction in soft clay, vol 5. In: Proceeding 12th world conf, Earthquake engineering. New Zealand, pp 915
10. ASTM D854–2005, Standard test method for specific gravity of soil solids by water pycnometer. American society for testing and materials
11. ASTM D 422–2001, Standard test method for particle size-analysis of soils. American society for testing and materials
12. ASTM D 4254–2000, Standard test method for minimum index density and unit weight of soils and calculation of relative density. American society for testing and materials
13. ASTM D 3080–1998, Standard test method for direct shear test of soils under consolidated drained conditions. American society for testing and materials
14. Chen G, Chen S, Chen W (2015) Shaking table tests on a three-arch type subway station structure in a liquefiable soil. Bull Earthq Eng 13(6):1675–1701
15. Almashhadany OY, Albusoda BS (2019) Assessment of soil tunnel interaction in sand soil. In: IOP conference series, Materials science and engineering 584(1):012057

16. Asheghababdi MS, Rahgozar MA (2018) Finite element seismic analysis of soil tunnel interactions in clay soils. J Sci Technol, Trans Civ Eng, Iranian 43(4):835–849
17. ABAQUS/CAE user manual (2012) United State of American, Version 6.14

Critical Tensile Strength for Reinforced Slope Under Seismic Loading

Mohamed H. Hammed, Suhail I. Khattab, and Qutayba N. Al-Saffar

Abstract Earthquakes expose slopes to dynamic load which increases shear stress and leads to a decrease in the factor of safety, when the factor of safety is less than 1.0; the slope becomes unstable and fails, which results in a loss in life and economic damage. The main purpose of this study is to illustrate the effect of seismic load on a reinforced cohesive slope. The first part of this study is to analyze the relationship between cohesion of soil and the factor of safety of slope under different combinations of horizontal and vertical seismic coefficients for a homogeneous cohesive slope model. The limit equilibrium method module of Geoslope software has been used in the analysis and the results are obtained from the previous study. The cohesion of the slope was taken from 20 to 40 kPa with an increment of 5 kPa, where the ratio of (k_v/k_h) is taken as (0.25, 0.5, 0.75 and 1.0) and k_h was fixed at (0.1, 0.2, 0.3 and 0.40). The second part of this study is to illustrate the effect of the horizontal coefficient on a reinforced cohesive slope and make a parametric study considering the length of reinforcement, several layers and tensile strength. Finally, the critical tensile strength (CTS) value for each length and number of reinforcement layers were computed. It was observed that the factor of safety and yield acceleration increase with the reinforcement length and number of reinforcement layers. Also, as the yield acceleration increased, the tensile strength of geogrid reinforcement increased until it reached critical values called critical tensile strength (CTS).

Keywords Seismic loading · Reinforced slope · Critical tensile strength · Geogrid

M. H. Hammed (✉) · S. I. Khattab · Q. N. Al-Saffar
Civil Engineering Department, University of Mosul, Mosul, Iraq
e-mail: eng.mohamedaljbure@yahoo.com

S. I. Khattab
e-mail: suhailkhattab@yahoo.com

Q. N. Al-Saffar
e-mail: dr.qutayba@gmail.com

M. O. Karkush and D. Choudhury (eds), *Modern Applications of Geotechnical Engineering and Construction*, Lecture Notes in Civil Engineering 112,
https://doi.org/10.1007/978-981-15-9399-4_26

1 Introduction

Slopes stability is one of the most important problems that civil engineers and geologists face in many projects, such as cutting, embankments, and retaining walls. Studying the stability of slope is required for the geometry of slope, soil properties, and the effecting forces to the slope. One of these forces is an earthquake which reduced the factor of safety and cause failure. In recent years north of Iraq was exposed to slight earthquakes that caused simple damage to some slopes. There are different methods used to protect slopes and one of them is reinforcing it. There are lots of researches done on reinforced slope which deals with different parameters. Mandal and Labhane [1] presented a procedure for the stability analysis and design of geosynthetic-reinforced soil slopes. Varadarajan et al. [2] studied the effects of the type of reinforcement, the type of clay, the depth of foundation, and the drainage condition on the behavior of a reinforced embankment-foundation system by using finite element analysis.

The effect of the distribution of the reinforcement forces on the stability of reinforced slopes (embankments) is studied by Tandjiria et al. [3] using the extended generalized slices method and the results are compared with the finite element method results. Madhavi and Varman [4] described shaking table model studies conducted on unreinforced and reinforced soil slopes and compared the results with the numerical model conducted by using Geoslope software. Song et al. [5] investigated the effects of the length, location, and the tensile strength of the reinforcement layers on the safety factor. In this research, the effect of seismic load on the reinforced cohesive slope has been studied considering the length of reinforcement, several layers, and tensile strength.

2 Methodology and Numerical Model

The geometric model used in the present study is described by Hossain et al. [6] and shown in Fig. 1 with a slope angle of 45°. Limiting equilibrium methods are widely used in calculating the factor of safety for the stability of slopes because of their simplicity and accuracy. These methods consist of cutting the mass of soil above the slip surface into fine slices and applying appropriate equilibrium equations (equilibrium of the forces and/or moments) [7]. In this paper, Morgenstern–Price [8] method was used to calculate the factor of safety. One of the earliest procedures to study seismic stability is the pseudo-static procedure, in which the earthquake loading is simulated by static forces F_h and F_v, which are equal to the soil weight multiplied by a seismic coefficient, k_h and k_v [9].

$$F_h = k_h \times W \tag{1}$$

$$F_v = k_v \times W \tag{2}$$

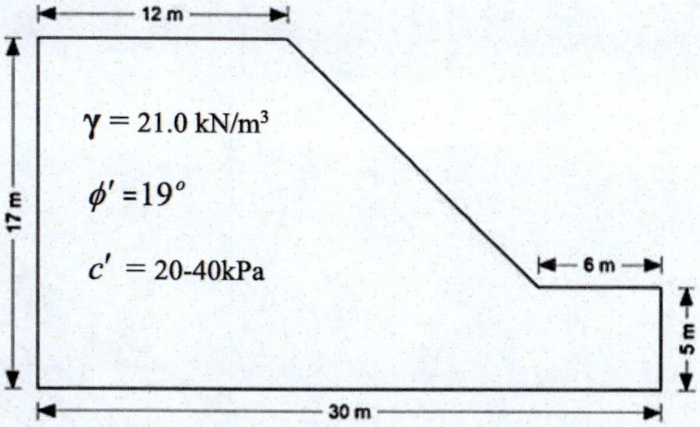

Fig. 1 Geometry of numerical model

where k_h and k_v are the coefficients of acceleration in horizontal and vertical directions, respectively.

3 Results and Discussion

The stability analysis in this study is conducted using Geoslope software, as shown in Fig. 2. The verification of the factor of safety with cohesion under different values of k_h and (k_v/k_h ratio) was verified, as shown in Figs. 3 and 4. The acceptable results obtained matched with the results of Hossain et al. [6]. The small differences in results may be contributed to the different programs used, a different number of slices used, and the location of the center of the slip surface.

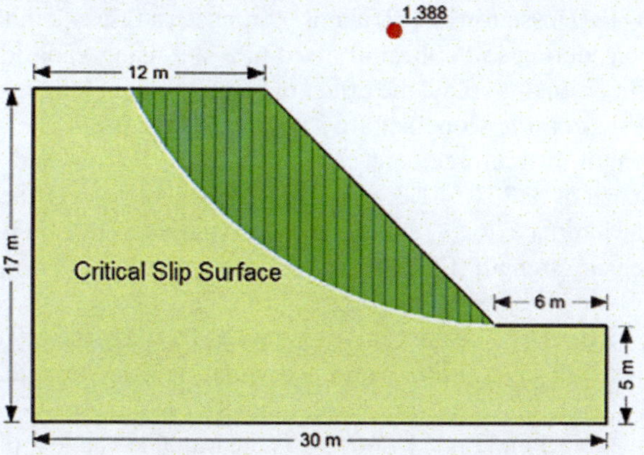

Fig. 2 Geoslope software simulation

Fig. 3 Variation of FOS with cohesion [6]

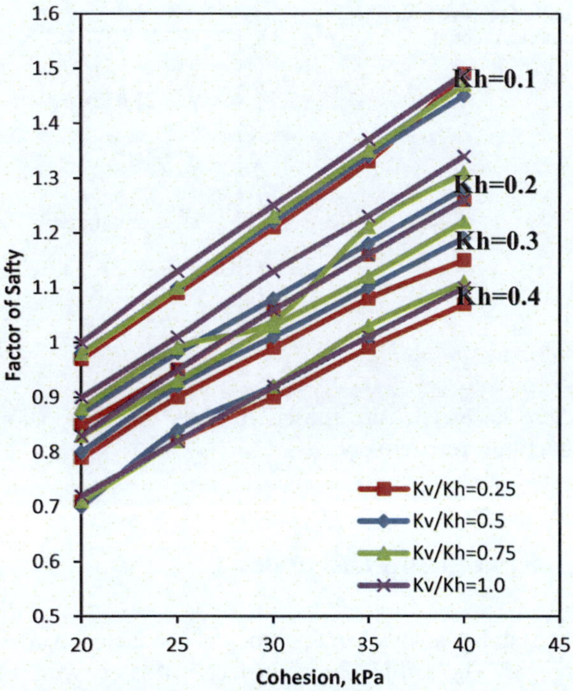

After the upper verification, the variation of yield acceleration[1] with cohesion is obtained, as shown in Figs. 5 and 6, which indicate increasing yield acceleration as cohesion increased.

Reinforced cohesive slope. A parametric study was conducted on the previous work (Hossain et al. [6]) using Morgenstern Price method for safety factor calculation considering soil cohesion of 20 kPa and using geogrid reinforcement, as shown in Fig. 7. In this study, the effect of reinforcement length and number of reinforcement layers on the slope stability were computed. For the study, 0, 6, 8, 10, and 12 m length of reinforcement is used. Also, 3, 5, 7, 9, and 11 layers were taken with a space of 3, 2, 1.5, 1.2, and 1 m, respectively, using the strength reduction factor in the interface (R_{inter}) equal to 0.67, as suggested by Song et al. [10]. Finally, the critical tensile strength (CTS) values for each length were computed and the critical length of reinforcement corresponding to CTS value for this case study was computed.

Effect of Reinforcement Length. To study the effect of the length of reinforcement, 11 layers were examined with a tensile strength of 80 kN. The safety factor and yield acceleration were computed for a length of 0, 6, 8, 10, and 12 m, as shown in Table 1 and Figs. 8 and 9. It is clear that as the length of reinforcement increased, both safety factors and corresponding yield acceleration also increased.

[1]The yield acceleration is the seismic coefficient that produces a factor of safety of unity in a pseudo-static slope stability analysis [9].

Fig. 4 Variation of FOS with cohesion (present study)

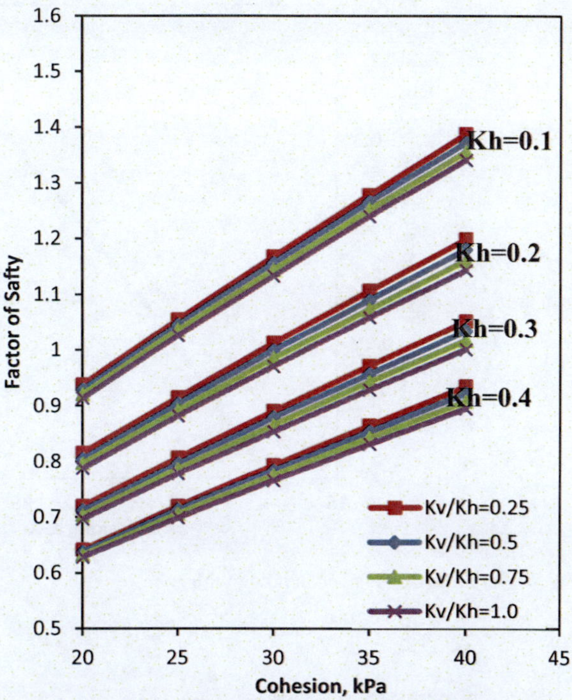

Fig. 5 Variation of FOS with k_h

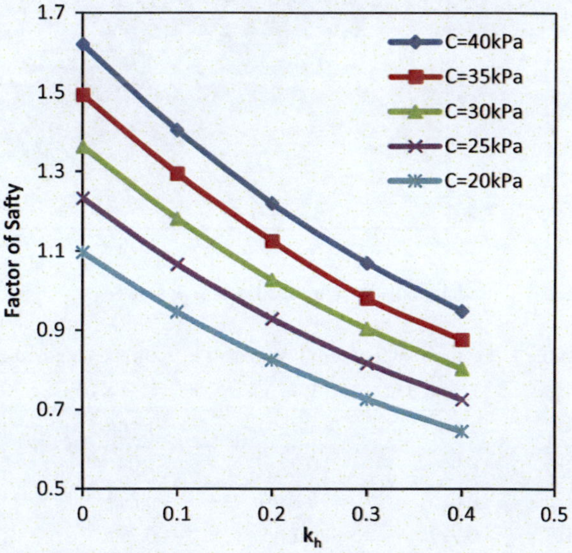

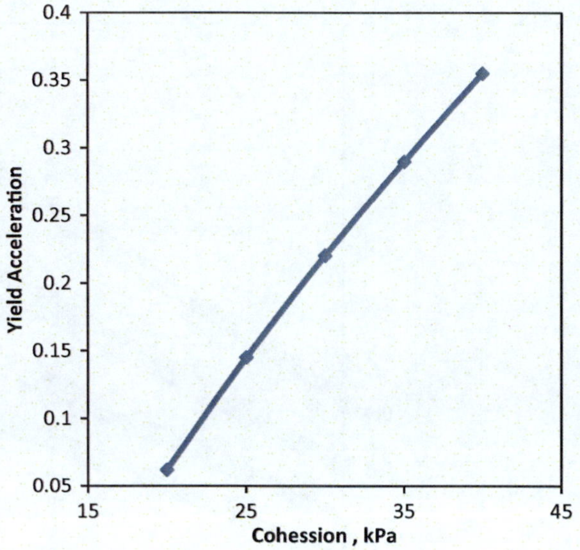

Fig. 6 Variation of yield acceleration with cohesion

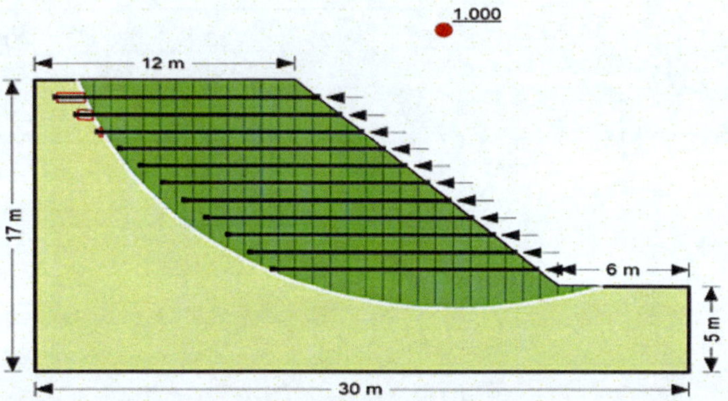

Fig. 7 Simulation of the reinforced cohesive slope

Table 1 Factor of safety for all values of k_h with the length of reinforcement

k_h	L = 0.0 m	L = 6.0 m	L = 8.0 m	L = 10 m	L = 12 m
0	1.097	1.129	1.203	1.289	1.377
0.1	0.946	0.977	1.03	1.095	1.159
0.2	0.825	0.856	0.897	0.939	1
0.3	0.727	0.749	0.777	0.812	0.877
0.4	0.645	0.662	0.684	0.713	0.777

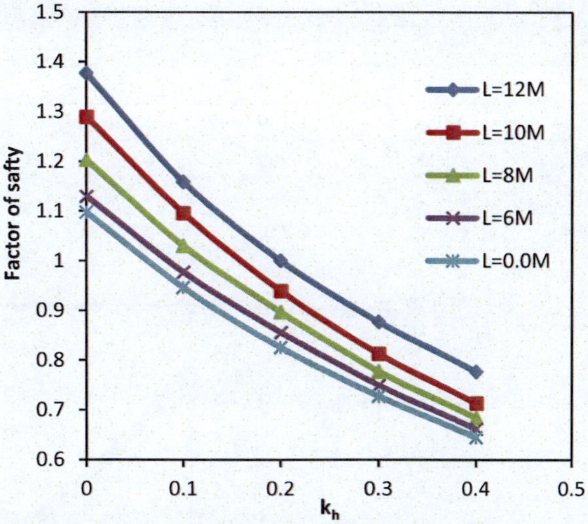

Fig. 8 Variation factor of safety with k_h

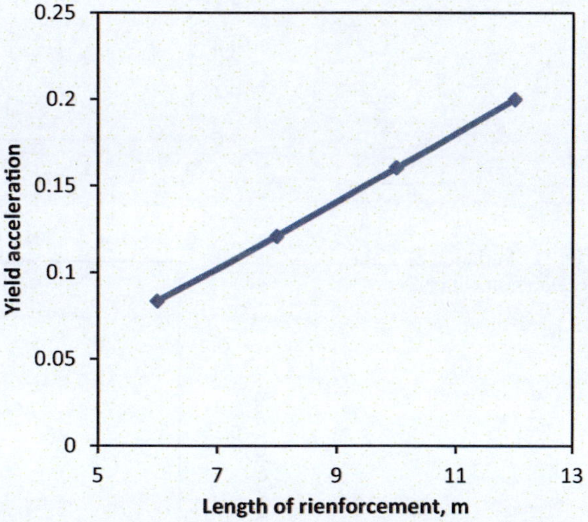

Fig. 9 Variation of yield acceleration with L

Effect of Number of Reinforcement Layers. To study the effect of the number of layers, the length of reinforcement for all layers was fixed at 12 m and tensile strength was fixed at 80 kN. The safety factor and yield acceleration for various number of layers (3, 5, 7, 9, and 11) were computed, as shown in Table 2 and Figs. 10 and 11. It has been shown that as the number of layers increased, both safety factors and corresponding yield acceleration increased until the number of layers reached nine and then remain constant.

Effect of the Tensile Strength. In this study, the tensile strength ranged from 8.0 to 80 kN for all length of reinforcement and different number of layers. Figure 12, as

Table 2 Factor of safety for all values of k_h with the number of layers

k_h	N = 3	N = 5	N = 7	N = 9	N = 11
0	1.321	1.344	1.365	1.373	1.377
0.1	1.088	1.139	1.146	1.156	1.159
0.2	0.918	0.969	0.97	0.995	1
0.3	0.793	0.837	0.838	0.86	0.877
0.4	0.696	0.736	0.737	0.757	0.777

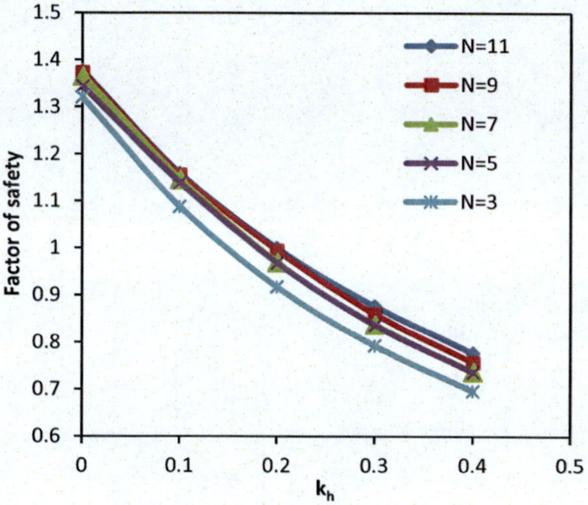

Fig. 10 Variation factor of safety with k_h

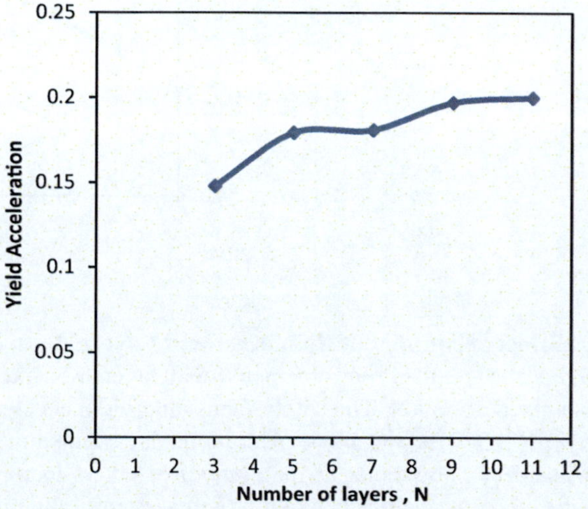

Fig. 11 Variation of yield acceleration with N

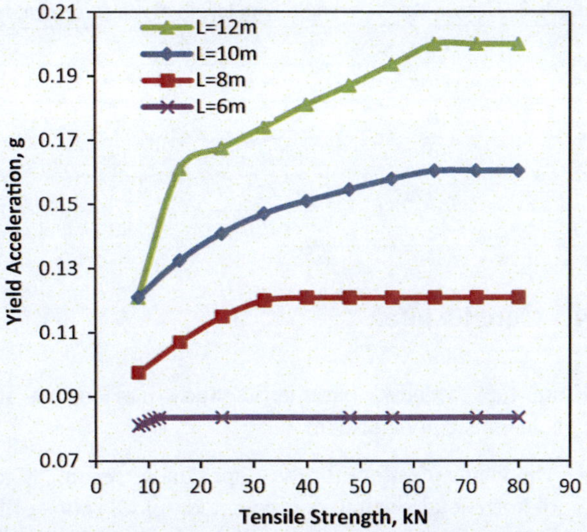

Fig. 12 Variation of yield acceleration with tensile strength

a sample of calculation, shows the variation of yield acceleration with tensile strength for several layers equal to 11 layers of reinforcement, which indicates that as the yield acceleration increased the tensile strength of reinforcement geogrid increased until critical value called critical tensile strength (CTS) is reached. The value of critical tensile strength increased with length of reinforcement until peak value is reached, as shown in Fig. 13. Table 3 summarizes the effect of the number of reinforcement layers and length of reinforcement on the critical tensile strength.

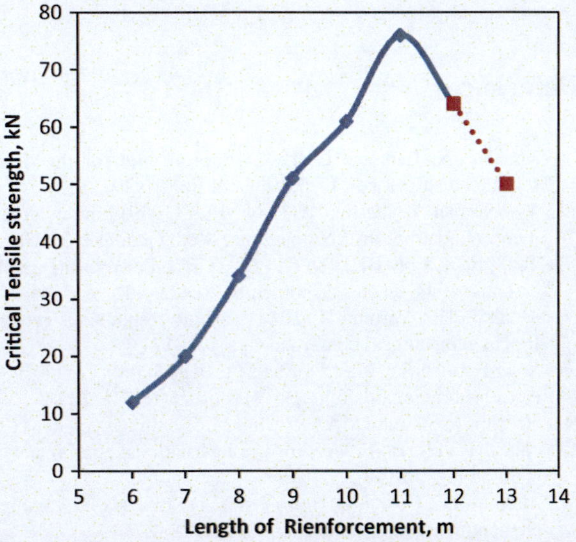

Fig. 13 Variation of critical tensile strength with length of reinforcement

Table 3 Critical tensile strength with length of reinforcement for all number of layers

L, m	CTS, kN				
	N = 3	N = 5	N = 7	N = 9	N = 11
6	10	16	26	17	12
8	19	35	35	38	34
10	30	42	63	34	61
12	45	79	80	80	64

4 Conclusions

From the extended parametric study carried out in this paper, the following conclusions can be drawn:

- The factor of safety for the slopes under seismic loads decreases with an increase of horizontal seismic coefficient k_h for all ratios of k_v/k_h.
- Factor of safety increases with increasing of soil cohesion for all ratios of k_v/k_h.
- The value yield acceleration increases with an increase of soil cohesion.
- Vertical seismic coefficient k_v is slightly affected by the factor of safety.
- The factor of safety and yield acceleration increase with increases in the reinforcement length.
- As the number of layers increased, both safety factors and corresponding yield acceleration increased until the number of layers reached critical values and then remain constant.
- As the yield acceleration increased the tensile strength of geogrid reinforcement increased until critical value called critical tensile strength (CTS) is reached.

References

1. Mandal JN, Labhane L (1992) A procedure for the design and analysis of geosynthetic reinforced soil slopes. Geotech Geol Eng 10:291–319
2. Varadarajan A, Sharma KG, Aly MAA (1999) Finite element analysis of reinforced embankment foundation. Int J Numer Anal Meth Geomech 23:103–114
3. Tandjiria V, Low BK, Teh CI (2002) Effect of reinforcement force distribution on stability of embankments. Geotext Geomembr 20:423–443
4. Madhavi GL, Varman N (2010) Seismic response of geosynthetic reinforced soil slopes. In: Indian geotechnical conference, pp 171–174
5. Song F, Chen RY, Ma LQ, Zhao J (2018) Stability analysis of reinforced slope based on limit equilibrium method. Tehnički Vjesnik 25(1):224–229
6. Hossain A, Sadman MAA, Rashid MM, Ashikuzzaman M (2019) Seismic stability of slopes in cohesive soils. In: Proceedings of international conference on planning, architecture and civil engineering, pp 1–7
7. Matthews C, Farook Z, Helm P (2014) Slope stability analysis–limit equilibrium or the finite element method. Ground Eng 48(5):22–28

8. Morgenstern NR, Price VE (1965) The analysis of the stability of general slip surfaces. Géotechnique 15:79–93
9. Duncan JM, Wright SG, Brandon TL (2014) Soil strength and slope stability. Published by John Wiley & Sons Inc., Hoboken, New Jersey, USA, pp 179–181
10. Song F, Cao GR, Zhang LY, Tan XM (2013) Numerical analysis of failure mode of geocell flexible retaining wall. In: IACGE 2013: challenges and recent advances in geotechnical and seismic research and practices, pp 136–145

Mathematical Analysis of Vertical Vibration for Circular Rigid Foundation Resting on Soil Surface

Ahmed I. Mohammed, Bayar J. Al-Sulayvani, and Mohammed N. Jaro

Abstract In this study, a dynamic analysis of a circular rigid foundation resting on the soil surface is carried out. An elastic half-space theory is used to compute the response of a circular foundation subjected to a vertical oscillation. The work deals essentially with the effect of Poisson's ratio μ. Besides, the work aims to study the effect of the dimension of rigid circular foundation (foundation radius r_o) and dimensionless frequency a_0 on the maximum vertical displacement of the foundation under vertical dynamic load. Two relationships have been conducted, one for dynamic stiffness versus dimensionless frequency a_0 and the other between damping coefficients versus dimensionless frequency a_0 for various new values of Poisson's ratio which have not been studied earlier. The dynamic stiffness (K_{dyn}) and dynamic damping (C_{dyn}) were used to estimate the maximum vertical displacement of the foundation. Results showed that the vertical dynamic stiffness coefficient is more affected by a variation of Poisson's ratio values than the vertical damping coefficient. Consequently, the dynamic stiffness plays the main role of estimating the maximum displacement of the circular foundation for low values of dimensionless frequency a_0. In conclusion, the effect of Poisson's ratio μ appears and increases gradually with increasing of dimensionless frequency a_0.

Keywords Dynamic analysis · Circular foundation · Dynamic stiffness · Dynamic damping · Maximum vertical displacement

---

A. I. Mohammed (✉) · B. J. Al-Sulayvani · M. N. Jaro
Civil Engineering Department, University of Mosul, Mosul, Iraq
e-mail: ahmedobadee1975@gmail.com

B. J. Al-Sulayvani
e-mail: dr.bayar.alsulayvani@uomosul.edu.iq

M. N. Jaro
e-mail: m.jaro@uomosul.edu.iq

M. O. Karkush and D. Choudhury (eds), *Modern Applications of Geotechnical Engineering and Construction*, Lecture Notes in Civil Engineering 112,
https://doi.org/10.1007/978-981-15-9399-4_27

1 Introduction

The study of vibration of foundation due to machine oscillation is important from a practical point of view. When dynamic load applied, foundation vibrates in a manner that depends on many parameters, and one of these important parameters which has a little attention from researchers is Poisson's ratio μ of the soil under the foundation. The variation of Poisson's ratio μ has a direct effect on the dynamic stiffness and damping coefficients and consequently on the displacement of foundation. Over the years, several methods have been conducted and developed to estimate each of the dynamic stiffness and damping coefficients of the soil under the foundation subjected to vertical oscillation. Gazetas [1] suggested a set of formulas and dimensionless charts to compute the dynamic stiffness and damping coefficients. All possible degrees of vibration, Poisson's ratio μ, and a practical range of oscillation frequencies are considered. Also, the effect of each of the shapes of the foundation for realistic geometric shape and embedded of the foundation under the ground surface on the dynamic stiffness and damping coefficients were analyzed.

Wolf and Deeks [2] presented the concept of the cone model to calculate the dynamic stiffness and damping coefficients of the surface circular rigid foundation of radius (r_o) resting on a homogenous half-space media with shear modulus G, Poisson's ratio μ and mass density ρ. The dynamic stiffness and damping coefficients for the vertical degree of freedom were processed with $\mu = 0.33$, 0.45 and 0.5. The results were presented as a function of the dimensionless frequency a_0. In another study, Bertha et al. [3] concluded a series of expressions for each of the effective damping and natural frequencies of rigid surface foundation subjected to vertical and coupled horizontal-rocking harmonic loading. They concluded that for vertical vibration the expression is only a function of the mass ratio (ratio of the mass of the foundation to an effective mass of soil) and Poisson's ratio μ. AL-Azawi et al. [4] studied the effect of embedment and geometry of the foundation on the stiffness and damping of soil. They found that increasing the depth of embedment leads to an increase in the resonant frequency and decreases the amplitude of vibration. Each of dynamic stiffness K_{dyn} and damping C_{dyn} can be estimated using the following expression [5]:

$$K_{dyn} = \eta \cdot k, \quad C_{dyn} = \lambda \cdot C \tag{1}$$

where:

η: vertical stiffness coefficient.
λ: vertical damping coefficient.

$$\text{k: static stiffness} = \frac{4Gr_o}{1 - \mu} \tag{2}$$

$$\text{C: static damping} = \frac{3.4r_o^2}{1 - \mu}\sqrt{\rho \cdot G} \tag{3}$$

G: shear modulus.

2 Vertical Vibration of a Rigid Circular Foundations Resting on Soil Surface

The contact pressure distribution q for rigid foundation resting on elastic half-space can be expressed as follows [6]:

$$q = \frac{Q_o e^{i(\&t+\alpha)}}{\pi r_o^2} \tag{4}$$

Figure 1 shows the pressure distribution under the rigid foundation resting on the soil surface.

The motion amplitude can be expressed as follows:

$$A_z = \frac{Q_o}{G r_o} Z \tag{5}$$

where

Az: the amplitude of the vibration.

Q_o: amplitude of the exciting force.

Z: dimensionless amplitude,

$$Z = \sqrt{\frac{f_1^2 + f_2^2}{\left(1 - b\, a_o^2\, f_1\right)^2 + \left(b\, a_o^2\, f_2\right)^2}} \tag{6}$$

$$\text{b: dimensionless mass ratio} = m/\rho r_o^3 = W/\gamma r_o^3 \tag{7}$$

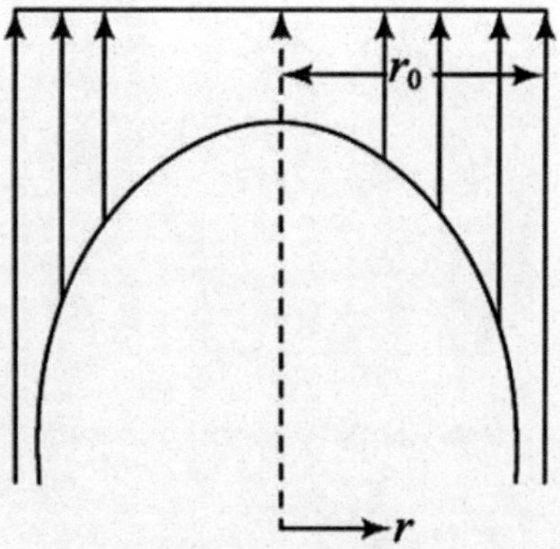

Fig. 1 Pressure distribution under rigid foundation resting on elastic half-space [6]

ρ: density of the material (soil)
γ: the unit weight of the soil
r_o: radius of the foundation

$$a_0\text{: dimensionless frequency} = \omega r_o \sqrt{\frac{\rho}{G}} = \frac{\omega \cdot r_o}{v_s} \tag{8}$$

G: shear modulus of the soil
v_s: shear wave velocity
f_1, f_2: displacement functions.

The displacement function (f_1, f_2) is related to the dimensionless frequency a_0 and Poisson's ratio μ and can be obtained from Fig. 2.

The expression of mass ratio was modified by Lysmer and Richart [7] as:

$$B_Z = \left(\frac{1-\mu}{4}\right) b = \left(\frac{1-\mu}{4}\right)\left(\frac{m}{\rho r_o^3}\right) \tag{9}$$

B_z: modified mass ratio.

Now, it is easy to obtain the resonant frequency and amplitude of vibration using the mathematical expression as:

A. **Resonant Frequency**:

1. Calculation of Natural Frequency.

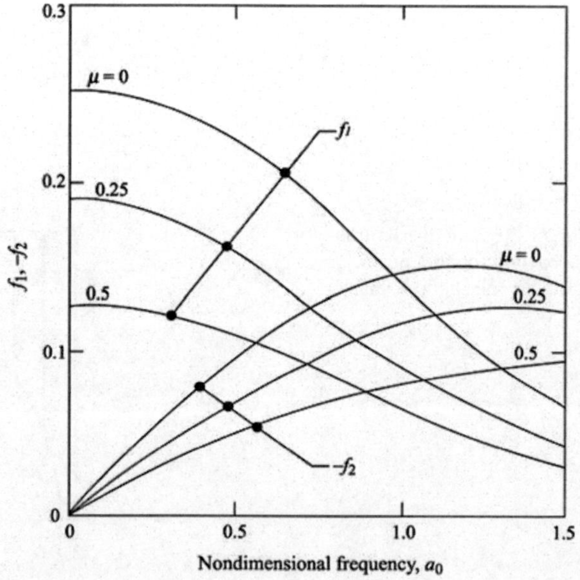

Fig. 2 Relationship between (f_1, f_2) and a_0 and for various μ [6]

$$f_n = \frac{1}{2\pi}\sqrt{k_z/m} \tag{10}$$

k_z: static spring constant (stiffness) for a rigid circular foundation.

$$k_z = \frac{4Gr_o}{1-\mu}$$

Then:

$$f_n = \frac{1}{2\pi}\sqrt{k_z/m} = \frac{1}{2\pi}\sqrt{\left(\frac{4Gr_o}{1-\mu}\right)\frac{1}{m}} \tag{11}$$

2. Calculation of Damping Ratio (Dz).

$$D_z = \frac{C}{C_{cz}} \tag{12}$$

where:
Cz: Damping constant $= \frac{3.4r_o^2}{1-\mu}\sqrt{\rho.G}$
C_{cz}: Critical damping

$$C_{cz} = 2\sqrt{k_z m} = 2\sqrt{\left(\frac{4Gr_o}{1-\mu}\right)(m)}$$

$$C_{cz} = 8\sqrt{\left(\frac{4Gr_o}{1-\mu}\right)\left(\frac{B_z\rho r_o^3}{1-\mu}\right)} = \frac{8r_o^2}{1-\mu}\sqrt{GB_z\rho} \tag{13}$$

Then:

$$D_z = \frac{C}{C_{c.z}} = \frac{\frac{3.4r_o^2}{1-\mu}\sqrt{\rho.G}}{\frac{8r_o^2}{1-\mu}\sqrt{GB_z\rho}} = \frac{0.425}{\sqrt{B_z}} \tag{14}$$

3. Calculation of the Frequency at Maximum Displacement (Resonance Frequency) is as Follows:

$$f_m = f_n\sqrt{1-2D_z^2} = \left[\frac{1}{2\pi}\sqrt{\left(\frac{4Gr_o}{1-\mu}\right)\frac{1}{m}}\right]\sqrt{1-2\left(\frac{0.425}{\sqrt{B_z}}\right)^2} \tag{15}$$

B. **The amplitude of Vibration**: The vibration amplitude can be estimated at any time (amplitude–time relationship) using the following formula:

$$A_z = \frac{Q_o/K_z}{\sqrt{[1-(\frac{\omega^2}{\omega_n^2})]^2 + 4D_z^2(\frac{\omega^2}{\omega_n^2})}} \tag{16}$$

and amplitude at resonance as:

$$A_{z(resonance)} = \left(\frac{Q_o}{K_z}\right)\left(1/2D_z\sqrt{1-D_z^2}\right) \tag{17}$$

where ω is the operation frequency and ω_n is the natural frequency. Equation (18) is commonly known as the lumped mass formula for estimating maximum displacement for a rigid circular foundation which is related to the elastic half-space theory proposed by Lysmer and Richart [7]. It can be used as an alternative to the previous equation and it is expressed as follows:

$$A_z = \left[\left(\frac{Q_o}{\omega^2}\right)/(\rho r_o^3)\right] \tag{18}$$

Jasmani and Kamalzar [8] compared lumped mass theory and elastic half-space theory methods for estimating the amount of displacement in a circular foundation. They illustrated that the difference in displacement between the two methods decreased by increasing the diameter of the foundation. The current research is conducted mainly to study the effect of Poisson's ratio μ on the dynamic stiffness and damping coefficients of soil, and consequently on the maximum displacement occurred in the circular rigid foundation resting on the surface of the soil and subjected to vertical dynamic oscillation.

The effect of foundation radius r_o and dimensionless frequency a_0 on the maximum displacement was also analyzed. From the above, it can be concluded that there are two stages in this study: The first stage includes creating relationships between each of dynamic stiffness and damping coefficients versus dimensionless frequency a_0 for various Poisson's ratio μ, including new values of Poisson's ratio which have not been studied before. After that, both dynamic stiffness K_{dyn} and dynamic damping C_{dyn} can be estimated for certain Poisson's ratio μ, dimensionless frequency a_0, and radius of foundation r_o. The second stage includes the estimation of the maximum vertical displacement for the foundation by using the dynamic stiffness K_{dyn} and dynamic damping C_{dyn} for certain realistic amplitude load Qo.

3 Results and Discussion

Figure 3 shows the relationship between dimensionless frequency a_0 and vertical stiffness coefficient (k) for the surface circular rigid foundation. The first upper curve (for $\mu = 0.33$) was proposed by Gazetas [1], while the lower two curves (for $\mu = 0.45$ and $\mu = 0.5$) were proposed by Wolf [2] using cone method approach. In this work and

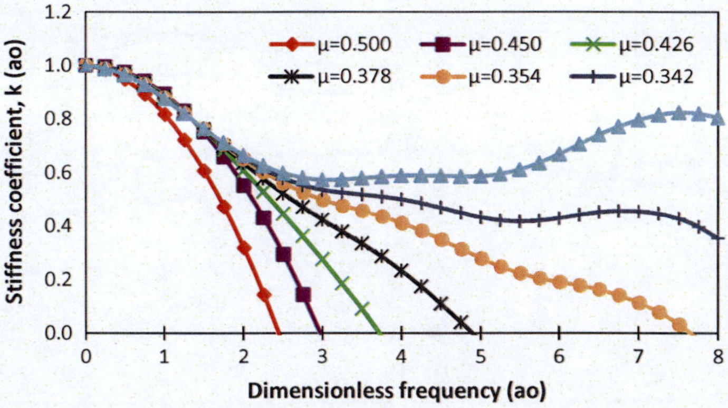

Fig. 3 Vertical dynamic stiffness coefficient of surface circular foundation for various Poisson's ratios

to study the effect of Poisson's ratio μ on the stiffness (k), four additional curves were conducted for μ = 0.342, 0.354, 0.378 and 0.426. These new curves were conducted by taking advantage of each of Gazetas and Wolf approaches. It is important here to mention that these additional curves for new values of Poisson's ratio μ represent the most common range for realistic soils, while the value conducted by Wolf (μ = 0.5) represents a special case of saturated clay with no volume change [5]. It is clear from Fig. 3 that the vertical stiffness coefficient k decreased significantly with an increment of Poisson's ratio μ. It is also clear that the vertical stiffness coefficient k decreased with the increase of dimensionless frequency a_0 for each value of Poisson's ratio μ.

The effect of both of Poisson's ratio μ and dimensionless frequency a_0 on the vertical damping coefficient is shown in Fig. 4. Again, the upper curve (for μ = 0.33) was conducted by Gazetas [1], while the lower curve (for μ = 0.5) was proposed by Wu et al. [9] using polynomial-fraction approximation. The additional curves were conducted in this study for μ = 0.370, 0.342, 0.350, 0.415 and 0.450. It can be seen that the value of vertical damping coefficient C depends on the value of Poisson's ratio μ. The value of the vertical damping coefficient for μ = 0.33 increased from 0.76 to about 1.072 at a_0 = 5.5, and then decreased. The same behavior can be noted for μ = 0.37, 0.41, 0.41 and 0.45. On the other hand, it can be observed that the vertical damping coefficient for μ = 0.5 decreases gradually with increasing of dimensionless frequency a_0. Also, it can be seen that all curves with different values of Poisson's ratio μ are meeting at a_0 = 1.75. This means that the vertical damping coefficient at this value of dimensionless frequency a_0 remains constant and is equal to 0. 9 regardless of the value of Poisson's ratio μ.

Each of dynamic stiffness (K_{dyn}) and dynamic damping (C_{dyn}) estimated from Eq. (1) was used to estimate the maximum displacement for various values of Poisson's ratios μ and different radiuses of foundation or according to Eq. (17) which was mentioned before. It is important here to point out that the damping ratio

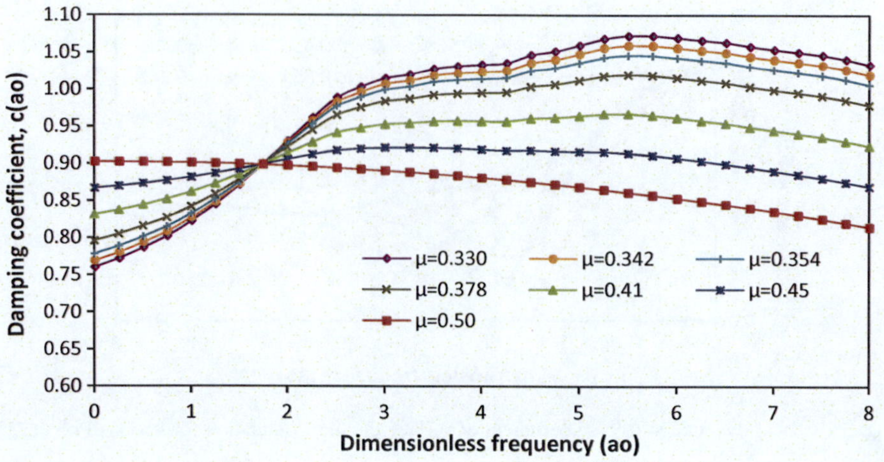

Fig. 4 Vertical dynamic damping coefficient of the surface circular foundation for various Poisson's ratios

D in Eq. (17) was estimated as ($D = C_{dyn}/C_c$). In this study, it is assumed that the soil under the circular rigid foundation has a shear modulus (G) of 20 MPa, the density of 1800 kg/m^3, and it is assumed that the amplitude of the existing force Q_o is constant and equal to 100 kN. The relationship between maximum displacement A_z estimated in millimeters and dimensionless frequency a_0 for $\mu = 0.33$, 0.342 and 0.35 are represented in Figs. 5, 6 and 7, respectively. The effect of the dimension of the foundation was also studied in this work by taking different values of radius r_o into account ($r_o = 2$, 2.75, 3.5 and 4.25 m). In general, these figures indicated that the maximum displacement (A_z) is decreasing with increasing radius. The decrement became small for high values of the radius ($r_o > 3.5$ m). It can also be seen that the shapes of curves in Fig. 5 for $\mu = 0.33$ are different from those in Figs. 6 and 7 for

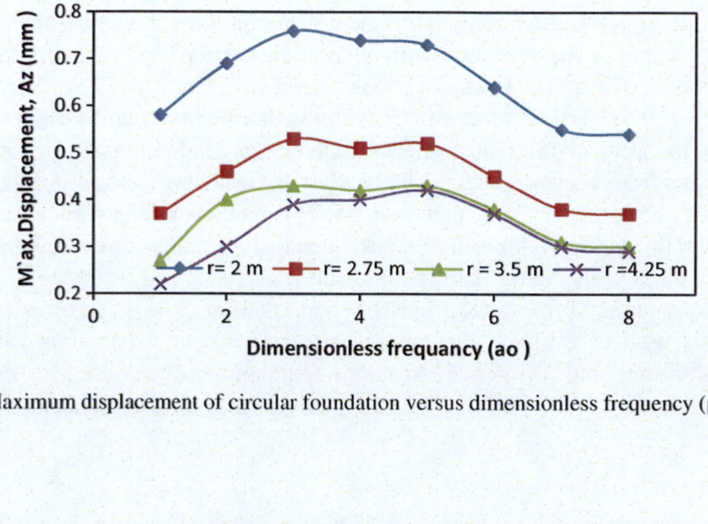

Fig. 5 Maximum displacement of circular foundation versus dimensionless frequency ($\mu = 0.33$)

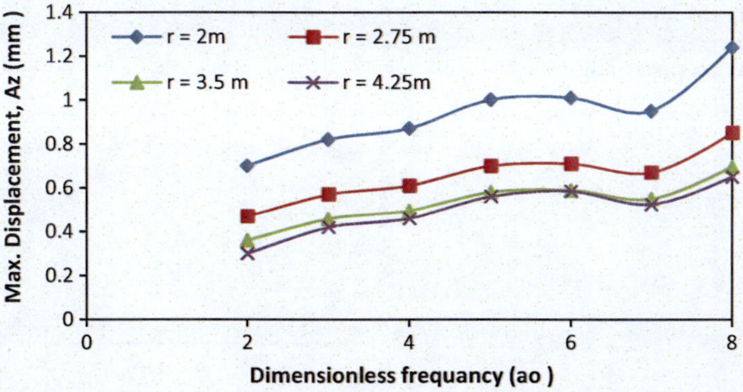

Fig. 6 Maximum displacement of circular foundation versus dimensionless frequency ($\mu = 0.342$)

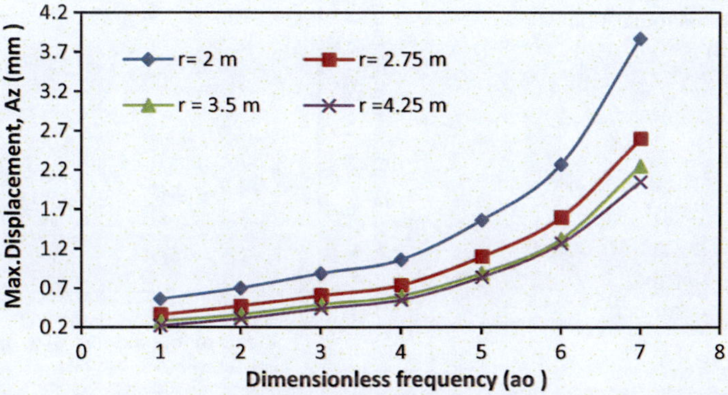

Fig. 7 Maximum displacement of circular foundation versus dimensionless frequency ($\mu = 0.35$)

$\mu = 0.342$ and 0.35, respectively. This difference is due to the different behavior of the stiffness coefficient concerning Poisson's ratios μ as shown in Fig. 3.

Figures 8, 9, 10, 11, 12, 13 and 14 show the maximum displacement of foundation versus foundation radiuses (r_o) for dimensionless frequency ($a_0 = 1$ to 7), respectively. It is obvious that for low values of dimensionless frequency ($a_0 = 1$ and 2), there is no effect of Poisson's ratios μ on the maximum displacement of foundation. While the effect of Poisson's ratios μ is appearing gradually with increasing of dimensionless frequency ($a_0 = 3$ to 7).

4 Conclusions

Based on the results of the analysis of circular foundation subjected to vertical load, the following points can be concluded:

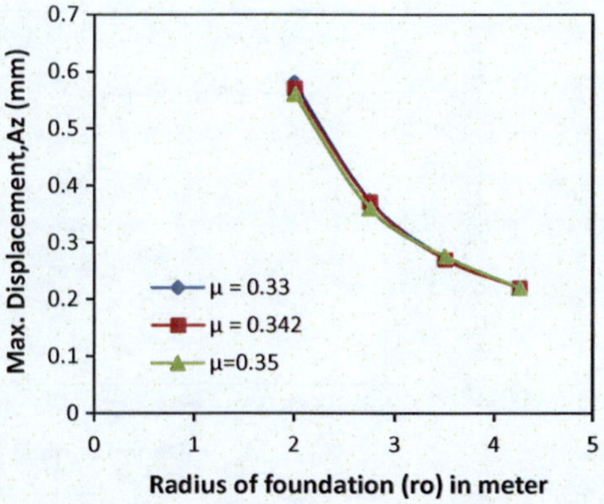

Fig. 8 Maximum displacement of circular foundation versus foundation radius ($a_0 = 1$)

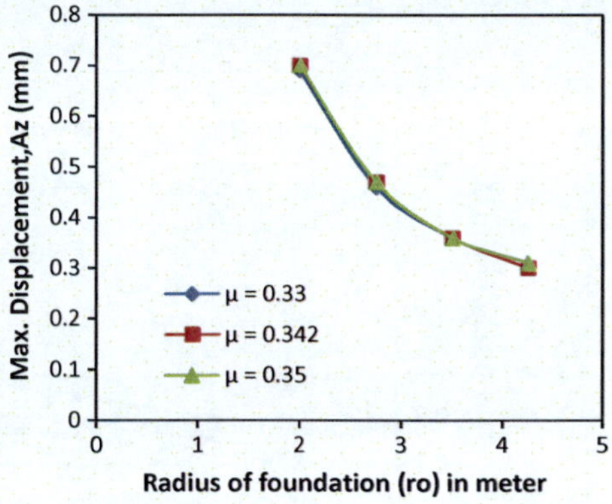

Fig. 9 Maximum displacement of circular foundation versus foundation radius ($a_0 = 2$)

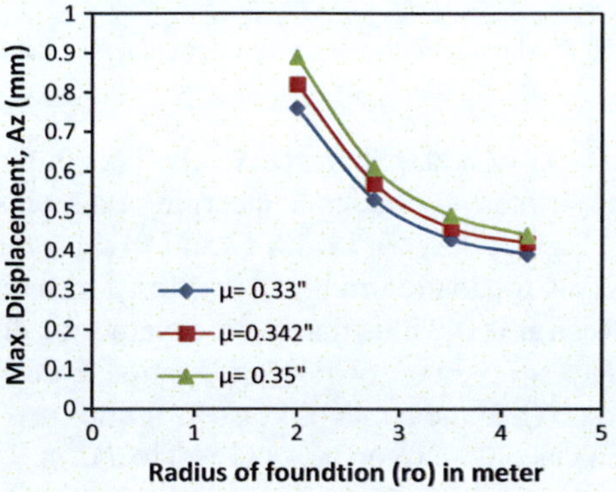

Fig. 10 Maximum displacement of circular foundation versus foundation radius ($a_0 = 3$)

Fig. 11 Maximum displacement of circular foundation versus foundation radius ($a_0 = 4$)

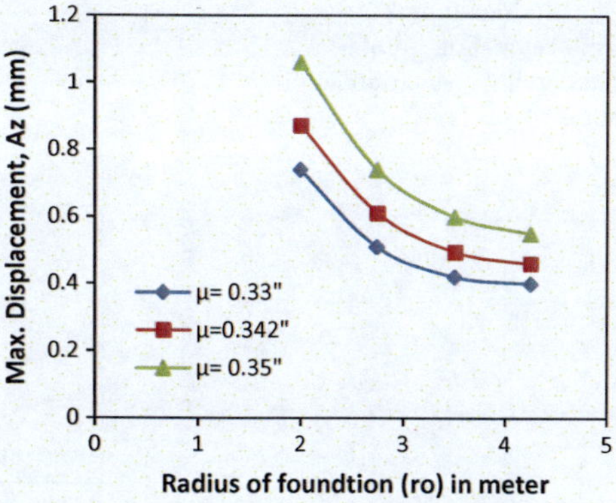

Fig. 12 Maximum displacement of circular foundation versus foundation radius ($a_0 = 5$)

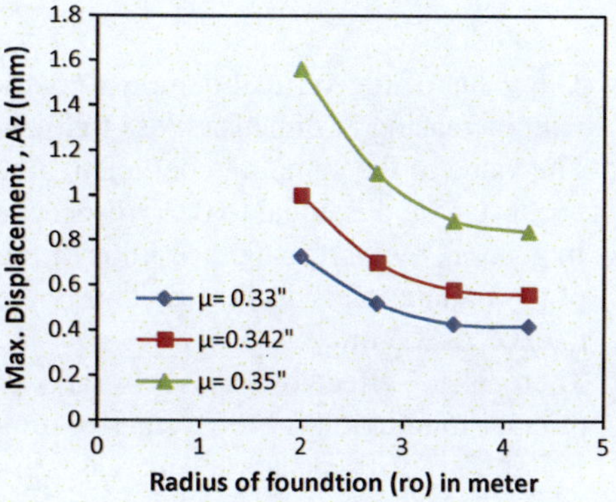

Fig. 13 Maximum displacement of circular foundation versus foundation radius ($a_0 = 6$)

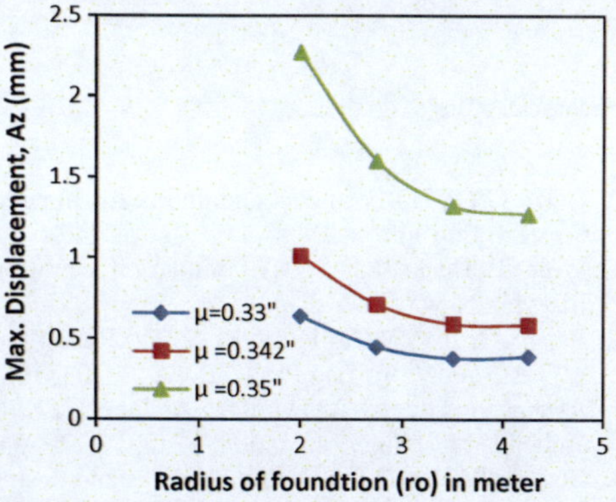

- The value of the vertical stiffness coefficient decreases with increment of both Poisson's ratio and dimensionless frequency a_0 for various values of Poisson's ratio.

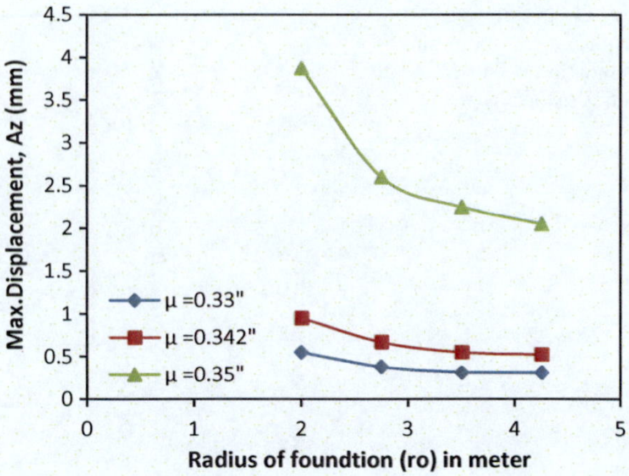

Fig. 14 Maximum displacement of circular foundation versus foundation radius ($a_0 = 7$)

- The value of the vertical damping coefficient is increasing and then decreasing with increasing of dimensionless frequency a_0 for $\mu = 0.33, 0.37, 0.41$ and 0.45. The value of the damping coefficient at the dimensionless frequency ($a_0 = 1.75$) is constant and is equal to 0.9 irrespective of the value of Poisson's ratio μ.
- In general, by increasing the radius of foundation r_o, the maximum displacement of the foundation is decreased. This decrement becomes lower when the value of r_o exceeds 3.5 m.
- There is no effect of Poisson's ratio μ on the maximum displacement of a circular foundation for low values of dimensionless frequency (a_0). The effect of Poisson's ratio μ appears and increases gradually with increasing of dimensionless frequency a_0.

References

1. Gatze G (1991) Formulas and charts for impedances of surface and embedded foundation. J Geotech Eng 117
2. Wolf JP, Deeks AJ (2004) Foundation vibration analysis: a strength of materials approach. Elsevier, NewYork
3. Bertha O, José J, Guillermo M (2019) Radiation damping for rigid foundations. Approximate expressions. Vibro Eng PROCEDIA 27:103–108
4. Al-Azawi TK, Al-Azawi RK, Al-Jaberi ZK (2006) Stiffness and damping properties of embedded machine foundation. J Eng 12:429–444
5. Bowles JE (2001) Foundation analysis and design. McGraw-Hill, New York
6. Das BM, Ramana GV (2010) Principles of soil dynamics. CL Engineering Publisher
7. Lysmer J, Richart FE Jr (2002) Dynamic response to footings to vertical loading. J Soil Mech Found Div ASCE 92:1091–1117
8. Jesmani M, Kamalzare M (2010) Comparison between numerical and analytical solution of dynamic response of circular shallow footing. Electron J Geotech Eng 15:1768–1781
9. Wu WH, Lee WH (2002) Systematic lumped-parameter models for foundations based on polynomial-fraction approximation. Earthq Eng Struct Dynam 31:1383–1412

Seismic Risk Assessment of Reinforced Concrete Frames at Al-Najaf City-Iraq Using Geotechnical Parameters

Sohaib K. Al-Mamoori, Ali N. Attiyah, Laheab A. Al-Maliki, Ahmed H. Al-Sulttani, Khaled El-Tawil, and Hussain M. Hussain

Abstract Seismic vulnerability is very important for buildings, especially old and dilapidated buildings that did not take into account the impact of earthquakes. In the past years, the seismic effect has not been taken into account, and recently Iraq has experienced a growing seismic activity. For this reason, it is crucial to create a seismic database for all cities. The present paper used SPT-N test results to assess seismic vulnerability for Al-Najaf city by producing five geotechnical maps using the GIS program for all seismic design categories. The data utilized in this research are obtained from 130 boreholes at 30 m depth distributed in the study area. The results show that the new RC frames must be designed as moment-resisting frames to resist future earthquakes. All the existing RC buildings are vulnerable to earthquake effects and need to be strengthened to act as seismic-resistant systems. The strengthened structure should act as ordinary reinforced concrete moment frame (ORCMF), intermediate reinforced concrete moment frame (IRCMF), or special reinforced concrete moment frame (SRCMF), depending on the building location and its risk category

S. K. Al-Mamoori (✉) · A. H. Al-Sulttani
Department of Environmental Planning, Faculty of Physical Planning, University of Kufa, Najaf, Iraq
e-mail: sohaib.almamoori@uokufa.edu.iq

A. H. Al-Sulttani
e-mail: ahmedh.alsulttani@uokufa.edu.iq

A. N. Attiyah
Department of Civil Engineering, Faculty of Engineering, University of Kufa, Najaf, Iraq
e-mail: alin.diebil@uokufa.edu.iq

L. A. Al-Maliki
Department of Hydraulic Engineering Structures, University of Al-Qasim Green, Babylon, Iraq
e-mail: laheab.almaliki@wrec.uoqasim.edu.iq

K. El-Tawil
Faculty of Engineering, Lebanese University, Beirut, Lebanon
e-mail: khaled_tawil@ul.edu.lb

H. M. Hussain
Department of Geology, University of Kufa, Najaf, Iraq
e-mail: hussainm.alshimmary@uokufa.edu.iq

M. O. Karkush and D. Choudhury (eds), *Modern Applications of Geotechnical Engineering and Construction*, Lecture Notes in Civil Engineering 112,
https://doi.org/10.1007/978-981-15-9399-4_28

type. The use of soil investigation data to assess the site class of soil (instead of assuming D for unknown soil) gave more economical choices for the reinforced concrete moment-resisting frames. Hence, for buildings types I, II, and III in some regions of the city, ORCMF can be used instead of IRCMF. In the same areas, IRCMF can be used instead of SRCMF for building type IV.

Keywords Seismic vulnerability · Geotechnical maps · GIS · SPT test · Al-Najaf city · Seismic risk assessment

1 Introduction

The design and construction of human-made facilities and every part of it require resistance to the impact of earthquakes. This resistance helps to protect them from collapse, minimize the damage of property and buildings, avoid human losses, and ensure the continuation of critical building services and vital installations as possible [1]. Loads produced by earthquakes are characterized by dissipating the construction energy beyond the flexibility limit; thus, the building design requirements and its structural details must take these influences into account. Owing to its location on the north part of the Arabian tectonic plate (Fig. 1), Iraq has recently experienced a growing seismic activity, which was felt clearly by people. Such events raised an interest in the structural strength of the existing buildings to resist seismic forces.

Moreover, the Iraqi seismic code was approved recently, which mentioned that the seismic design requirements should be considered in any structural design process. The utmost aim of seismic hazard assessment is to intensify the knowledge of the seismic parameters. This knowledge can help to predict earthquakes, which in turn can be used to design earthquake-resistant construction.

The seismic risks have to be taken into account for every structure, depending on its importance [2]. The structural system should decide according to the seismic design category (SDC) to resist the seismic force [3]. Depending on the construction industry, the earthquake-resistant system may be built using steel, concrete, and masonry, or composition of them. RC frames are the conventional structural system used in Iraq to build multistory buildings. There is a need to decide what type of reinforcement should be used for the new constructions. According to design codes, ordinary reinforced concrete moment frame (ORCMF), intermediate reinforced concrete moment frame (IRCMF), or special reinforced concrete moment frame (SRCMF) must be used. One of the crucial factors that increase seismic risks during an earthquake is the geotechnical features of the area, which is termed as site class in design codes. According to the ASCE Standards (ASCE/SEI 7–16), the soil is classified according to its type, which is based on the upper 100 ft (30 m) of the site profile [3].

The main purpose of the research is to assess the seismic risk of used RC frames in multistory buildings at Al Najaf city-Iraq based on the site class and the risk category (i.e., building type). This research considered the first Al Najaf city for the

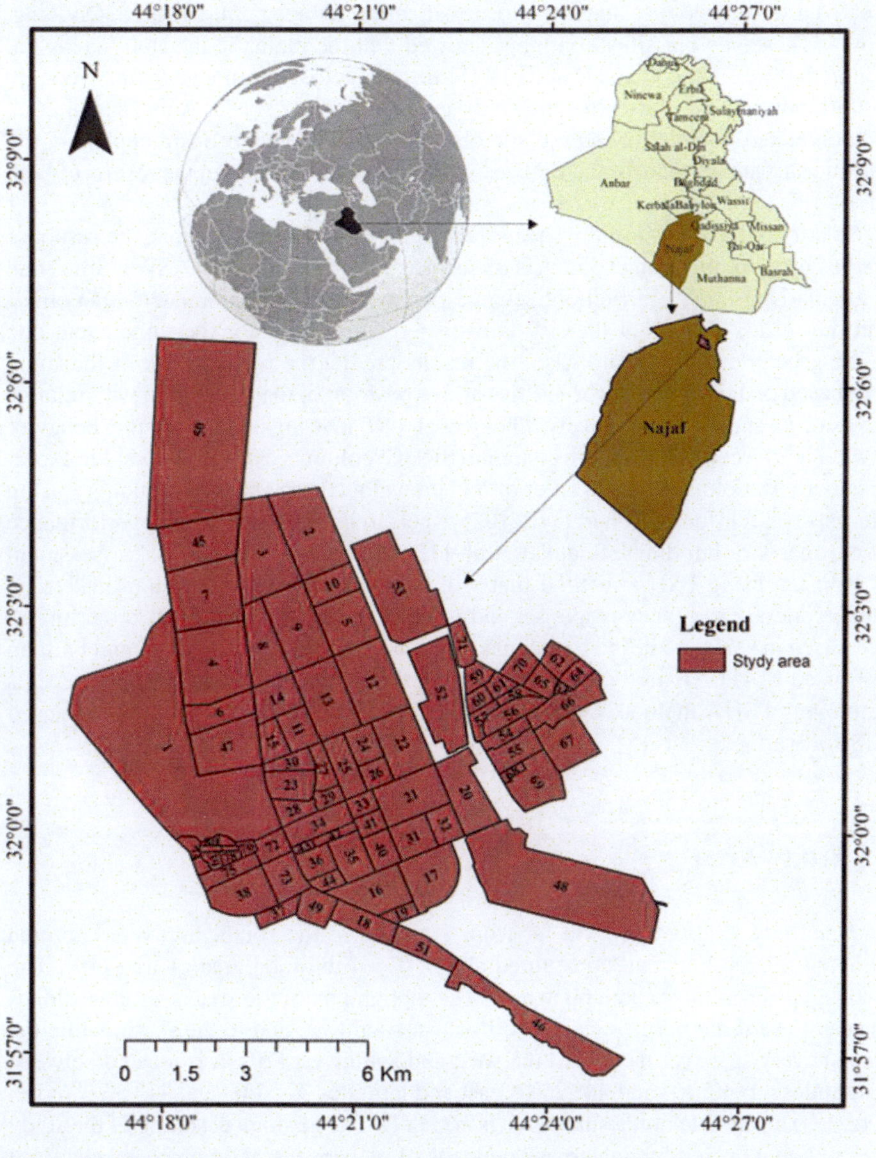

Fig. 1 The study area

study area but many studies have been conducted to assess seismic vulnerability using different approaches around the world; for example, Formisano et al. [4] examined old masonry buildings vulnerability in the historical center of Torre del Greco to help inhibit any susceptive damaging. The study shows that aggregated grouped buildings are less vulnerable to one level than these units are isolated. Sun [5, 6] created a seismic zoning map for site classification for a small urban area, Ongoing, Korea. These site classification maps showed that the plains in the study areas fall within the site classes C and D and in 2011 he conducted a similar study in Gyeongju to find that the site classes for the study area were also C or D. Anbazhagan et al. [7] conducted a study to find the site class for Indo-Gangetic Basin depending on seismic surface wave tests and SPT tests, and then they compared the results of both cases.

Shakir et al. [8] proposed a shear-wave velocity model for seismic site response studies using multichannel analysis of surface wave technique (MASW). Šipoš and Hadzima-Nyarko [9] carried out seismic risk assessments in some Croatian cities, and the results show that the risk is expected to increase for the same event due to the growth of population. Also, the results predict the realistic risk of the study area based on buildings vulnerabilities and exposition of the population with relative errors of 12 and 8%, respectively. Zheng et al. [10] investigated the seismic behavior of steel reinforced high strength concrete (SRHC) columns using pseudo-static experiments of 16 columns. Also, Li et al. [11] developed retrofitting technology using precast steel reinforced concrete (PSRC) panels to improve the seismic performance of old masonry buildings. Shaukat et al. [12] conducted a seismic risk assessment in Pakistan using PSHA method that will help engineers and decision-makers on preparedness, emergency response, and mitigation actions. MATLAB program has been used in some studies to re-evaluate the seismic hazard in Iraq. Said and Farman [13] used MATLAB in their research to get the net catalog to independent events employing PSHA method. Other studies related to seismic design assessment used GIS software [e.g. 14–18].

2 Study Area

Al-Najaf city is about 160 km far from Baghdad to the south. It covered an area of about 29,000 km^2 and constituted about 7% of the total area of Iraq [19]. The study area elevation is about 60 m a.s.l. The topography of the area is: gradual slopes plateau toward the north, western north, east, southeast, and to the south, while the slope is very steep (cliff) toward the west and southwest [20]. It is located between longitudes 44° 17′00″ and 44° 25′0″ east and latitudes 32° 7′0″ and 31° 56′0″ north (Fig. 1). The climate of the study area is arid [21]. The geological setting of the study area mainly covered by quaternary deposits of pleistocene and holocene because it is located in Mesopotamia which is tectonically classified as an unstable shelf; only a small area in the west is covered by pliocene–pleistocene (Dibdiba Formation) [22–24] (see Fig. 2).

The soil of Al-Najaf city mainly consists of the classes SP, SM, SP-SM with different ratios at deferent depths [25]. The geotechnical properties ranges of the study area are as follows: the internal friction angle is 26° and 42° [26]. The bearing capacity is 5 and 20 ton/m^2 [19], the soil sulfate content is 0.360 and 14.0%, and groundwater sulfate content is 84–239% [26]. The gypsum content is 10 and 25% [27]. The liquid

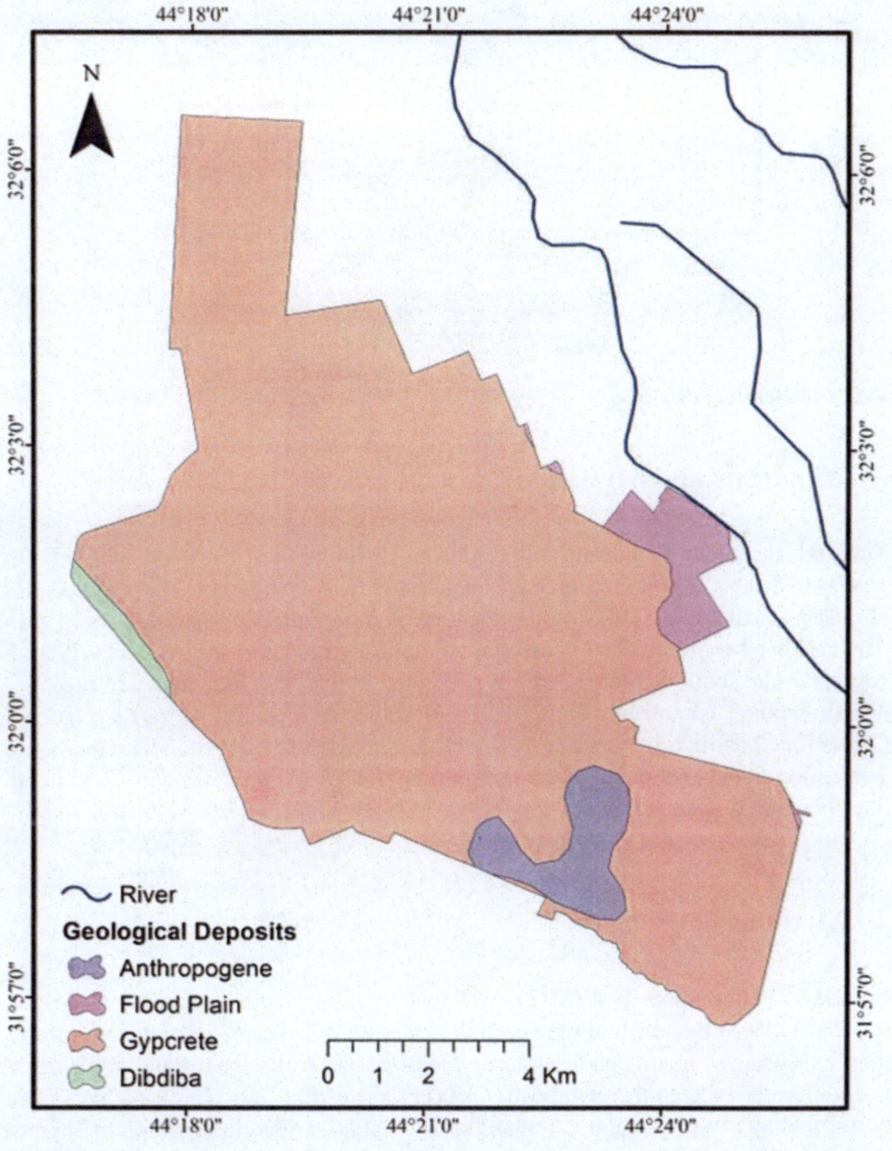

Fig. 2 Geological deposits in the study area. *Source* After [25, 26]

limit (LL) and plastic limit (PL) are 21–29% and 11–15%, respectively [28]. The maximum dry density is 17–19 kN/m^3. Finally, the optimum moisture content is 8–14% [25].

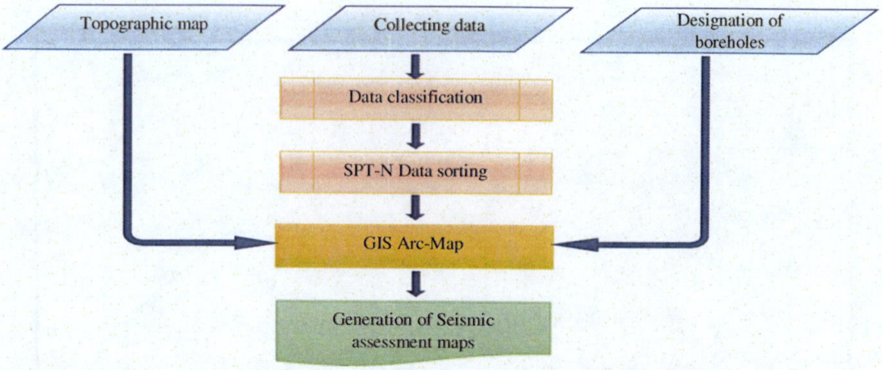

Fig. 3 Flowchart of the work

3 Materials and Methods

The study depended on the data collected from 130 boreholes. The depth of boreholes was 0–36 m, and 13 soil samples have been tested for each borehole. Two approaches are used to calculate SPT-N value; the first is data collecting, and secondly, GIS database has been built. The geotechnical maps are created using the ArcGIS 10.5 software. The scope of the research is presented in the flowchart shown in Fig. 3.

Materials. The soil lab. Data of SPT-N value have been done in the National Center for Construction Laboratories and Researches (NCCLR)/Babylon laboratory. The collected data belong to 464 boreholes, but only 130 boreholes with 30 m depth have been used (Fig. 4).

4 Methods

Standard Penetration Test (SPT). SPT had been performed according to ASTM (D-1586) [29] While drilling the boreholes, samples have been collected at uniform durations using standard penetration sampler to assess soil's density and also consistency in order to keep the samples for laboratory testing [30]. The tests have been performed by driving a 5 cm external diameter, solid-walled tube sampler [31] into the borehole by a 63.5 kg hammer freely falling from a 76 cm elevation (automatic falling trip-hammer). The sampler is driven 15 cm at first to break through the loose soil and "seat" the sampler. Afterward, the sampler drove an additional 30 cm gradually, where the results registered as the congruous number of blows needed (N value) to drive the sampler 30 cm [32]. Soil samples obtained from the bores were packaged and sealed in the field to prevent disturbance and keep moisture, then the samples are brought to the laboratory for testing. Blows number was referred to as N values and comprises the resistance of the test [29]. The results were given as a blow count value

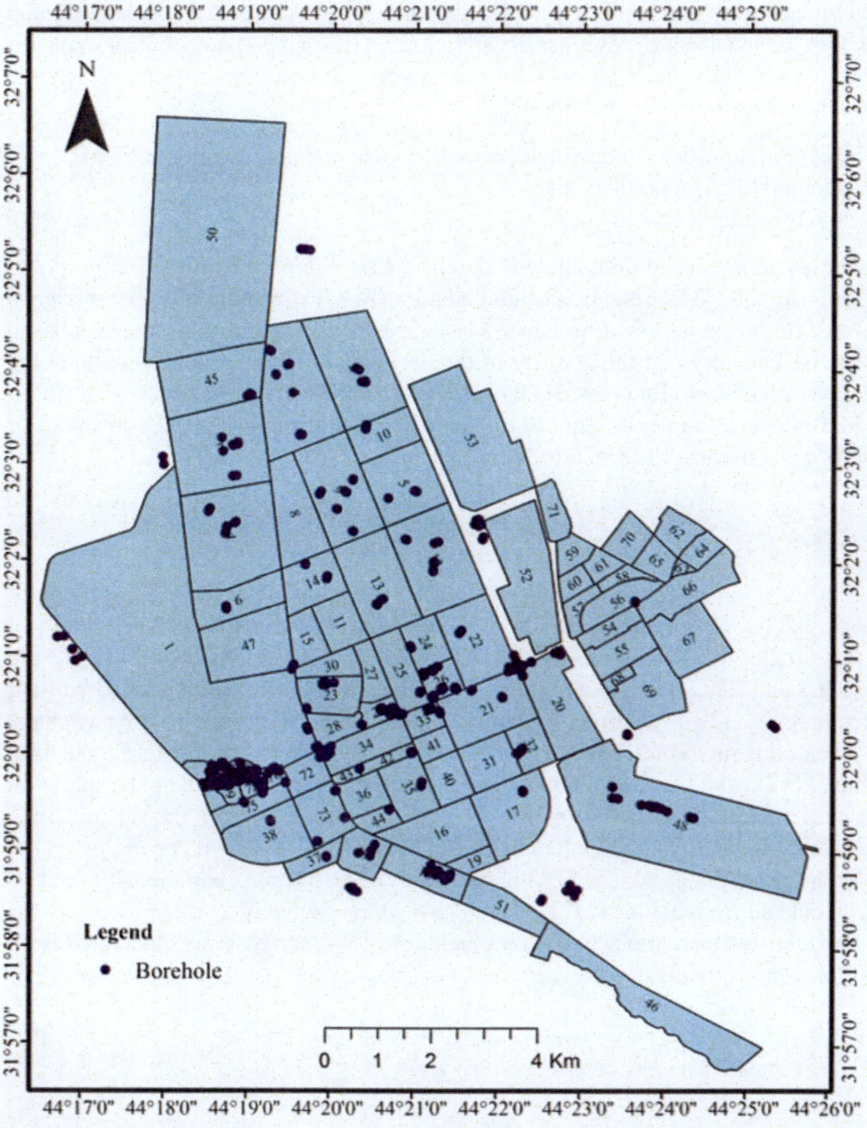

Fig. 4 Boreholes location

N which differs from the small values (1–5) if the soil is loose, to most significant values (over 50) when the soil is dense [33].

Standard Penetration Test (SPT) and Seismic Assessment. SPT is performed to assess soil density and also consistency in order to keep the samples for laboratory testing [30]. SPT-N values can indicate soil compactness and seismic risk as they used to identify site class type. For every seismic risk assessment or design, there are two cases to be considered: the long-period spectral response acceleration and the short-period spectral response acceleration. In this paper, geotechnical maps for the two cases have been produced. For categorizing buildings to achieve seismic resistance at economical cost, three factors are significant:

(a) Seismic intensity zone, which depends on where the building is located,
(b) the building importance, and
(c) the foundation soil stiffness.

These parameters combinedly will determine the extent of building's appropriate seismic strength [34]. Seismic design category (SDC) is another vital parameter in the seismic risk assessment, in which classification assigned to a structure is based on its risk category and the severity of the design earthquake ground motion at the site. The calculations for assessing the site class and SDC for the study area are based on SPT-N over 30 m depth. Equation 1 can calculate the equivalent SPT-N values of soil for any other depth from 0 to 30 m:

$$N_{30} = \frac{\sum_{i=1}^{n} d_i}{\sum_{i=1}^{n} (\frac{d_i}{N_i})} \tag{1}$$

where $\sum_{i=1}^{n} d_i$ = cumulative depth in meters, and d_i, N_i = the thickness (m) [7]. In this article, the ASCE STANDARD ASCE/SEI 7–10 is relied on to figure the site class and SDC for the study area. The tables mentioned in the manuscript are listed in ASCE/SEI 7–10. First, the building classification (i.e. type of category) must be assessed from Table (1.5-1). Then the importance factor I_e is determined from Table (1.5-2), which is dependent on category type. Next, we calculate the spectrum response accelerations at short period S_s and one second S_1 from the seismic map. The site class (A, B, C, D, E, or F) from Table (20.3-1) is assessed depending on the above information. To assess SDC for the study area, the site coefficients F_a and F_v are calculated from Tables (11-4.1) and (11-4.2), respectively.

The adjusted response spectrum accelerations (S_{MS} and S_{MI}) are calculated from the following equations:

$$S_{MS} = F_a S_S \tag{2}$$

$$S_{MI} = F_v S_1 \tag{3}$$

Then, the design response spectrum accelerations (S_{DS} and S_{DI}) are calculated from the following equations:

$$S_{DS} = \frac{2}{3} S_{MS} \tag{4}$$

$$S_{DI} = \frac{2}{3} S_{MI} \tag{5}$$

Finally, the seismic design category (SDC) is assessed depending on Tables (11.6-1) and (11.6-2), respectively. The flowchart in Fig. 5 illustrates the steps of preparing the data and producing the maps.

Seismic history. The seismology data for the last 60 years were obtained from the United States Geological Survey Agency (USGA) to clarify the seismic action background in Iraq. Figure 6 shows the seismic magnitude data (according to the Richter scale) collected by the USGS mapping agency for the period 1950–2019 [35].

Based on the earthquake historical data acquired from USGS for earthquakes active zone in Iraq and the surrounding area from 1950 till May 2019, 1707 earthquakes above 2.5 magnitudes have been recorded. Figures 7 and 8 show the interpolation maps in two methods (IDW and LPI). This magnitude intensity range (3.8–4.14) classified as intensity level (III to IV) according to modified Mercalli intensity scale (MMIS) [36, 37], which means it has light to moderate magnitude intensity. The

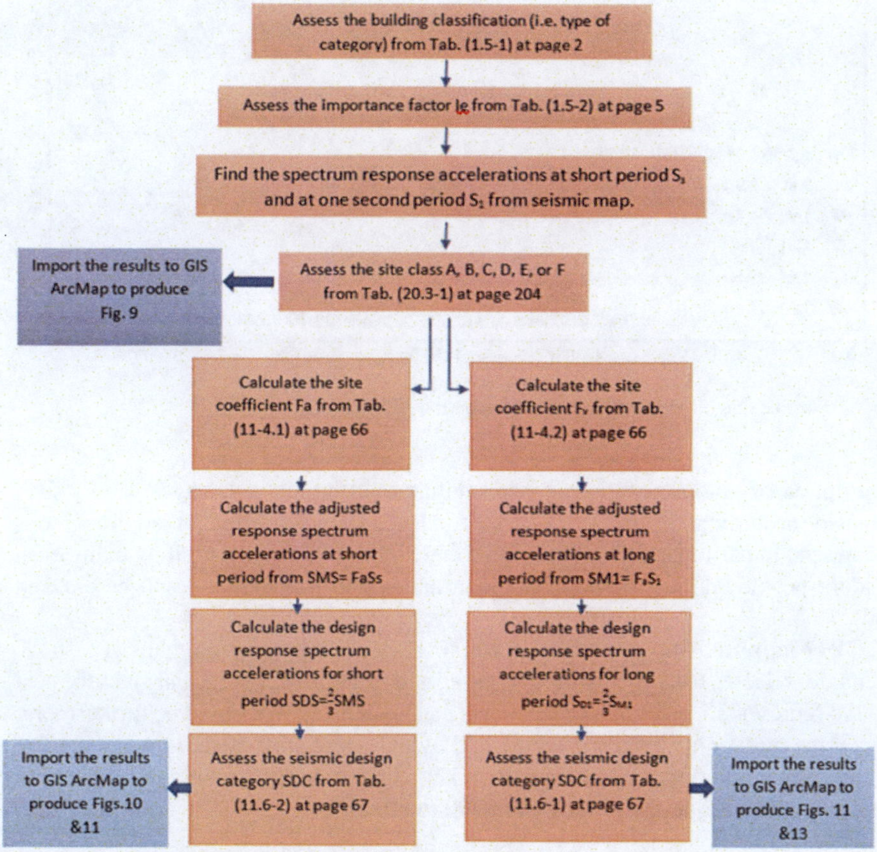

Fig. 5 The steps of preparing the data and producing the seismic risk maps

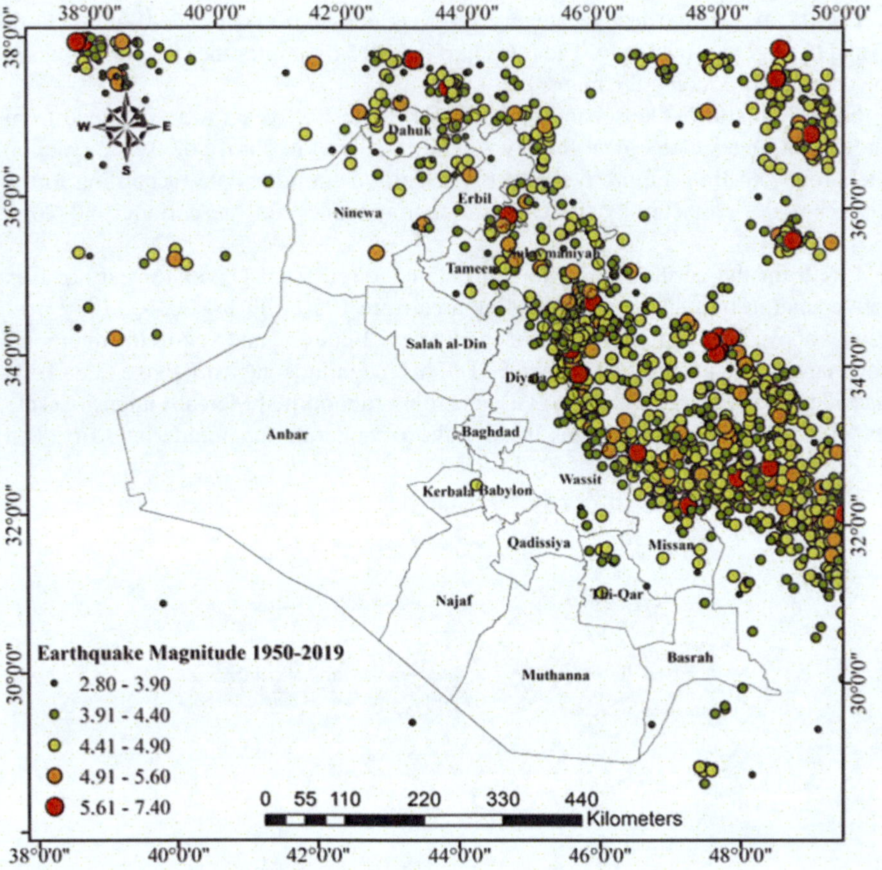

Fig. 6 Seismic magnitude data of Iraq for the period 1950–2019 [33]

description of human reaction, ground failure, and building damages in this such intensity according to the USGS 1995 for level III will be felt by some persons indoor and in particular in the second floor, while level IV will be felt by many even outdoor people and move windows, door, and sound like the wall will be cracked [36].

GIS Mapping. Arc-Map 10.5 is used to generate interpolation maps. The IDW method is selected because it is the most used and suitable technique that gives the best spatial representation for the SPT-N data [38, 39]. The IDW method is, furthermore, one of the most considerably used method by geoscientists to determine the models in spatial interpolation [40]. A cell size (pixel) 20 m is chosen because the residential land area in Al-Najaf city ranged from 200 to 400 m^2.

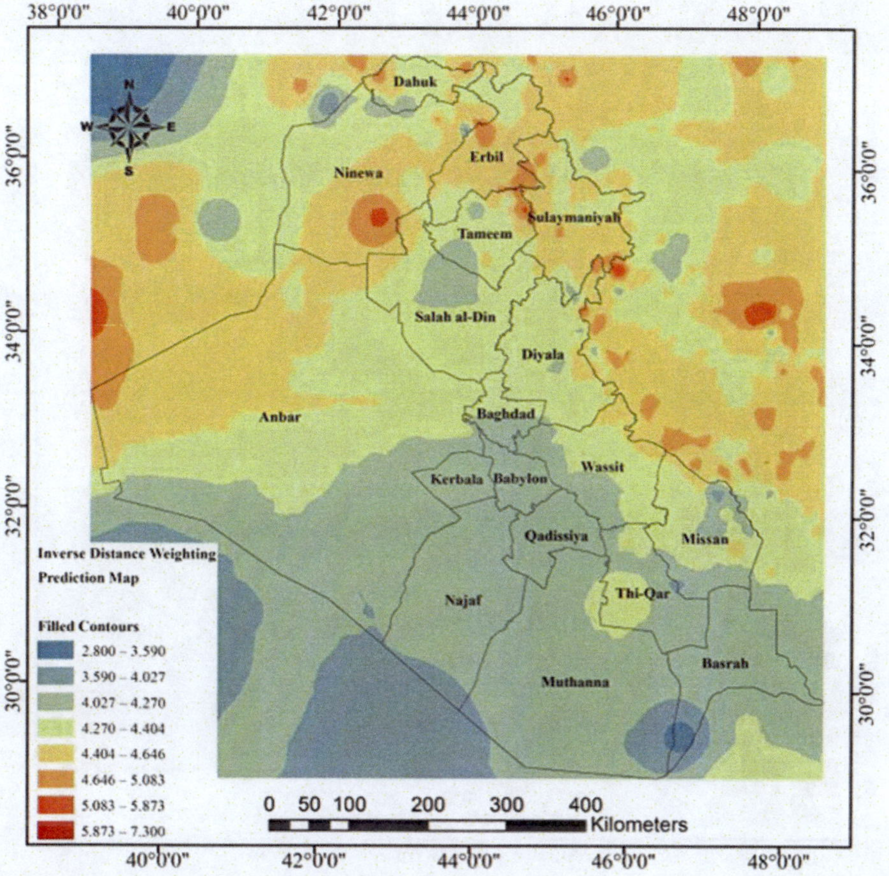

Fig. 7 Interpolated magnitude map by IDW method

5 Results and Discussion

According to the American standards ASCE_7–16 and the Iraqi Seismic Code, site class D shall use when the soil properties are unknown. It can be seen clearly from Fig. 9 that some of the zones in the city will consider as site class D instead of its real state (i.e., site class C). In other words, the soil in those zones is stiffer than it assumed according to design codes. The seismic vulnerability of structures is directly related to SDC, which is dependent on the spectrum acceleration response (i.e., earthquake intensity), site class (i.e., soil type), and risk category (i.e., building type). Figure 10 shows the SDC for buildings types I, II, and III for spectrum acceleration response calculated in a short period. Figure 12 exhibits SDC for the same building type for spectrum acceleration response calculated at long period. It can be seen clearly that different SDC (A, B, or C) should be considered in seismic design. In the case when the soil is unknown, all the cities will have SDC B and SDC C in Figs. 10 and 12,

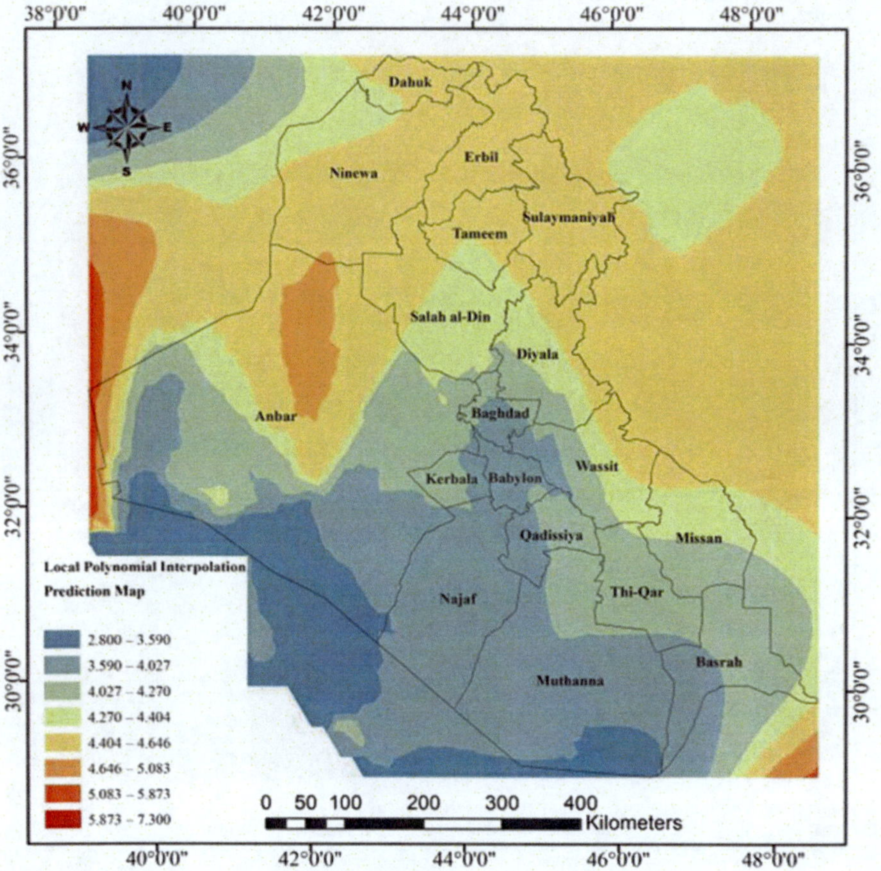

Fig. 8 Interpolated magnitude map by LPI method

respectively. The structural design would be more conservative in some areas of the city.

In the same way, for the other type of buildings (type IV), it can be seen from Figs. 11 and 13 that SDC is considered as A, C, or D for spectra accelerations responses calculated at short and long periods, respectively. However, the SDC for building type IV is considered as C instead of A, and D instead of C, if the soil properties are unknown. Conservative design is expected for building structures in such cases. According to the American standards ASCE_7–16 and the Iraqi Seismic Code, the worst SDC should consider from spectra accelerations responses calculated in short and long periods. Hence, SDC shown in Figs. 11 and 13 will govern the seismic structural design. The primary step in the design is to choose the structural system, which will be dependent on the SDC. Most of the multistory buildings in Al-Najaf city used RC frames as structural systems. If such a system is used to be seismic-resistant, three types can be used: ORCMF, IRCMF, or SRCMF.

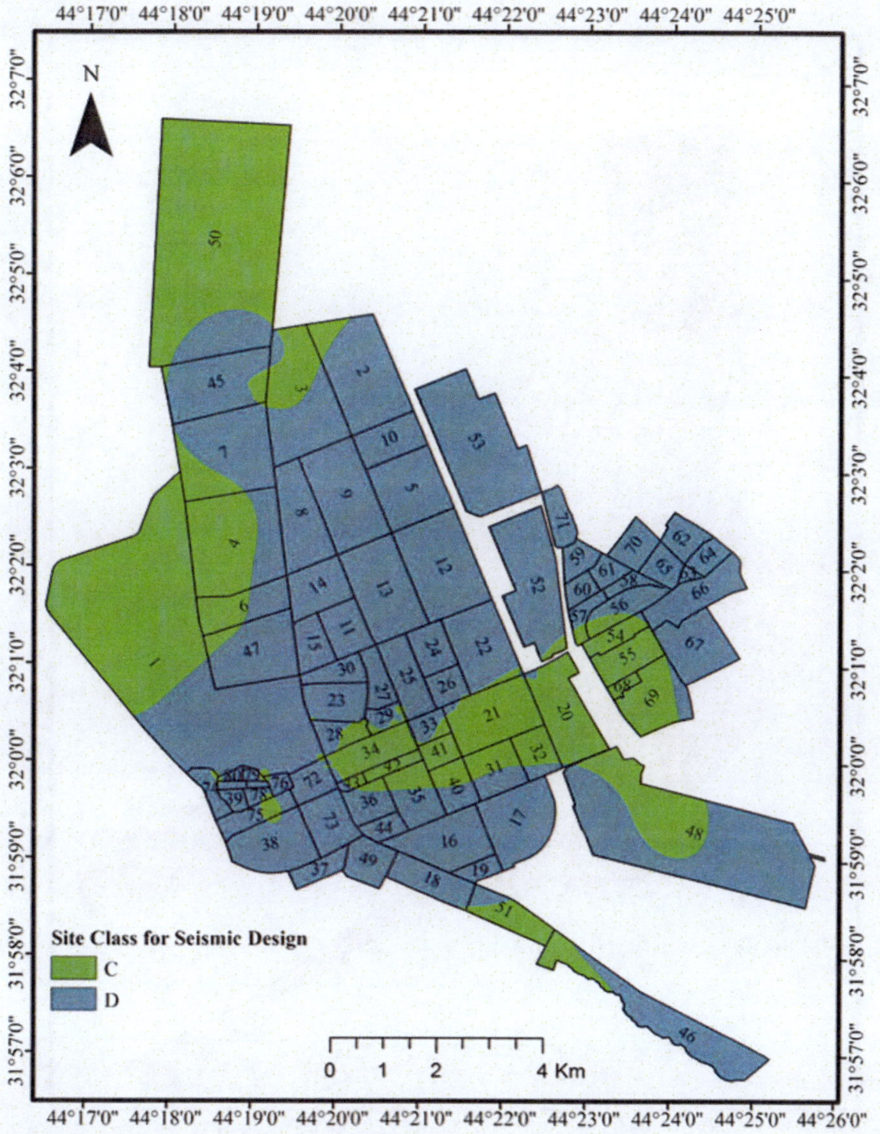

Fig. 9 Site class for seismic design

According to the American standards ASCE_7–16 and the Iraqi Seismic Code, only SRCMF can be used in SDC D, either IRCMF or SRCMF in SDC C, and the three systems are permitted in SDC B zones. From an economic point of view and ease of construction, the designer will choose ORCMF, IRCMF, and SRCMF for SDC B, C, and D, respectively. Figures 10 and 11 show the system type can be used to resist earthquakes, where the first map is prepared for buildings types I, II, and

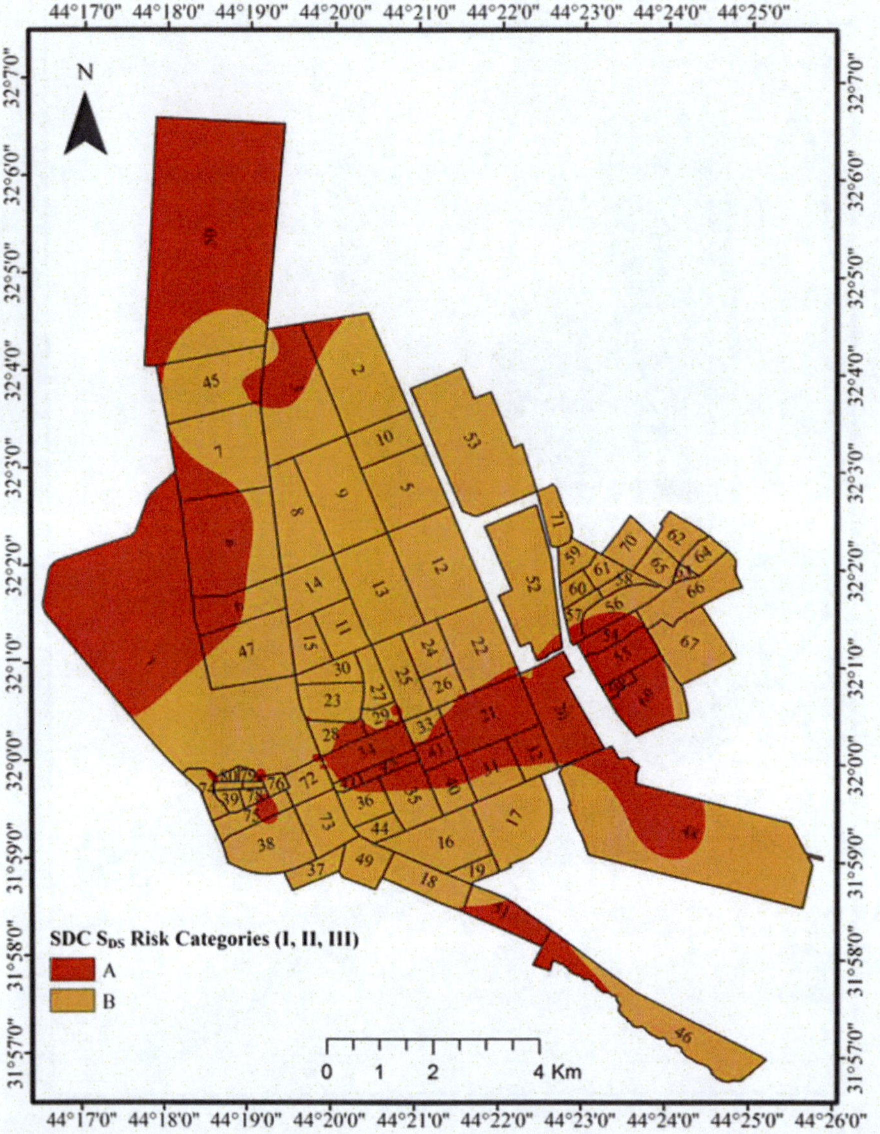

Fig. 10 Seismic design category (SDC) for building types I, II, and III for spectrum acceleration response—short period

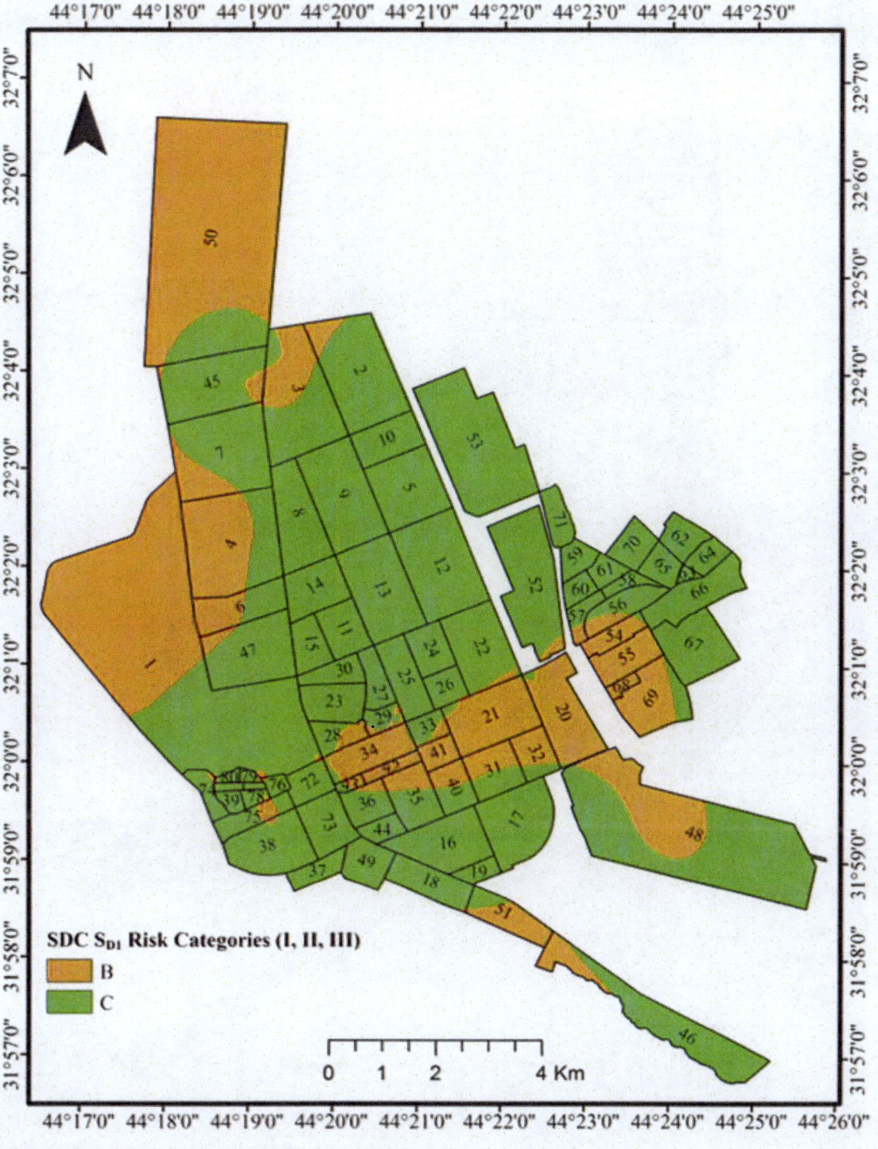

Fig. 11 Seismic design category (SDC) for building types I, II, and III for spectrum acceleration response—long period

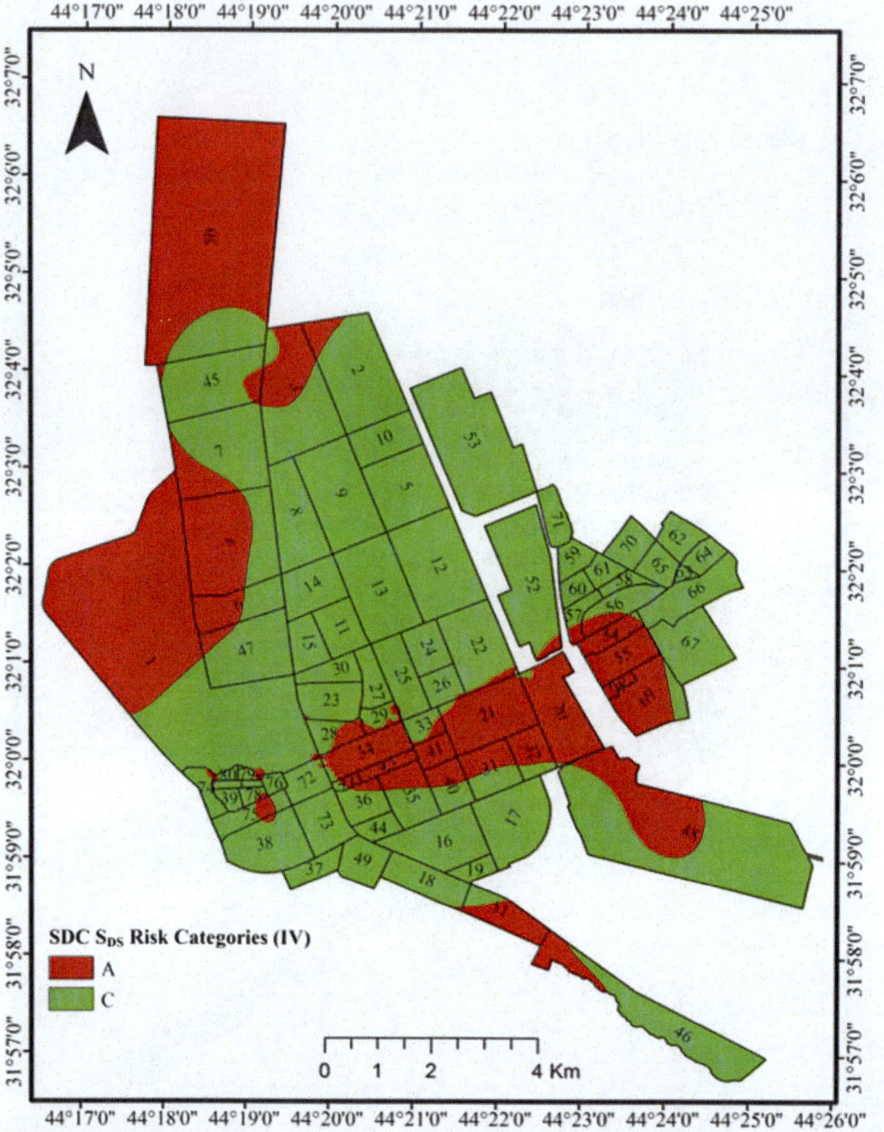

Fig. 12 Seismic design category (SDC) for building type IV for spectrum acceleration response calculated for a short period

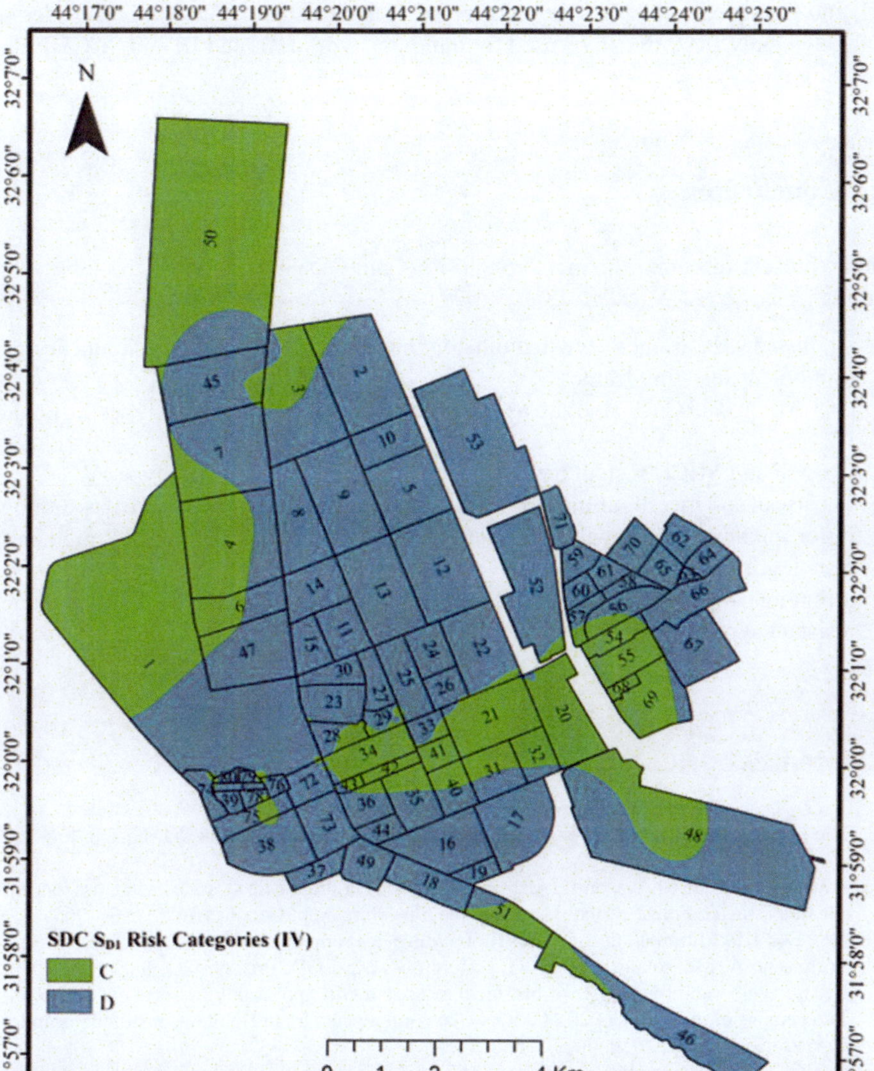

Fig. 13 Seismic design category (SDC) for building type IV for spectrum acceleration response calculated for a long period

III, and the second one for building type IV. It is clear that if the soil is considered unknown, only IRCMF can be used for buildings types I, II, and III and SRCMF for building type IV.

6 Conclusions

In the present research, the main conclusions can be summarized by the following points:

- All the new RC frames used in multistory buildings at Al Najaf city should design as moment-resisting frames.
- ORCMF and IRCMF shall be used for buildings of risk category types I, II, and III.
- IRCMF and SRCMF shall be used for buildings of risk category type IV.
- The use of soil investigation data to assess the site class of soil (instead of assuming D for unknown soils) will give more economical choices for the RC moment-resisting frames.
- Strengthening requires for all RC frames built at Al Najaf city to act as seismic resistant systems.

References

1. Xie L, Qu Z (2018) On civil engineering disasters and their mitigation. Earthq Eng Eng Vib 17(1):1–10
2. Ameer A, Sharma M, Wason H, Alsinawi SA (2005) Probabilistic seismic hazard assessment for Iraq using complete earthquake catalogue files. Pure Appl Geophys 162(5):951–966
3. ASCE (2010) Minimum design loads for buildings and other structures. Reston, VA
4. Formisano A, Mazzolani F, Florio G, Landolfo R, Masi GD, Priscoli GD, Indirli M (2010) Seismic vulnerability analysis of historical centres: a GIS application in Torre del Greco. In: Proceedings of the final COST ACTION C26 conference: Urban Habitat Constructions under Catastrophic Events, 2010, pp 583–588
5. Sun CG (2010) Methodical assessment of geotechnical characteristics related to earthquake motion in a small urban area using a GIS-based tool. System 2:10
6. Sun CG (2012) Applications of a GIS-based geotechnical tool to assess spatial earthquake hazards in an urban area. Environ Earth Sci 65(7):1987–2001
7. Anbazhagan P, Kumar A, Sitharam TJP (2013) Seismic site classification and correlation between standard penetration test N value and shear wave velocity for Lucknow City in Indo-Gangetic Basin. Pure Appl Geophys 170(3):299–318
8. Shakir AM, Foti S, Garofalo F, Hijab BR, Laftah AA (2013) Laterally constrained inversion of surface wave data at Najaf City (IRAQ). Soil Dyn Earthq Eng 45:89–95
9. Šipoš TK, Hadzima-Nyarko M (2017) Rapid seismic risk assessment. Int J Disaster Risk Reduct 24:348–360
10. Zheng S, Qin Q, Zhang Y, Zhang L, Yang W (2017) Research on seismic behavior and shear strength of SRHC frame columns. Earthq Eng Eng Vib 16(2):349–364

11. Li W, Wang T, Chen X, Zhong X, Pan P (2017) Pseudo-dynamic tests on masonry residential buildings seismically retrofitted by precast steel reinforced concrete walls. Earthq Eng Eng Vib 16(3):587–597
12. Khan SA, Pilakoutas K, Hajirasouliha I, Garcia R, Guadagnini M (2018) Seismic risk assessment for developing countries: Pakistan as a case study. Earthq Eng Eng Vib 17(4):787–804
13. Said AI, Farman MS (2018) Re-evaluations of seismic hazard of Iraq. Arab J Geosci 11(11):279
14. Kheradmand M, Jahangiri K, Sohrabizadeh S, Safarpour H, Jazani RK (2019) Physical seismic vulnerability assessment of neighborhood emphasising on critical land use. Int J Struct Integr 10(2):176–187
15. Mohamad I, Yunus MM, Harith N (2019) Assessment of building vulnerability by integrating rapid visual screening and geographic information system: a case study of Ranau township. In: IOP conference. Materials science and engineering. IOP Publishing, 012042
16. Liu Y, Li Z, Wei B, Li X, Fu B (2019) Seismic vulnerability assessment at urban scale using data mining and GIScience technology: application to Urumqi (China). Geomat Nat Hazards Risk 10(1):958–985
17. Zhai Y, Chen S, Ouyang Q (2019) GIS-based seismic hazard prediction system for urban earthquake disaster prevention planning. Sustainability 11(9):2620
18. Mazumder RK, Salman AM (2019) Seismic damage assessment using RADIUS and GIS: a case study of Sylhet City, Bangladesh. Int J Disaster Risk Reduct 34:243–254
19. Al-Maliki LAJ, Al-Mamoori SK, El-Tawel K, Hussain HM, Al-Ansari N, Al Ali MJ (2018) Bearing capacity map for An-Najaf and Kufa Cities using GIS. Engineering 10(05):262–269
20. Al-Mamoori SK, Al-Maliki LAJ, El-Tawel K, Hussain HM, Al-Ansari N, Al Ali MJ (2018) Chloride, calcium carbonate and total soluble salts contents distribution for An-Najaf and Al-Kufa Cities' soil by using GIS. Geotech Geol Eng 1–19
21. Farhan SL, Jasim IA, Al-Mamoori SK (2019) The transformation of the city of Najaf, Iraq: Analysis, reality and future prospects. J Urban Regen Renew 13(2):160–171
22. Al-Mubarak M, Amin R (1983) Report on the regional geological mapping of the eastern part of the Western Desert and western part of the Southern Desert, GEOSURV. Int. Rep. (1380) (1983)
23. Buday T (1980) The regional geology of Iraq: stratigraphy and paleogeography. State Organization
24. Jassim SZ, Goff JC (2006) Geology of Iraq, Published by Dolin, Prague and Moravian Museum Brno, Czech Republic
25. Ali TS, Fakhraldin MK (2016) Soil parameters analysis of Al-Najaf city in Iraq: case study. J Geotech Eng 3(1):56–62
26. Al-Mamoori SK, Al-Maliki LA, Hussain HM, Al-Ali MJ (2018) Distribution of sulfate content and organic matter in An-Najaf and Al-Kufa cities'soil using GIS. Kufa J Eng 9(3):92–111
27. Al-Mamoori SK (2017) Gypsum content horizontal and vertical distribution of An-Najaf and Al-Kufa Cities' soil by using GIS. Basra J Eng Sci 17(1):48–60
28. Al-Mamoori SK, Jasem Al-Maliki LA, Al-Sulttani AH, El-Tawil K, Hussain HM, Al-Ansari NJG (2019) Horizontal and vertical geotechnical variations of soils according to USCS classification for the City of An-Najaf, Iraq using GIS. Geotech Geol Eng
29. ASTM (2000) Standard test method for standard penetration test (SPT) and split barrel sampling of soil
30. ASTM (2009) Standard practice for description and identification of soils (visual-manual procedure). ASTM International West Conshohocken, PA
31. Governorate of Babylon (2016) Investigation Reports, in N.C.f.C.L.R. (NCCLR) (Ed.)
32. ASTM (2011) Standard test method for direct shear test of soils under consolidated drained conditions, D3080/D3080M
33. Atkinson J (2007) The mechanics of soils and foundations. CRC Press
34. Arya AS, Boen T, Ishiyama Y (2014) Guidelines for earthquake resistant non-engineered construction. UNESCO
35. USGS, Earthquake Data (2019. https://earthquake.usgs.gov. Accessed May 2019

36. Dewey JW, Reagor BG, Dengler L, Moley K (1995) Intensity distribution and isoseismal maps for the Northridge, California, earthquake of January 17, 1994, US Geological Survey
37. USGS, Magnitude/Intensity Comparison (2019). https://earthquake.usgs.gov/learn/topics/mag_vs_int.php. Accessed August 2019
38. Al-Ani H, Oh E, Chai G (2013) Characteristics of embedded peat in coastal environments. Int J GEOMATE 5(1):609–618
39. Al-Ani HO, Chai E, Al-Uzairy G, Nader B (2014) GIS-interpolated geotechnical zonation maps in surfers paradise, Australia. In: 6th international conference on advanced geographic information systems, applications, and services (GEO-processing 2014), Spain, Barcelona, Citeseer, 2014, p 148
40. Lu GY, Wong DW (2008) An adaptive inverse-distance weighting spatial interpolation technique. Comput Geosci 34(9):1044–1055

Geometric Nonlinear Synthetic Earthquake Analysis of Base Isolated Tall Steel Buildings Under Site-Specific Seismic Loading

Rafaa M. Abbas and Ameer J. Abdulkareem

Abstract The incorporation of base isolation in building construction in the regions of medium to high seismicity is investigated with special emphasis on the geometric nonlinearity response due to the second-order effect. A thorough investigation is carried out by using time history analysis for different building heights located in different seismic design categories with site soil class D. A synthetic earthquake is developed by using spectral matching to adjust reference time series that consisted of available readings from the past earthquake of the 1940 El Centro earthquake adopted as a reference time series to generate time series compatible with the target response spectrum of the selected regions. The superstructure of the adopted tall buildings, with lead rubber bearings, is represented by finite element models. The study reveals that for high-rise building construction, isolation can significantly reduce seismic response in stiff soil (i.e., site class D). Results presented reflect the potential of synthetic time history analysis to assess the dynamic response of base-isolated tall buildings. The results show that site seismic intensity has a negligible effect on the percentage variation in the story shear, overturning moment, and building acceleration values due to the p-delta effect. Generally, story shear and acceleration values reduced, while overturning moment values increased when the p-delta effect is accounted for. The study has shown that geometric nonlinearity due to p-delta increases building flexibility and reduces the pseudo acceleration of the superstructure and hence the earthquake-induced forces in the structure.

Keywords Synthetic earthquake · Time history analysis · p-delta analysis · Response spectrum · Base isolation · High-rise building

R. M. Abbas (✉) · A. J. Abdulkareem
Department of Civil Engineering, University of Baghdad, Baghdad, Iraq
e-mail: dr.rafaa@coeng.uobaghdad.edu.iq

A. J. Abdulkareem
e-mail: a.abdulkarim1001@coeng.uobaghdad.edu.iq

M. O. Karkush and D. Choudhury (eds), *Modern Applications of Geotechnical Engineering and Construction*, Lecture Notes in Civil Engineering 112,
https://doi.org/10.1007/978-981-15-9399-4_29

1 Introduction

The effect of earthquakes is one of the most dangerous lateral force effects on buildings because of the location of their effect. Unlike other lateral forces, the impact of earthquakes moves from the base to the entire building and in most cases, it causes severe damage and can lead to the collapse of the building [1]. By increasing the height of buildings with the same lateral dimensions means more slender and thus less lateral stiffness and more flexible [2]. Because of this flexibility, when exposed to lateral forces such as earthquakes, it generates lateral displacement that increases directly with the height and inversely with the stiffness of the building. This resulting displacement as a response of the building to the lateral forces works with the weight of the building to generate additional moment called (p-delta) in addition to the initial moment [3]. Many solutions and techniques have been developed to resist and minimize the impact of seismic forces. One of the most important techniques that have been used has proven to be effective in reducing the damage caused by earthquakes and keeps the building usable during and after the earthquake is base isolation [4]. One of the most important strategies of isolators is to convert the energy transferred from the base to the displacement through the expansion of the isolator to prevent the transfer of this kinetic energy to the body of the building [5]. This displacement generated in the isolator is added to the displacement of the building and therefore the total displacement of the isolated building could be higher than in the case of a fixed-base building [6]. The method of analysis of the fixed and isolated building depends on several considerations, and the most important of which is the height of the building [7].

Since the study is concerned with tall isolated buildings, as well as for knowing the linear and geometric nonlinear behavior of isolated buildings, it is necessary to rely on the dynamic methods for the analysis of response spectrum and time history. The response spectrum is an expression of the amount of seismic activity at a specific location, which depends heavily on the characteristics of the soil being able to modify the properties of the seismic waves, which can be maximized or reduced depending on the nature of the soil, whether stiff, loose or others [8]. In general, most of the specifications such as ASCE/SEI 7-10 recommend the use site class D when the characteristics of the soil of the site are unavailable [9]. However, many of the codes and specifications dictate the use of dynamic analysis methods, especially time history records to evaluate the response of base-isolated buildings because they have the potential to sense the properties of these isolators and capability to show their effect on the results of the analysis [10]. On the other hand, real-time history data may be limited in many countries and the available time history records of ground motion gives a response spectrum not matched to target response spectrum of the site in which the building is to be constructed; thus, synthetic and artificial time histories are widely used instead [11].

Although a great deal of previous research has been carried out on base isolation technique for multistory buildings and to analyze the effect of p-delta analysis on high-rise building response, there is no reported work on the effect of p-delta analysis on the seismic response of base-isolated tall buildings. This study aims to investigate the effect of p-delta on the dynamic response of base-isolated high-rise buildings under site-specific earthquake excitation. Tall buildings with different heights located in three different seismic design categories with site soil class "D" will be analyzed using spectral matching in the time domain to simulate the response spectrum of the selected site-specific base excitation.

2 Time-Domain Spectral Matching

The seismic design of structures is generally based on a design response spectrum obtained from hazard analysis for a specified site. For many engineering applications, such as the design of base-isolated buildings, a more complex dynamic analysis is often conducted. Such analysis requires inputs in the form of design time series where the response spectrum for this time series must be close to and consistent with the design response spectrum of the site in which the building is to be constructed [12]. Since the possibility of using real earthquake data is limited, hence design time series is developed by modifying initial time series that consists of empirical recordings from past earthquakes representative of the design event or numerical simulations of the ground motion for the design event. To achieve this goal spectral matching is commonly used in engineering practice. The time-domain method is generally considered as a better approach for spectral matching since this method adjusts the acceleration time histories in the time domain by adding wavelets. A wavelet is a mathematical function that defines a waveform of effectively limited duration which has a zero average.

The development of spectral matching using wavelet transforms was started by Lilhanand and Tseng [13]. The original proposed wavelet consists of the true acceleration impulse response function in reverse time order. The proposed algorithm by Lilhanand and Tseng suffers from two shortages. Firstly, the modifying wavelets adversely alter the resulting velocity and displacement time series by introducing drift into the computed accelerogram results. Secondly, when matching closely spaced periods the proposed algorithm yields unstable results. To correct the resulting velocity and displacement time histories after adjustment, an improved tapered cosine wavelet was developed by Alatik and Abrahamson [12]. The corrected tapered cosine wavelet can be described by:

$$\psi_j(t) = e^{-\left[\frac{(t-t_j+\Delta t_j)}{\gamma_j}\right]^2} \cos\left(\acute{\omega}_j\left(t - t_j + \Delta t_j\right)\right). \quad (1)$$

where Δt_j is the differences between the time of peak response with the reference time of the wavelet (t_j). The correction factor γ_j is a frequency-dependent coefficient used to adjust the duration of the adjustment function. The $\gamma(f)$ factor was developed to ensure a smooth taper and leads to zero velocity and displacement at the end of the wavelet. The adjustment at all frequencies, $\gamma(f)$, is given by:

$$\gamma(f) = 1.178 f^{-0.93}. \quad (2)$$

The corrected tapered cosine wavelet leads to zero velocity and displacement results when the wavelet has sufficient time duration to start and end with zero value. The minimum t_j values required to prevent drift is described by

$$t_{j\min} = 3.9223 f^{-0.845}. \quad (3)$$

3 Numerical Study

In this study, five multistory steel building models shown in Fig. 1 with different number of stories and height ranging from 10 stories (35 m height) and up to 30 stories (105 m height) were analyzed using ETABS software. In all building models, the height of each story is 3.5 m including the ground floor with 25 m in the x and y directions for the plane view as shown in Fig. 2. The height to width ratio for the buildings range from 1.4 up to 4.2 for both directions. The buildings were analyzed assuming a fixed base (FB) for the first time and later using base isolation with lead rubber bearing (LRB) technique.

To perform seismic analysis using the time history method, a synthetic time series will be employed to conduct linear and geometric nonlinear seismic analysis for the fixed and base-isolated models to assess the p-delta effect. Synthetic time series compatible with the target response spectrum is generated by adjusting reference time series that consists of real records from past earthquakes. In this study reference

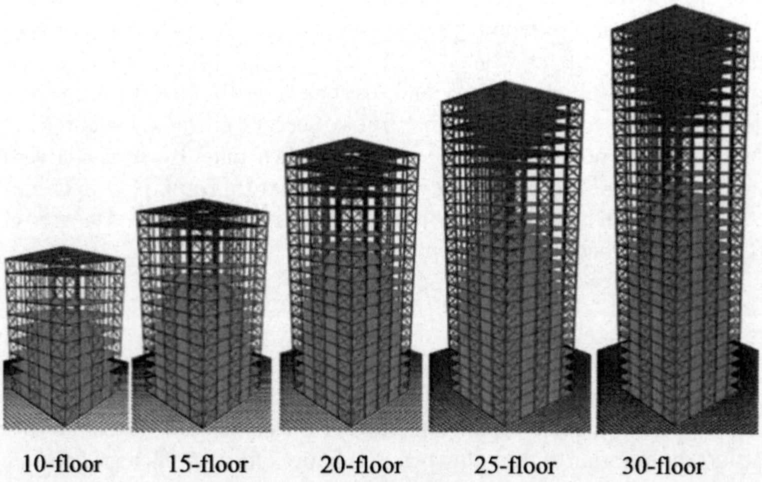

Fig. 1 Three-dimensional building models

Fig. 2 Plan dimensions

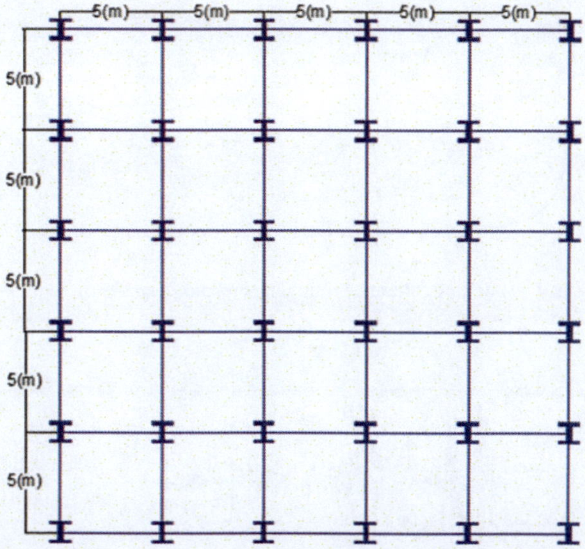

time history of the N-S component of the 1940 El Centro earthquake is adopted as an initial time series. Tall building models located in three different seismic design categories (SDC C, CDC D, and SDC E) with site soil class "D" and design "target" response spectra shown in Fig. 3 will be adopted. Spectral matching is based on the technique of adjusting the frequency content of the reference time series so that its response spectra match the design spectrum at all periods of interest. This is obvious from the comparison presented in Fig. 4 for the target response spectrum and the generated time series spectra, which reveals that both spectra match each other over

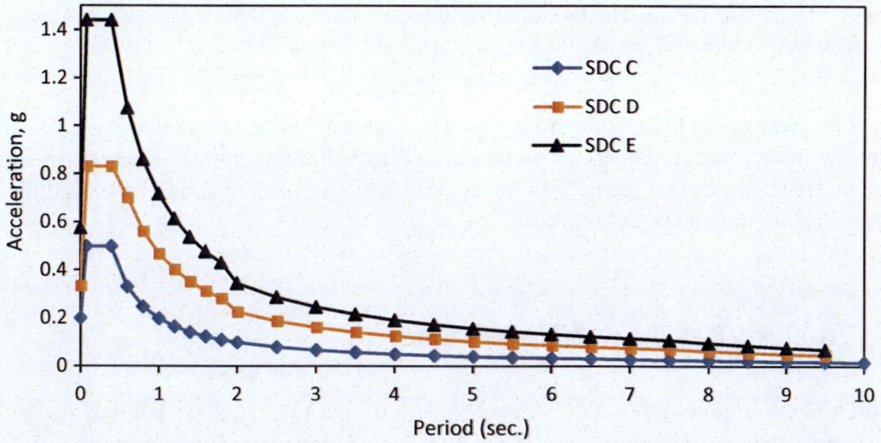

Fig. 3 Target "design" response spectra for SDC C, SDC D, and SDC E

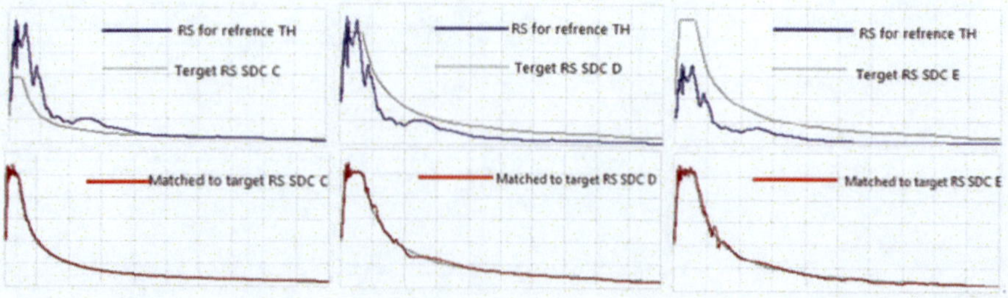

Fig. 4 Matching reference time series response spectrum with the design response spectra for SDC C, SDC D, and SDC E

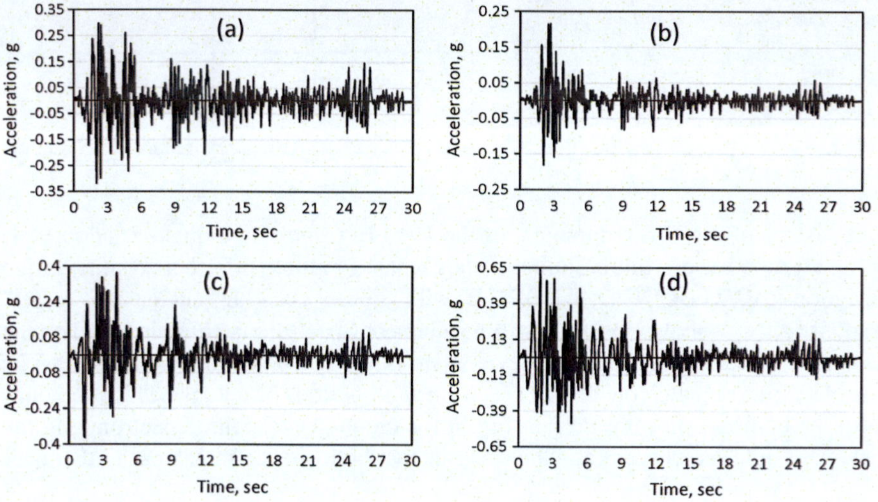

Fig. 5 Comparison for acceleration histories between **a** reference time series and generated time series, **b** SDC C, **c** SDC D, and **d** SDC E

the complete range of periods of interest. Comparison between initial and modified time series acceleration histories presented in Fig. 5 shows that the characteristics of the reference time series have been maintained while frequency content adjusted for each SDC to match target spectra.

4 Analysis Results and Conclusions

Story Shear. Figure 6 shows the story shear variation for the 10-story building due to different methods of analyses and SDC D when analyzed using synthetic earthquake time history. Comparison for maximum story shear variation for the different building heights, methods of analyses, and SDC D is shown in Fig. 7. Results presented in

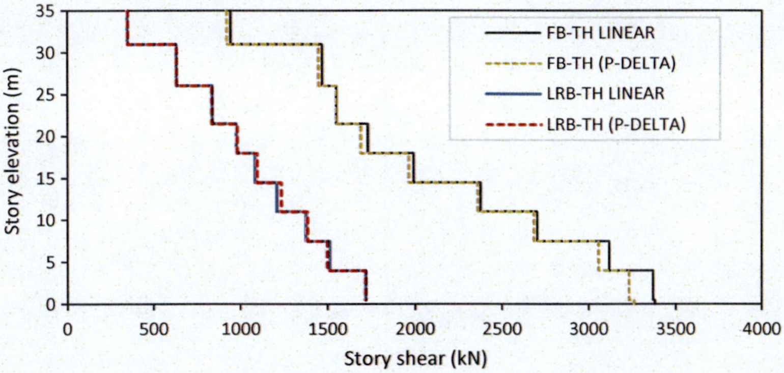

Fig. 6 Story shear variation with height for the 10-story building (SDC D)

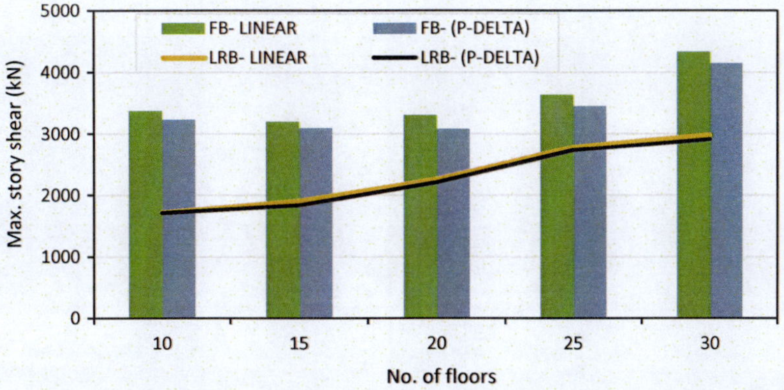

Fig. 7 Comparison for maximum story shear variation for the different building heights, methods of analyses, and SDC D

Figs. 6 and 7 reveal that story shear values significantly reduced and the difference due to nonlinear p-delta effect is minimized when the base isolation technique is utilized. This result reflects the significance of the damping and reduced stiffness introduce at the building base due to base isolation on the dynamic response. The p-delta effect on the maximum story shears for the different building heights and different site-specific excitations is summarized in Table 1. Results compiled in this table indicate that maximum story shear values reduce due to the p-delta effect. It is observed that increasing site seismic excitation does not contribute to the general trend of p-delta effect. Generally, the maximum reduction in the story shears for base-isolated buildings is about 4%, while a maximum value of about 7% is observed for fixed-base buildings.

Overturning Moment. The variation of the overturning moment with the height for the 10-story building due to different methods of analyses and SDC D is presented

Table 1 Percentage (%) reduction in the maximum story shear due to p-delta effect

Seismic design category	SDC C					SDC D					SDC E				
No. of floors	10	15	20	25	30	10	15	20	25	30	10	15	20	25	30
Fixed-base (FB)	0.4	3.6	4.0	5.6	2.2	4.2	3.1	6.8	5.1	4.3	1.1	2.2	1.0	3.7	1.2
Isolated-base (LRB)	1.7	2.8	2.1	4.0	2.1	0.3	2.7	2.2	2.0	2.4	0.5	1.3	0.7	2.7	1.1

in Fig. 8. However, Fig. 9 shows base moment variation for the different building heights, methods of analyses, and SDC D. These figures indicate that buildings overturning moments significantly reduced when base isolation is employed and that moment values are generally increased when p-delta effect is included. It is noted that base moment values increase as building height is increased. Results for the percentage increase in the base moments due to p-delta effect for different building heights and different site-specific excitations are compiled in Table 2. It is observed that site-specific base excitation has a negligible effect on the variation of the percentage increase in the maximum overturning moment. Generally, the maximum increase in the maximum overturning moment for the base-isolated buildings due to the p-delta effect is about 4.6%.

Story Drift. Results presented in Figs. 10 and 11 show the variation of the story drift ratios for the 10-story building and other building heights due to different

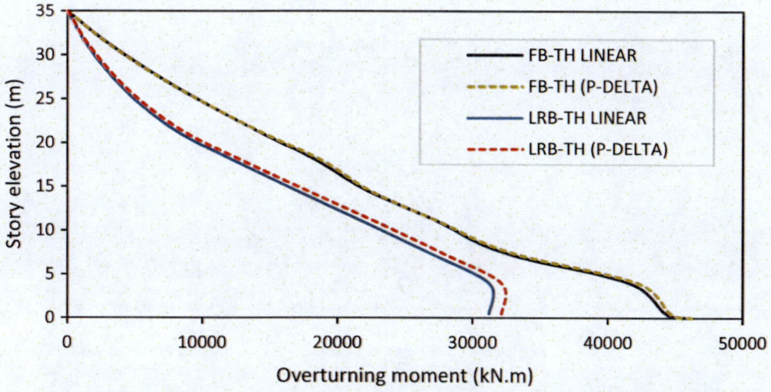

Fig. 8 Overturning moment variation with height for the 10-story building (SDC D)

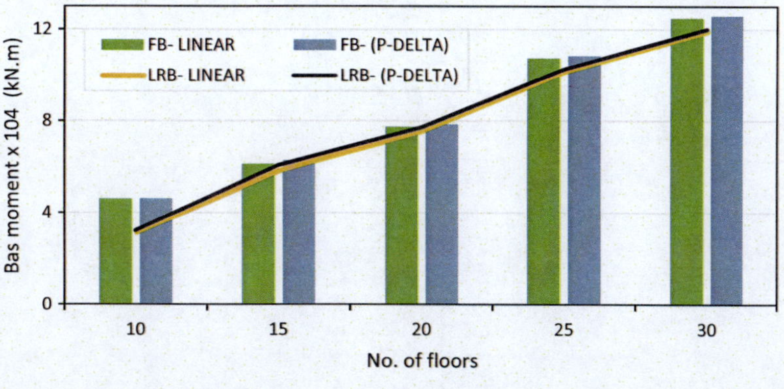

Fig. 9 Comparison for the maximum overturning moment variation for the different building heights, methods of analyses, and SDC D

Table 2 Percentage (%) increase in the buildings maximum overturning moments due to p-delta effect

Seismic design category	SDC C					SDC D					SDC E				
No. of floors	10	15	20	25	30	10	15	20	25	30	10	15	20	25	30
Fixed-base (FB)	1.8	3.5	2.6	2.6	1.2	0.6	2.5	1.4	0.9	0.8	2.4	3.5	2.2	2.2	2.3
Isolated-base (LRB)	2.3	3.7	1.3	0.5	0.5	2.8	4.2	2.0	0.9	0.7	2.9	3.0	4.6	3.5	3.1

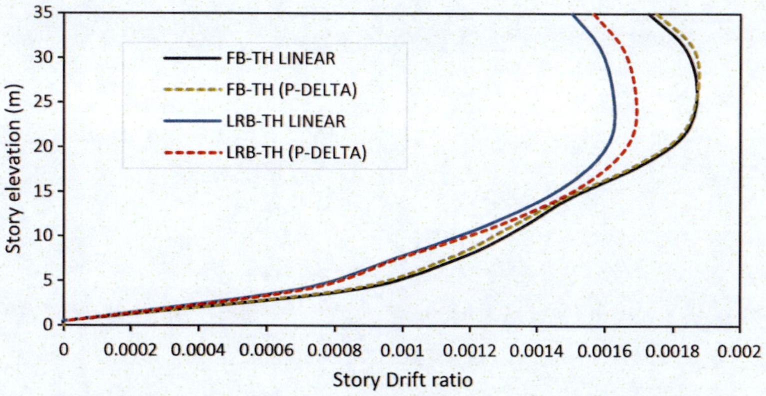

Fig. 10 Story drift ratio variation with height for the 10-story building (SDC D)

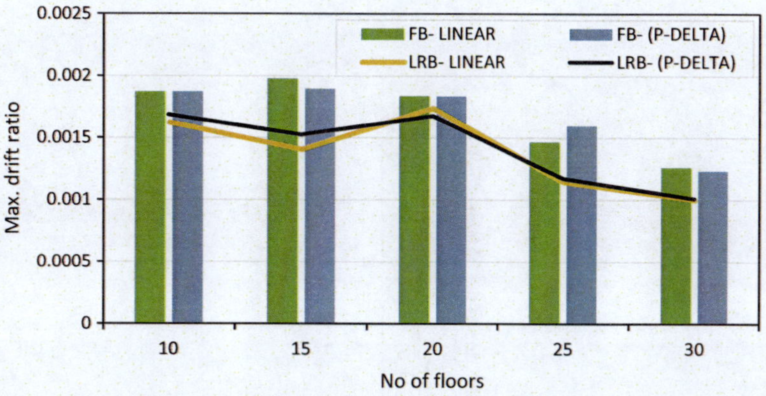

Fig. 11 Comparison for maximum drift ratios variation for the different building heights, methods of analyses, and SDC D

methods of analyses and SDC D. Comparisons presented in these figures indicate that buildings story drift ratios significantly reduced due to base isolation. Also, Fig. 10 shows building story drift ratio increases when the p-delta effect is included in the analysis. Results for the percentage change in the story drift ratios due to p-delta effect for different building heights and different site-specific excitations are compiled in Table 3. It is observed that site-specific base excitation has a negligible effect on the trend of the variation in the drift ratios. Generally, the variation in the drift ratios for the base-isolated buildings due to the p-delta effect range between −6.6% and +8.3%. The percentage variation of the story drift ratios presented in Table 3 is mainly attributed to the interaction of the dynamic characteristics for the different buildings and the site-specific excitation frequencies.

Table 3 Percentage (%) variation in the maximum drift ratios due to the p-delta effect

Seismic design category	SDC C					SDC D					SDC E				
No. of floors	10	15	20	25	30	10	15	20	25	30	10	15	20	25	30
Fixed-base (FB)	3.1	3.1	3.3	−5.5	−6.6	−0.1	−4.1	−0.2	8.9	−2.2	4.4	4.8	5.6	0.6	3.2
Isolated-base (LRB)	−0.5	2.3	2.4	−4.4	−6.8	3.9	8.3	−3.6	2.3	0.8	−0.4	0.3	2.0	−0.4	−0.9

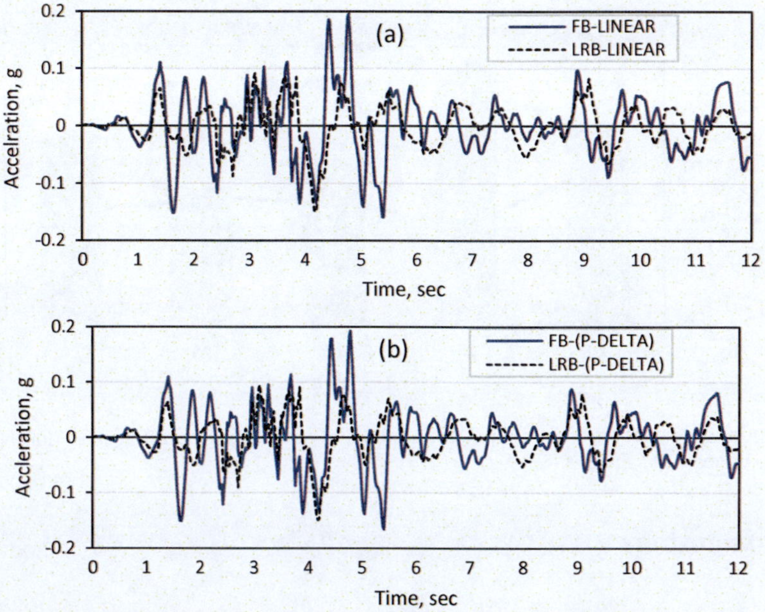

Fig. 12 Top story acceleration time history for the 10-story building and SDC D, **a** linear analysis, **b** nonlinear analysis

Story Acceleration. One of the most important tasks of base isolators is to reduce the acceleration generated as a result of the building's response to the ground acceleration. Hence, emphasis is placed on the acceleration values generated at the highest point of the models. The results shown in Fig. 12 for top story acceleration time history for the 10-story building reveal that the base isolators significantly reduce building acceleration values. The same effect is observed in Fig. 13 when considering different building heights, methods of analyses, and SDC D. Moreover, Fig. 13 reveals that the top story acceleration is reduced due to p-delta effect especially for fixed-base buildings and to a lesser extent for base-isolated once. Table 4 summarizes the effect of p-delta analysis on the reduction in the maximum top story acceleration for different building heights and site-specific base excitations. It is interesting to note that the maximum reduction in the top story acceleration values for base-isolated buildings is about 3.1% in comparison with 9.2% for fixed-base buildings, which indicated that base isolation reduces p-delta effect on building's acceleration.

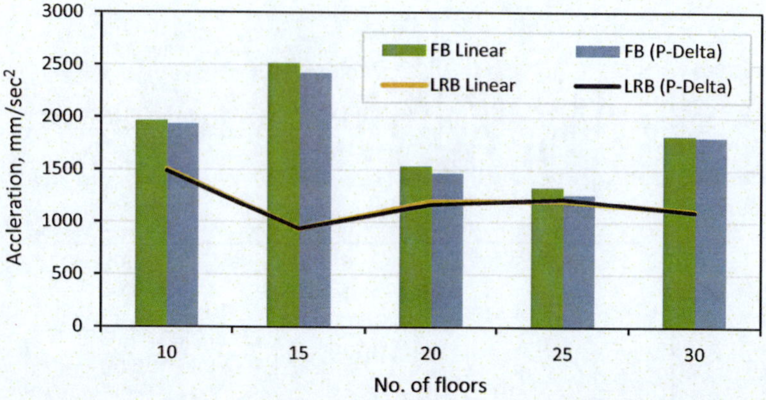

Fig. 13 Comparison for maximum top story acceleration variation for the different building heights, methods of analyses, and SDC D

5 Conclusions

The seismic response of base-isolated high-rise steel buildings with different heights subjected to synthetic earthquake records matched to different site-specific response spectra in the time domain has been analyzed to investigate the effect of geometric nonlinearity due to p-delta on the dynamic response. It has been observed that the dynamic response of tall buildings significantly reduced due to base isolation. The study revealed that for high-rise building construction, base isolation could significantly reduce seismic response in stiff soil (i.e., site class D) recommended by the ASCE 7 when the characteristics of the soil of the site are unavailable. The results presented have shown that site seismic intensity has a negligible effect on the percentage variation in the story shear, overturning moment, and building acceleration values when considering the p-delta effect due to different site-specific base excitations. A maximum reduction of less than 4% for the story shear and acceleration values and an increase of about 5% for the overturning moment values due to p-delta effect on the seismic response of base-isolated buildings have been found. Moreover, a variation in the range of −6.6% to +8.3% in the inter-story drift ratios has been observed, which could be attributed to the interaction between different building periods due to variation in the building height and base-isolation properties, and the site-specific excitation frequencies.

Table 4 Percentage (%) reduction in the maximum top story acceleration due to p-delta effect

Seismic design category	SDC C					SDC D					SDC E				
No. of floors	10	15	20	25	30	10	15	20	25	30	10	15	20	25	30
Fixed-base (FB)	3.5	0.8	6.9	9.2	1.1	1.6	3.9	4.6	5.2	0.8	0.7	0.9	4.2	0.3	1.1
Isolated-base (LRB)	0.8	0.2	2.0	1.0	0.7	0.9	0.2	2.5	1.2	1.1	3.1	0.3	2.3	1.4	1.2

References

1. Yuen YP, Kuang JS (2015) Nonlinear seismic responses and lateral force transfer mechanisms of RC frames with different infill configurations. Eng Struct 91:125–140. https://doi.org/10.1016/j.engstruct.2015.02.031
2. Adam P, Zembaty Z, Minch MY (2019) Seismic effects on leaning slender structures and tall buildings. Eng Struct 198:2. https://doi.org/10.1016/j.engstruct.2019.109518
3. Palermo M, Silvestri S, Landi L, Gasparini G, Trombetti T (2018) A "direct five-step procedure" for the preliminary seismic design of buildings with added viscous dampers. Eng Struct 173:933–950. https://doi.org/10.1016/j.engstruct.2018.06.103
4. Calvi PM, Calvi GM (2017) Historical development of friction-based seismic isolation systems. Soil Dyn Earthq Eng 106:14–30. https://doi.org/10.1016/j.soildyn.2017.12.003
5. Lago A, Trabucco D, Wood A (2018) Case studies of tall buildings with dynamic modification devices. In: Damping technologies for tall buildings. Butterworth-Heinemann, Elsevier Inc. https://doi.org/10.1016/B978-0-12-815963-7.00008-7
6. Providakis CP (2008) Effect of LRB isolators and supplemental viscous dampers on seismic isolated buildings under near-fault excitations. Eng Struct 30:1187–1198. https://doi.org/10.1016/j.engstruct.2007.07.020
7. Cancellara D, De Angelis F (2017) Assessment and dynamic nonlinear analysis of different base isolation systems for a multi-storey RC building irregular in plan. Comput Struct 180:74–88. https://doi.org/10.1016/j.compstruc.2016.02.012
8. Irfan M, El-Emam M, Khan Z, Abdalla J (2012) Local site effects on seismic ground response of Dubai-Sharjah metropolitan area 1. Geo Congress 2012, ASCE 1869–1878. https://doi.org/10.1061/9780784412121.192.
9. ASCE/SEI, ASCE 07-10 (2010) Minimum design loads for buildings and other structures. ASCE standard
10. Kircher C (2012) Seismically isolated structures, Fema P-751, NEHRP Recommended Provisions, pp 31–64
11. Shin G, Song O (2016) A time-domain method to generate artificial time history from a given reference response spectrum. Nucl Eng Technol 48:831–839. https://doi.org/10.1016/j.net.2016.01.023
12. Alatik L, Abrahamson N (2015) An improved method for nonstationary spectral matching. Earthquake Spectra 26:601–617. https://doi.org/10.1193/1.3459159
13. Lilhanand K, Tseng W (1987) Generation of synthetic time histories compatible with multiple damping response spectra. In: Proceedings of the international conference on structural mechanics in reactor technology, Lausanne, Switzerland

Experimental Investigation for Dynamic Response of Saturated Clay Under Machine Foundation

Ahmed S. Abdulrasool, Mohammed Y. Fattah, and Nahla M. Salim

Abstract A machine foundation is considered important and because of that the foundation design under dynamic loads must be governed by serviceability limit states performance considerations rather than strength requirements. The dynamic response is carefully calculated to ensure high-quality and efficient performing for machinery. This paper presents a model for estimating the dynamical response of a machine foundation on saturated clay by experimental work. The machine is produced to simulate forced vibrations by harmonic loading at various operating frequencies. The depth influence of embedding and load frequency on model footings on soft clay soil is investigated. As well as, stresses and excess pore water pressure get measured at different depths in the soil below the footing. It was concluded that the maximum displacement increases as the operating frequency increases, as well as, there is an increase in the excess pore water pressure when the dynamic load increases.

Keywords Dynamic analysis · Dynamic response · Saturated clay · Machine foundation · Pore water pressure

1 Introduction

Dynamic response is one of the primary parameters used to describe the soil behavior under dynamic loads. The machine foundation's behavior is affected by different factors such as soil properties, the weight of machine and foundations, the magnitude of the unbalanced force of machine, contacting zone between foundation and soils,

A. S. Abdulrasool (✉) · M. Y. Fattah · N. M. Salim
Civil Engineering Department, University of Technology, Baghdad, Iraq
e-mail: ahmeedsameerabd@gmail.com

M. Y. Fattah
e-mail: myf_1968@yahoo.com

N. M. Salim
e-mail: nahla_salim2007@yahoo.com

M. O. Karkush and D. Choudhury (eds), *Modern Applications of Geotechnical Engineering and Construction*, Lecture Notes in Civil Engineering 112,
https://doi.org/10.1007/978-981-15-9399-4_30

and static soil pressure [1]. For these reasons, the most critical step in the design to get a successful machine foundation can be relied on as an analyzing technique for dynamic responding and it should not surpass the limit given to a machine according to the manufacturing company. Anyway, before starting the acceptable limits of the settlement of the machine foundation, it is very important to explain that the failure of the vibrating foundations is reached when the motion exceeds a limiting value, which is usually expressed as displacement of the foundation at specified frequencies. From the data available so far, it appears that vibrations in neighboring structures will be negligible if the vibration of the foundation is less than 0.20 mm. For machine generating translation and torsion, the combined settlement should be estimated [2].

Many researchers have investigated the dynamic response under machinery; among them, the authors of [3–5] along with several researchers investigated the vertical vibrating manner for surface footing, whereas the authors of [6–8] along with many researchers investigated the vertical vibrating manner for embedding footings such as in [9, 10]. Through previous studies, it was proved that the experimental tests should take some considerations such as soil properties that are well documented before the test. The dynamic response measurement is not affected by wave reflections, as well as the dynamical loading represents a significant factor to interpret the behavior of the foundation soil.

Kumar and Reddy [11] examined the impact of the spring-mounting base on the responding of rigid footing and machine exposed to steady-state harmonious vertical load that had been expressly made via executing various block vibrating testing. The work of the spring-mounting pad, which possesses solidness a lot little compared to the one of soils strata, caused an extreme decrease in the resounding displacing amplitude for the footing. It additionally resulted in a critical decrease in the full frequency of the machine foundation. For every one during the examinations with the spring-mounting base, the resonant frequency of the machine base can be seen as constantly similar to the footing frequency. Mandal et al. [12] studied empirically the dynamical response of the foundation of layer soils over the rigid base. The relative strain level within the entire location of testing outcomes was in the order of 10^{-4}. The impacts of the last response of the strain level had been explained. Also, it was shown that to any applicable working, it was reasonable to use the analysis of direct elastically. In any case, finding comments could be discovered dependent on the testing outcomes as well as comparing the empirical working with the conclusive outcomes, and these comments can be given as follows:

1. The location along with thickness for a single layer affects the foundation's responding, for the most part. The increase in the thickness of the stiff layer causes an increase within the foundations' resonance frequency, while the increase in the thickness of the soft layers at the top causes a decrease within the foundation's resonance frequency.
2. Radiating dampness presumed as equal to zero to the foundation that rests upon two-layered soils that lie beneath rigid layers, whereas the entire dampness could be presumed along with the soil materials dampness to the entire applied working.

3. The responding of the foundation evaluated by soils of a multi-layer below the rigid layer by a mass-spring-dashpot example having to determine if the identical stiffness have been taken from Baidya and Muralikrishna [13], as the materials' dampness got proposed, it was concluded sensibly good in comparison with the hypothesis of comparable elastic half-space.

Based on the literature review and problems described above, this study aims at determining the dynamic response of soft clay by a two-mass oscillator machine.

2 Experimental Model and Material

Harmonic vibration tests are performed on a shallow foundation in the vertical directions to investigate the dynamic response on saturated clay. Tests are conducted on a machine foundation in a container box of dimensions (720 × 720 × 560 mm) and contains an absorbent layer to prevent reflected waves. These dimensions are found by finite element analysis carried out by [14]. The soil used from Al-Rashid camp embankment-Baghdad city was a relatively homogeneous brown clayey and it consists of 3% sand, 38% silt, and 59% clay. The Atterberg limits are 47 for liquid limit and 23 for plastic limit. The specific gravity is 2.67. The soil is classified as clay of low plasticity (CL) according to USCS (Unified Soil Classification System). Laboratory tests are implemented at the Soil Mechanics Lab of the Civil Engineering Department of the University of Technology.

Trial testing has been conducted to determine the undrained shear strength at variation water content as shown in Fig. 1. The soil bed is prepared at undrained shear strength (c_u) 20 kPa. The soil lumps are positioned in layers into the box container and each layer is leveled gently by using a wooden tamper to remove any entrapped air. A steel circular foundation of 110 mm diameter and thickness 25 mm

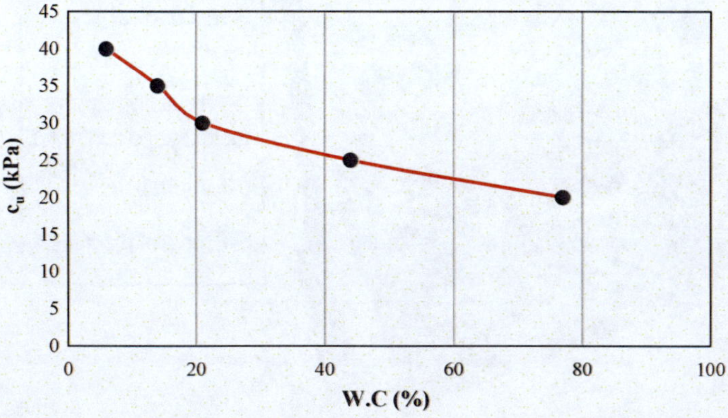

Fig. 1 Variation of undrained shear strength versus water content after 48 h curing

is positioned on the soil surface. The foundation is subjected to harmonic vibration in the vertical direction by a mechanical oscillator.

The oscillator consists of two AC motors operating together but rotates in opposite directions. Each motor comprises power 400 W, electric potential 240 V, alternating current motor speed controller (speed range 0–4500 rpm) and an eccentric weight 14.71 g and it has been placed to rotating disk with diameter 75 mm and thickness 11 mm. This weight is positioned at an eccentricity distance e_o of 27.5 mm from the axis of rotation. The horizontal component of force cancels out the corresponding component of the eccentric weight on the other motor, and the other way around. Just the vertical part of the force is remaining. Figure 2 shows the equipment for harmonic vibration and container box.

This type of machine was adapted by DEGEBO (German Research Society for Soil Mechanics) and different textbooks such as [2, 15]. The amplitude harmonic vibrating force (F_o) can be found in the vertical positions by:

$$F_o = 2m_e e_o \omega^2 \tag{1}$$

where $\omega =$ angular frequency of each counter-rotating machine, m_e refers to eccentricity mass and e_o is an eccentricity distance. The mechanical oscillator produces

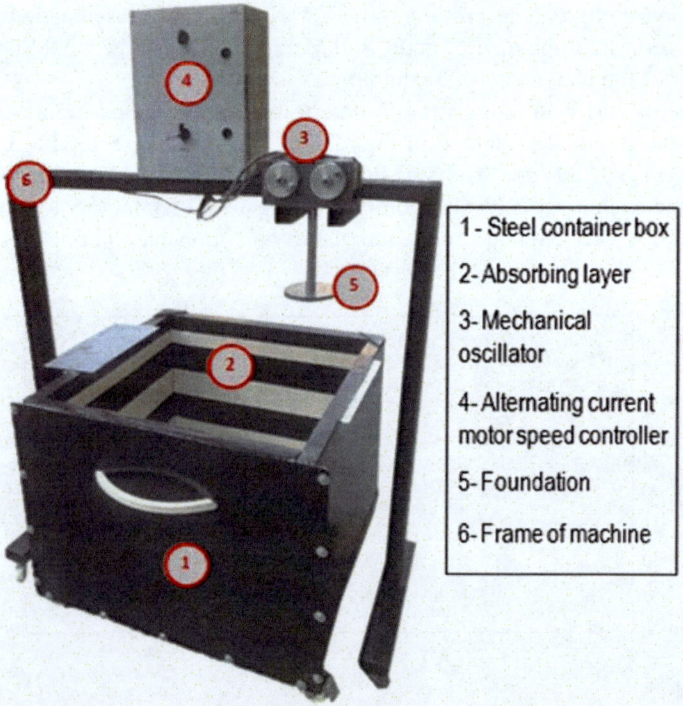

Fig. 2 Equipment for harmonic vibration and container box

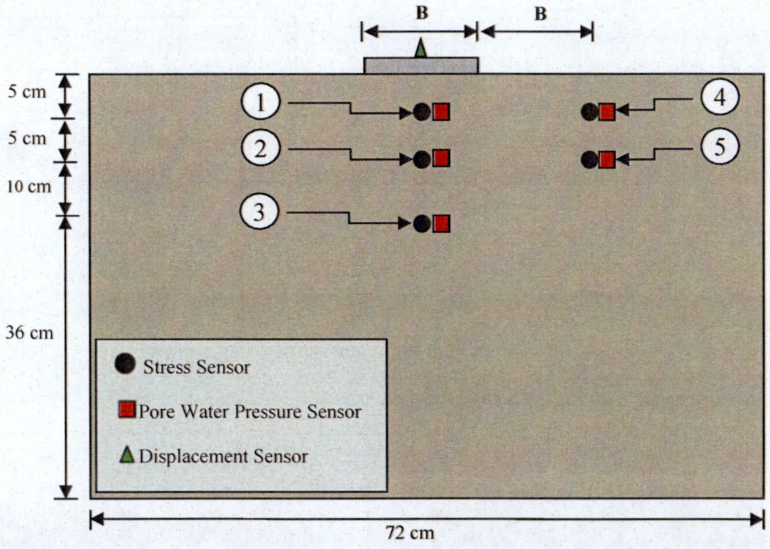

Fig. 3 Schematic view of sensors in model

a sinusoidal harmonic vibration; the vertical vibrating force of all times can be represented as:

$$F(t) = F_o \sin\omega t. \tag{2}$$

The vibration-measuring equipment consisted of the flexible ultra-thin tactile pressure sensors to measure the stresses under foundations (earth pressure transducers), pore water pressure gauge of range 0–5 MPa, and vibration meter is utilized for measuring displacement. The distribution sensors are shown in Fig. 3. Tests are carried out at different operating frequencies (1000, 2000, and 3000 rpm) and the applied dynamic load duration is 1000 s. Tests are summed up as clarified below due to the large quantity of data for stress and pore water pressure.

3 Results and Discussion

Figures 4, 5 and 6 present the displacement–time history for the footing at different depths and operating frequencies. From Figs. 4, 5 and 6, an increase within the amplitude displacement of the foundation with frequency can be noticed, and the amplitude displacement is changed from 0.865 to 2.60 mm with an increase in operating frequency from 1000 to 3000 rpm. On the other hand, there was a dramatic fall in amplitude displacement, as the foundation embedding increases from 50 mm and that the reduction in the amplitudes could either be on account of change in damping,

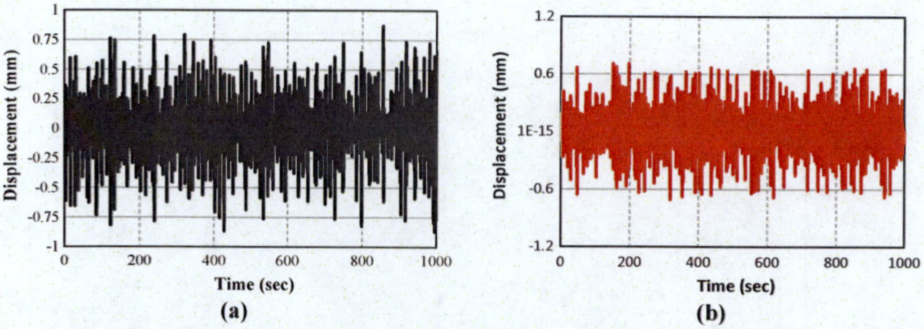

Fig. 4 Measured displacement at the edge of foundation with operating frequency 1000 rpm. **a** Foundation on the surface; **b** Foundation at depth 50 mm

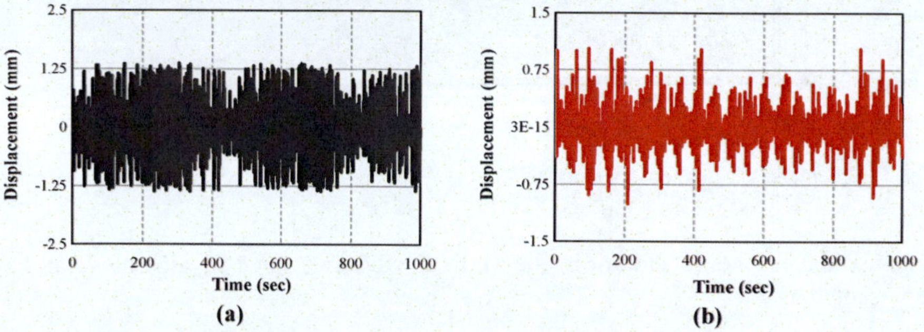

Fig. 5 Measured displacement at the edge of foundation with operating frequency 2000 rpm. **a** Foundation on the surface; **b** Foundation at depth 50 mm

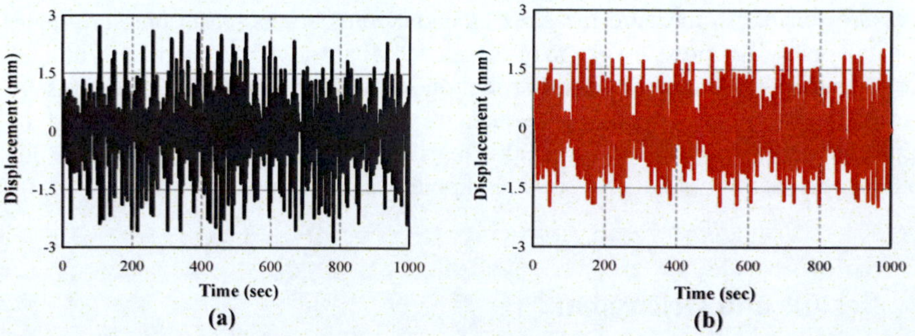

Fig. 6 Measured displacement at the edge of foundation with operating frequency 3000 rpm. **a** Foundation on the surface; **b** Foundation at depth 50 mm

as well as shear stresses caused by the extra inertia that provide by the embedding. Additionally, the overly soil restricts the vertical moving; as a result, decreases the foundation base settling via the increase within its vertical stiffness. Similarly, the amount of the practical loading gets transferred to the grounds by shear stressing alongside the vertical sides.

From the above results, it can be seen that the total stresses propagated to greater depth when operating frequency increased and maximum total stress occurred in point 1 at 50 mm under foundation. At the same point, the maximum total stress increases clearly by 17, 31, and 7% at operating frequency 1000, 2000, and 3000 rpm, respectively, when the embedment of foundation is 50 mm, but the sensor number 5 beside foundation at depth 100 mm, the total stress increases by 29, 16, and 31% when the embedment of foundation increase 50 mm. The embedment of foundation leads to an increase in dynamic stresses due to change in stiffness of system (soil–structure interaction) and increase the effective confining pressure due to overburden pressure of soil in the foundation (Fig. 7).

From Fig. 8 which shows the relationships between the excess pore water pressure at different points within the soil as well as operating frequency, the excess pore water pressure is increased via the increase within the embedding. It can be seen that the maxi excess pore water pressure occurs at operating frequency 3000 rpm for all cases. The water existence in the clay particles might need high frequency to dissociate water from particles.

4 Conclusions

- For circular foundation at the soil surface of soft clay, there was an increase in the amplitude displacement on foundation, total stress, and pore water pressure when there was an increase in the operating frequency.
- The increase of the embedment depth leads to a decreasing within the displacement response of the foundation.
- The total stress in soil is influenced by embedment depth; the effect will be increasing when the embedment depth is increased.
- As the embedment depth increases, the maximum pore water pressure decreases, and this effect is pronounced when increasing the operating frequency.

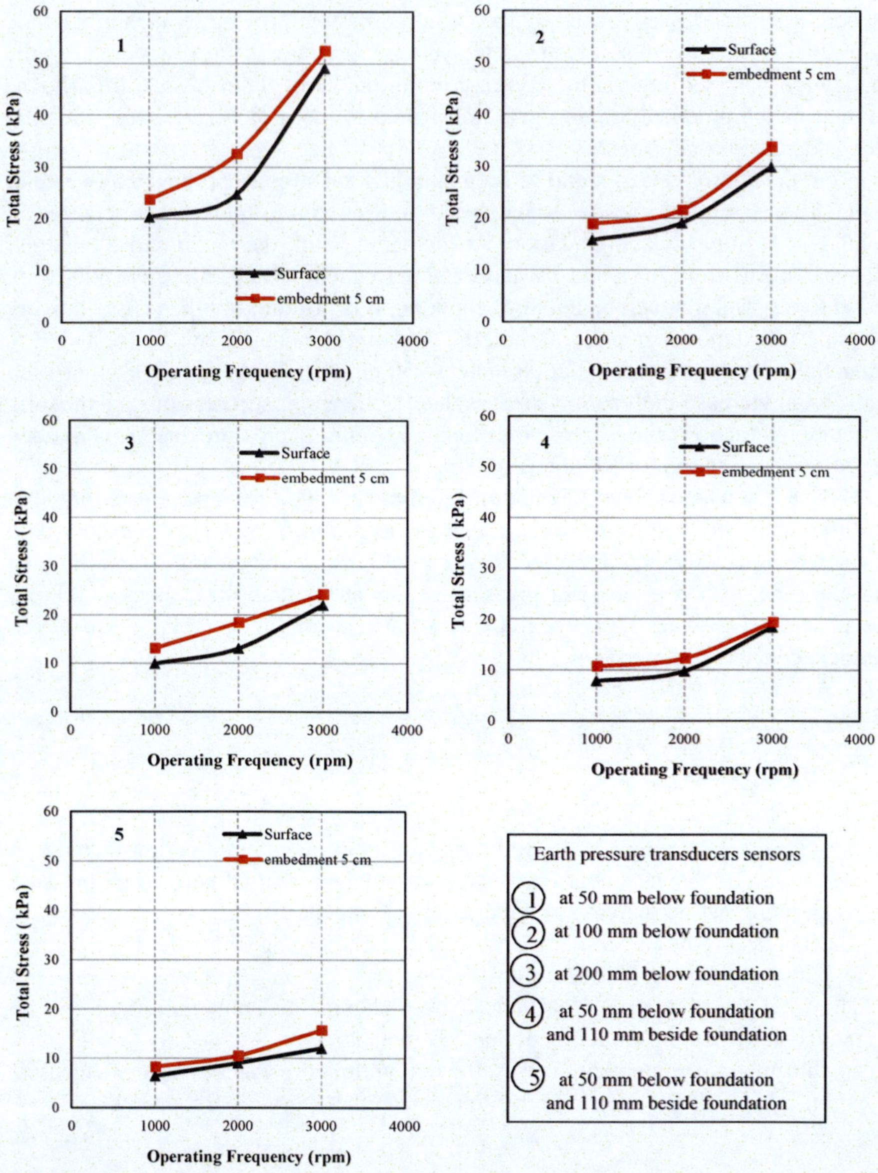

Fig. 7 Variance for the total maxi stress within the operating frequency of the machine at different points under the foundation

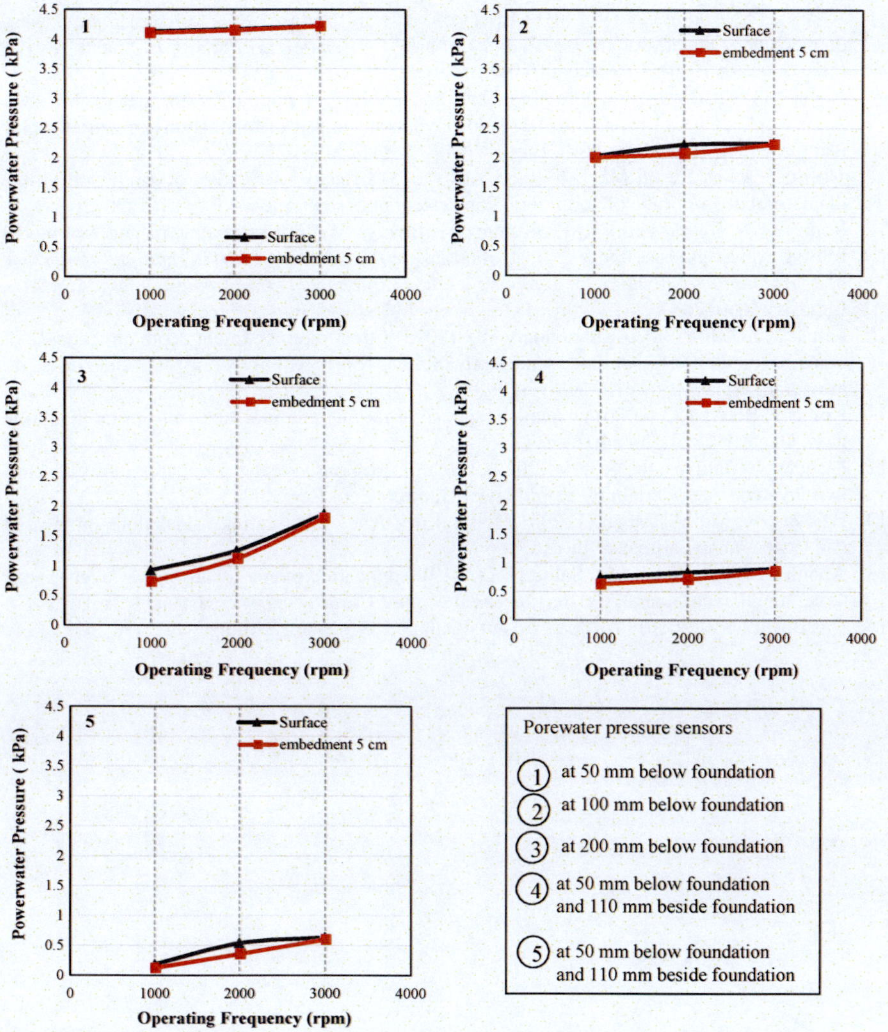

Fig. 8 Variance for the maxi excess pore water pressure within the operating frequency of the machine at various points under foundation

References

1. Ramesh HN, Kumar MT (2011) Concepts and problems in the design of foundations subjected to vibrations. Int J Geomech ASCE 312–321
2. Prakash S (1981) Soil dynamics. McGraw-Hill, New York
3. Gazetas G, Stokoe KH (1991) Free vibration of embedded foundation: theory versus experiment. J Geotech Eng ASCE 117(9):1382–1401
4. Baidya DK, Muralikrishna G, Pradhan PK (2006) Investigation of foundation vibrations resting on a layered soil system. J Geotech Geoenviron Eng ASCE 132:116–123

5. Al-Wakel S, Abdulrasool A (2018) Effect of soil stabilized by cement on dynamic response of machine foundations. In: *Proceedings, the BCEE3 conference*, Sharm El-Sheikh, Egypt. https://doi.org/10.1051/matecconf/201816201001
6. Novak M (1970) Prediction of footing vibrations. J Soil Mech Found Div ASCE 96(3):837–861
7. Al-Wakel SF, Fattah MY, Karim HH (2011) Dynamic analysis of foundations on saturated clay using an energy absorbing layer. Eng Technol J 29(11):2189–2201
8. Al-Mosawi MJ, Fattah MY, Al-Ameri AFI (2015) Effect of saturation of sandy soil on the displacement amplitude of soil foundation system under vibration. J Eng 21(2):20–36
9. Ramesh HN, Raghavendra RMV, Kumar MT, Bhavya M (2008) Dynamic response of model footing over a rigid base under vertical vibration. In: *Proceedings, 12th international conference of international association for computer methods and advances in geomechanics (IACMAG)*, India, pp 2680–2687
10. Fattah MY, Salim NM, Al-Shammary WT (2015) Effect of embedment depth on response of machine foundation on saturated sand. Arab J Sci Eng (Springer Publishing) 40(11):3075–3098. https://doi.org/10.1007/s13369-015-1793-8
11. Kumar J, Reddy CO (2006) Dynamic response of footing and machine with spring mounting base. Geotech Geol Eng 24:15–27
12. Mandal A, Baidya DK, Roy D (2012) Dynamic response of the foundations resting on a two-layered soil underlain by a rigid layer. Geotech Geol Eng 30:775–786
13. Baidya DK, Muralikrishna G (2001) Natural frequency of vibrating foundations on layered soil system-an experimental investigation
14. Abdulrasool AS, Fattah MY, Salim NM (2019) Application of energy absorbing layer to soil-structure interaction analysis. In: *Proceedings, the BCEE4 conference*, Istanbul, Turkey
15. Das BM, Luo Z (2016) Principles of soil dynamics. Cengage Learning

Estimation of the Settlement Components of Municipal Solid Waste

Bilal Muiassar M. Salih, Mohammed A. Ibrahim, and Raid R. Al-Omari

Abstract The results of the one-dimensional simulator represented the behavior of synthetic municipal solid waste (MSW), including biodegradation, physical, and mechanical properties, and main stages of settlement. The MSW settlement was divided into two main stages: immediate settlement and secondary settlement. Based on the MSW composition generated in Baghdad City, synthetic specimens of MSW were prepared. They were randomly collected by distributing bags to 30 houses at three different neighborhoods in Baghdad and transported to the laboratory in locked bags and were decomposed in anaerobic bioreactor cell with leachate recirculation. Many parameters were monitored simultaneously for 400 days, to permit for an overall description of biodegradation of MSW. The water content ($w = 57.9\%$) and the specific gravity ($G_s = 1.58$) of the entire specimen was measured. The values of the initial void ratio (e_0) and wet unit weight (γ_{bulk}) of the specimen were 2 and 8.12 kN/m^3, respectively. Measuring of all geotechnical, physical, microbial, and chemical parameters is important to generate a universal understanding of long-term biodegradation of MSW. The main parameters of the model, which are $C_c{}'$, $C_\alpha{}'$, k, and ε_{bt}, are important in the evaluation of settlement components and total settlement in landfill. The settlement and mass loss data obtained from this study could be used for the development of the settlement models.

Keywords Municipal solid waste (MSW) · Compression · Settlement · Creep · Biodegradation · Landfill · Anaerobic · Leachate recycling · Biogas · Bioreactor

B. M. M. Salih (✉)
Civil Engineering Department, College of Engineering, Al-Iraqia University, Baghdad, Iraq
e-mail: bilal.muasser@aliraqia.edu.iq

M. A. Ibrahim · R. R. Al-Omari
Civil Engineering Department, College of Engineering, Al-Nahrain University, Baghdad, Iraq
e-mail: moh_env@yahoo.com

R. R. Al-Omari
e-mail: tosharaid@yahoo.com

M. O. Karkush and D. Choudhury (eds), *Modern Applications of Geotechnical Engineering and Construction*, Lecture Notes in Civil Engineering 112,
https://doi.org/10.1007/978-981-15-9399-4_31

1 Introduction

The settlement of MSW can be divided into three main stages: immediate compression, mechanical creep, and biocompression [1–3]. The first stage of immediate compression is the initial fast strain accumulation that resulted from applying load. Here, the term immediate is used to differentiate it from the term primary in soil, which represents the consolidation of the soil as pore water pressure dissipates [4]. After the completion of immediate compression, the long-term secondary compression owing to mechanical creep and biodegradation starts. Mechanical creep occurred by bending, crushing, reorientation, and raveling of particles of MSW [5]. Strain due to biodegradation occurs as a result of loss in the strength and decomposition of organic matter. It was noted that the time when gas generation started in the bioreactor is the end of the mechanical creep stage and this indicates the beginning of biodegradation of organic matter. Biodegradation-induced settlement continues until complete decay of organic matter takes place. The magnitude of gas generation rates and the equivalent mass loss offers a good method to estimate the settlement [6].

Bioreactor landfills are modern landfills developed to distribute recycled leachate and additional liquids into the waste; however, this type of landfills aims to enhance decomposition of waste which leads to acceleration of production of biogas, treatment of leachate in site, and increase the waste settlement [7, 8].

MSW secondary settlement caused by mechanical creep and biodegradation is calculated by estimating each process independently [9, 10]. Many field and laboratory studies on MSW long-term settlement were carried out to study biodegradation-induced compression. The rate of compression due to biodegradation represents a challenge for researchers. Biodegradation-induced compression strain increases with increasing organic matter content [10]. References [11, 12] found a more appropriate simulation to calculate biodegradation-induced compression based on first-order kinetics.

A one-dimensional compression model was proposed by Chen et al. [10] and was applied by [13] and [14]. In this model, the long-term secondary compression (creep and biodegradation) was combined by two first-order kinetic equations to analyze storage capacity and settlement of MSW landfill. A modified model was proposed by Xiao et al. [15] separating the creep effect on secondary compression based on a one-dimensional model of Chen et al. (2010) to show the relationship between biodegradation process and biodegradation-induced compression. The main study objectives are to investigate the gas generation potential and leaching behavior of anaerobic bioreactor landfill, studying the biodegradation compression and behavior of MSW in anaerobic bioreactor landfill, and to evaluate the loss of mass versus the loss of volume in an anaerobic bioreactor for use in future settlement model development.

2 Model Description

Chen et al. [10] proposed a modified model that estimates the total MSW compression strain (ε) as a sum of immediate compression, mechanical creep compression, and biodegradation-induced compression, which are expressed as follows.

$$\varepsilon_t = \varepsilon_I + \varepsilon_c + \varepsilon_b \tag{1}$$

$$\varepsilon_I = C_c^{'}.\log\frac{\sigma}{\sigma_0} \tag{2}$$

$$\varepsilon_c = C_{\alpha e}^{'}.\log\frac{t}{t_I} = \frac{\Delta h}{h_I} \tag{3}$$

$$\varepsilon_b = \varepsilon_{bt}\left(1 - e^{-kt'}\right) \tag{4}$$

where ε_I is immediate compression strain of MSW due to load $\Delta\sigma$. ε_c and ε_b are creep and biodegradation-induced compression strains at time t since the application of $(\sigma_0 + \Delta\sigma)$, respectively. $C_c^{'}$ is the modified immediate compression index of fresh MSW; $C_{\alpha e}^{'}$ is the creep index of MSW; t_I (day) represents the time for the completion of immediate compression; ε_{bt} is the ultimate biodegradation-induced strain; t' (day) is the time from the start of biodegradation, and k (day^{-1}) is the rate coefficient of biodegradation-induced compression.

3 Materials and Methods

In this study, MSW was tested in a laboratory simulator to simulate the MSW biodegradation in landfills. Degradation of MSW is continued for approximately 400 days inside a simulator. Measuring of all geotechnical, physical, microbial, and chemical parameters is important to generate a universal understanding of long-term biodegradation of MSW.

Test apparatus design. The rate of MSW samples decomposition was promoted in the laboratory to create ideal conditions for biodegradation compression. Figure 1 shows the system used during the laboratory tests which had leachate collection and recirculation system, gas collection and analyzing system, and loading system that measure axial deformation. Some of the system components are described below.

Sample cell. PVC pipe is used as a specimen cell, with dimensions 600 mm height, inner diameter of 200 mm, and 10 mm wall thickness. Two stainless steel caps were clamped by steel rods on both ends of the waste columns of the simulator. The contact surface between the column and the caps was sealed by rubber gaskets (see Fig. 2). All inner surfaces were grease-treated to minimize the friction. A perforated leachate distribution plate is placed on the top of the waste sample to uniformly

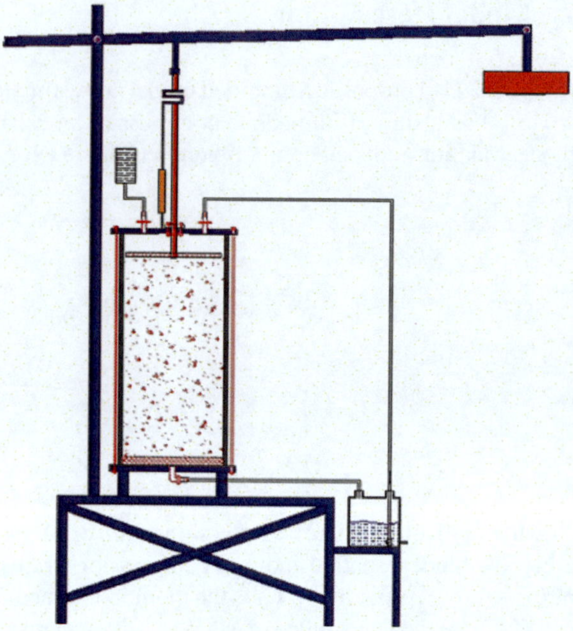

Fig. 1 The test system layout

distribute the recycling leachate. Below the specimen, a perforated plate is placed, and a non-woven geotextile was placed above the perforated plate to avoid losing of fine-grained material.

Compression system. A steel weight was utilized to make an axial load over the MSW sample. Digital weight indicator and load cell was used to determine vertical load. The specimen settlement was monitored using an LVDT and dial gauge.

Leachate Recirculation System. Leachate was collected from the drainage valve below the base of the sample cell. The goal of this system is to collect the leachate in a container and recycle it to the top of the sample cell and also apply alkali or acid for pH control. A submersible pump delivers leachate from the tank to the top of the waste sample through a perforated leachate distribution plate above the waste specimen to uniformly distribute the recycling leachate.

Gas Collection System. A biogas components analysis was used to study the biodegradation degree of MSW. The gas collection system consists of a tube at the top of the cell cap and a bag of gas sampling. The biogas was analyzed using two analyzing gas sensors fixed at top of 100 mm diameter PVC pipe. The concentrations of CH_4 and CO_2 for all biogas samples were recorded. The generated biogas volume was determined by the technique of water displacement at various periods (see Fig. 3). The volume of generated CH_4 and CO_2 was calculated as the product of the CH_4 and CO_2 concentrations and the volume of biogas.

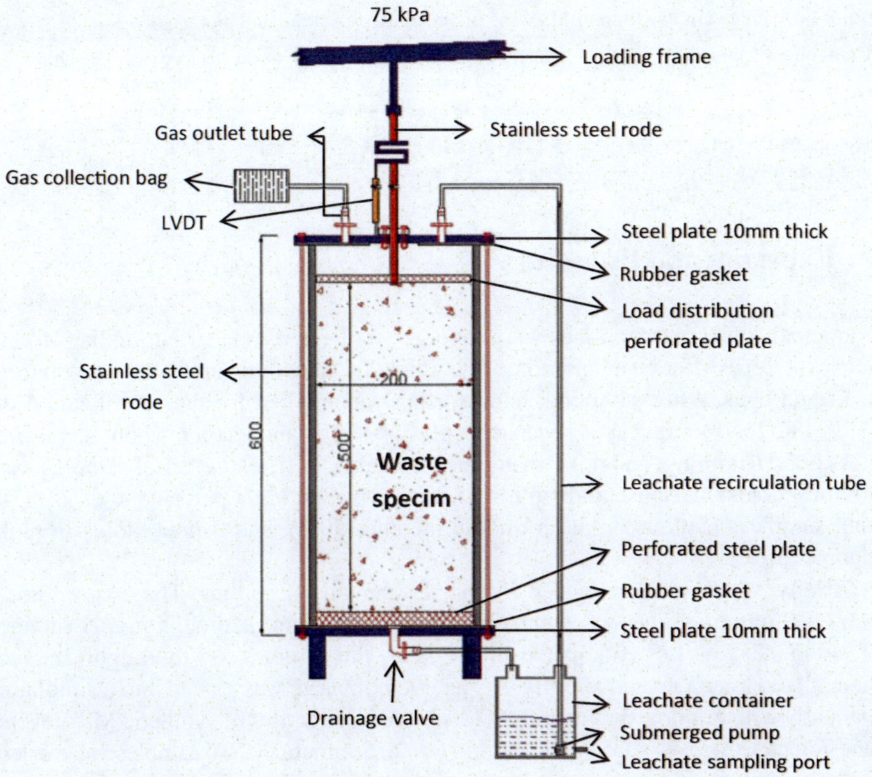

Fig. 2 Details of test apparatus (simulator)

Fig. 3 Schematic of biogas system

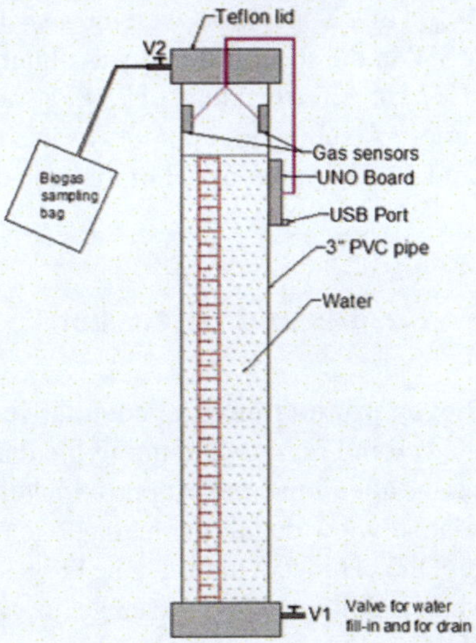

Table 1 Average composition of MSW of Baghdad used in simulated bioreactor

Component	Food scrap	Paper and carton	Plastic	Textile and leather	Metal and glass	Wood	Yard trimming and leaves	Others
Weight (%), wet basis	69.2	5.4	5.11	2.96	3.96	3.0	2.96	7.41

4 Experimental Procedure

Municipal solid waste samples were randomly collected by distributing bags to 30 houses at three different neighborhoods in Baghdad, and transported to the laboratory in locked bags. After sampling, hand sorting was applied for the classification of MSW into food scrap, paper and carton, plastic, textile and leather, wood, metal and glass, yard trimming, and other materials (see Table 1). The number of samples was 30 households; 12 households in low income representing a percentage of 40% of total samples, 15 households in middle income (50%), and 3 households in high income (10%).

Briefly, waste was segregated to coarser and finer fractions. The coarser fraction (>30 mm material) was manually cut to be <30 mm. Although proper mixing of components is not anticipated in the field, the sample was thoroughly mixed manually to assure reproducibility. A specimen of waste was placed into a simulator manually with minimal compaction. The physical properties of synthetic MSW were determined using two subsamples of this waste. Synthetic MSW samples were dried at a temperature of 60 °C in the oven for 48 h to avoid losing of organic matters. The water content ($w = 57.9\%$) and the specific gravity ($G_s = 1.58$) of the entire specimen was measured. The values of the initial void ratio (e_o) and wet unit weight (γ_{bulk}) of the specimen were 2 and 8.12 kN/m^3, respectively. The initial height of the MSW in the specimen cell was about 500 mm and has a volume of about 0.0157 m^3 (15.7 L). A vertical load of 75 kPa was utilized over the MSW sample during the test period. A submerged pump was used for recycling the leachate solution at a rate of 6 mL/s. The value of pH of the added solution was at a range of 6–8.

5 Results and Discussion

Biogas production. The cumulative gas production of the anaerobic simulator is showed in Fig. 4 by summing the daily biogas production volume. Gas production begins after four weeks (approximately) of operation production of biogas which was relatively low due to the methanogenic bacteria acclimatization and leachate which kept the pH low. After that, the biogas generation increased and methanogenic status was established by the change in gas concentrations to about 50–60% CH_4 and 25–35% CO_2 by volume.

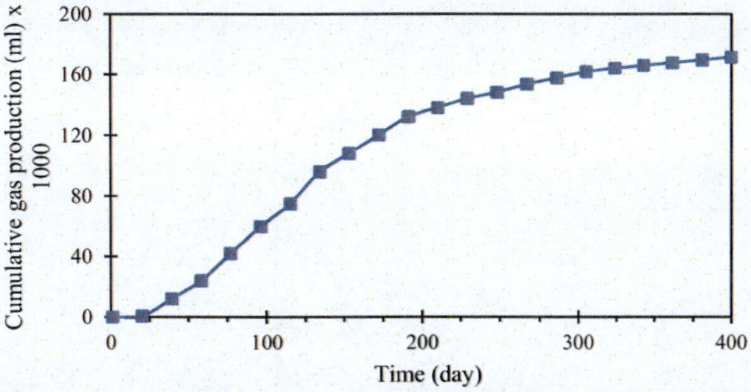

Fig. 4 Cumulative of biogas production

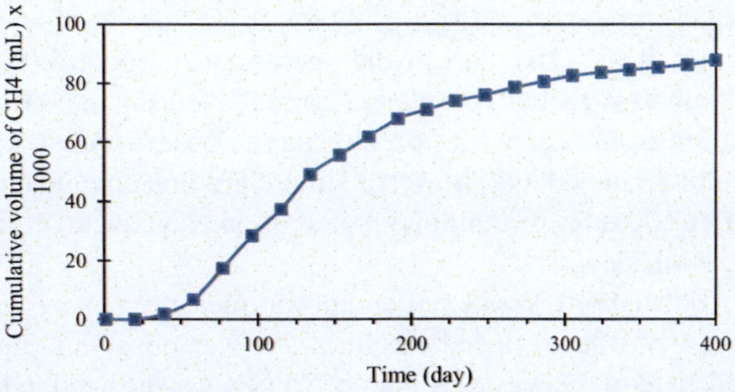

Fig. 5 Cumulative volume of CH_4

At the beginning of the waste decomposition, aerobic bacteria break down the complex carbohydrates, proteins, and fats that contain organic waste. During this process, aerobic bacteria consume the existing oxygen and convert it initially into carbon dioxide (CO_2). A total of 171,600 mL of biogas ($CH_4 + CO_2$) was generated. A total of 87,900 mL of CH_4 was generated as shown in Fig. 5 by day 400, which is equivalent to 6761 mL of CH_4/kg of initial total waste mass.

The volume of CH_4 increased rapidly from day 40 of operation to around day 190. CH_4 generation slowed down after day 190. A time-series biogas composition was recorded, as shown in Fig. 6. At the beginning, CO_2 concentrations increased from

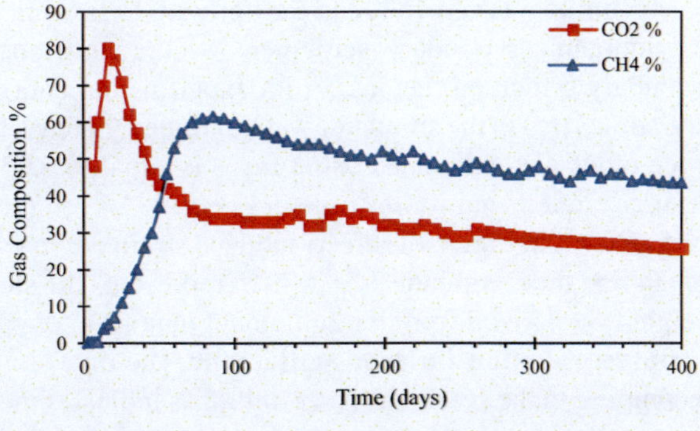

Fig. 6 Biogas composition

Fig. 7 Immediate settlement

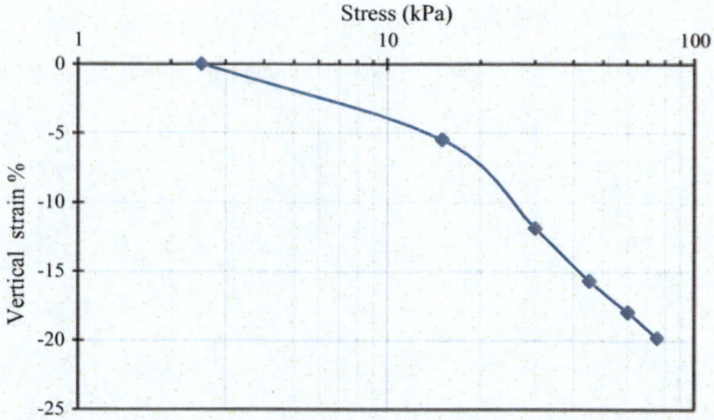

about 50% at the start of the operation to around 80% and it decreases quickly to about 30–40% between days 25 and 55. During the same period, CH_4 concentrations raised from nearly 0 to 62% through 75 days. After 150 days, the compositions of both CH_4 and CO_2 showed a simple gradual change until the operation has stopped after 400 days of continuous operation. The average gas production rate was about 429 mL/day.

Settlement. Waste settlement was measured for a period of 400 days. At the beginning (zero load), a settlement of 2% was observed due to the addition of synthetic leachate to the waste. A load of 75 kPa was utilized to the sample and continued for about 400 days.

Immediate (Primary) Settlement. This settlement was calculated as a percentage of the height of waste before and after the application of load (75 kPa) and was about 20%. The initial load σ_0 was determined to be 2 kPa. Figure 7 shows the relationship between vertical strain and stress. From Eq. 2, the value of $C_c^{'}$ for fresh waste was determined to be 0.125.

Secondary Settlement. The mechanical creep and biodegradation referred to as long-term secondary settlements were analyzed via simplified settlement models and then plotted against log-time. Secondary settlement was recorded after the immediate settlement which lasted around 72 h after the application of the load. The secondary settlements relevant model parameters were determined using the experimental results.

Mechanical creep-induced Settlement. Following primary settlement, mechanical inorganic secondary settlement occurs until such time that biodegradation secondary settlement begins. From experimental data, the creep-induced compression strains (ε_c) in the simulator was continued till day 91 and was equal to 3.21%. The creep index $C_{\alpha e}^{'}$ calculated using Eq. 3 is equal to 0.022. Creep-induced settlement strain is plotted against log-time in Fig. 8.

As shown in Fig. 8, the creep-induced strain curve was approximated as a straight line to find the creep index $C_{\alpha e}^{'}$ which represents the slope of this line. The bulk unit weight (γ_b) derived from the settlement data from day 3 to 91 was 10.77 kN/m^3.

Biodegradation-induced settlement. The time at which organic matter starts to decompose in the cell in the form of biogas indicates the end of the mechanical creep

Fig. 8 Creep compression with time

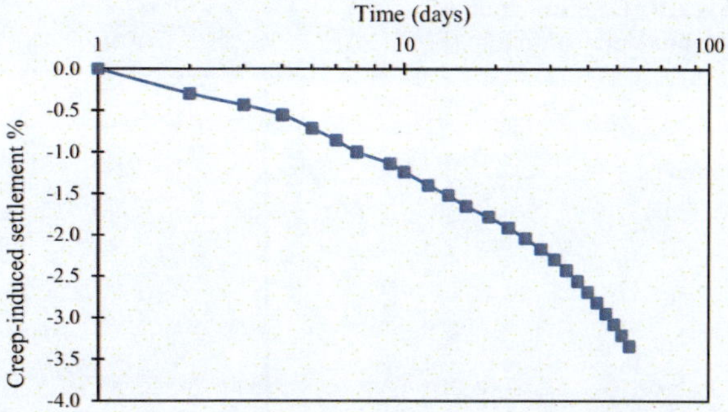

[16]. Biodegradation settlement continues until full degradation of organic matter. The value of this settlement can be estimated from the rates of gas generation and the equivalent mass loss [6]. Biodegradation-induced settlement was analyzed using Eq. 4. The initial organic content of the MSW measured according to ASTM D2974 (heated at 550 ± 50 °C for 6 h) was 70%. The organic matter content was measured at 75, 150, 225, and 400 days to determine the MSW biodegradation rate, *k*, as shown in Fig. 9 and Table 2.

Degradation in organic solid waste is the result of a combination of chemical and predominantly biological events. As shown below, an equation of organic matter degradation was derived with time [17], and expressed as:

$$C_t = C_o e^{-kt} \tag{5}$$

C_t = the organic matter concentration at time *t*.
C_o = the initial organic matter concentration.

Fig. 9 Losing of organic matter during degradation

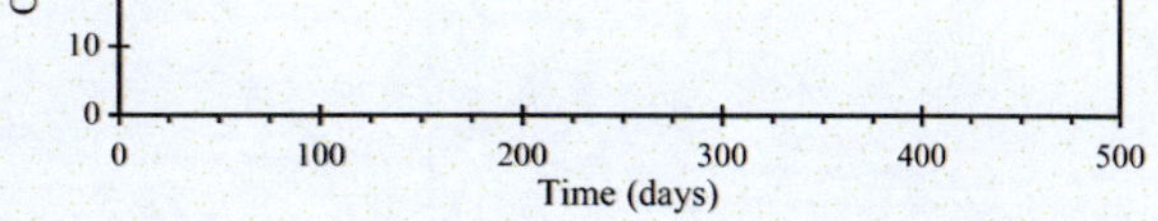

Table 2 MSW organic matter content

Time (Days)	0	75	150	225	400
OC (%)	70	65	55	50	40

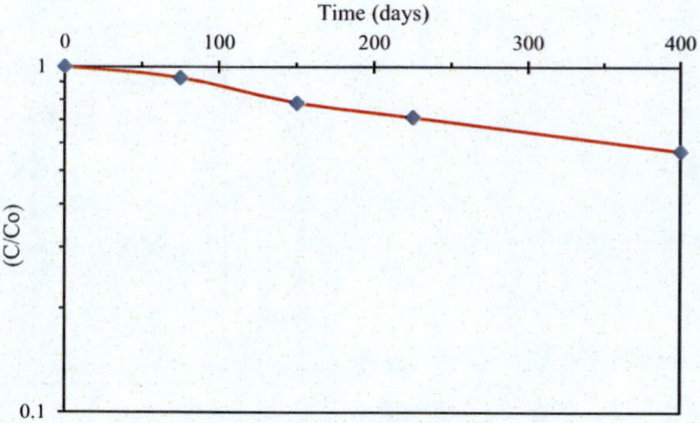

Fig. 10 Determination of the biodegradation rate (k)

k = overall biodegradation rate.

$$\frac{C_t}{C_o} = e^{-kt}$$

$$\ln\left(\frac{C_t}{C_0}\right) = -kt \quad (6)$$

From Eq. 6 the value of k was measured as the approximated fitted straight line's slope, as shown in Fig. 10. Knowing the values of ε_b and k, the values of ε_{bt} and t' were calculated by substituting in Eq. 4. The values of k, ε_{bt} and t' is 0.0014 day^{-1}, 34.4%, and 4900 days, respectively. Figure 11 shows the biodegradation-induced settlements with time. The starting time of biodegradation secondary settlement begins after mechanical creep settlement ended.

Total secondary settlement. Total long-term secondary settlement was determined from summing of mechanical creep and biodegradation, and can be shown by summing Eqs. 3 and 4 as:

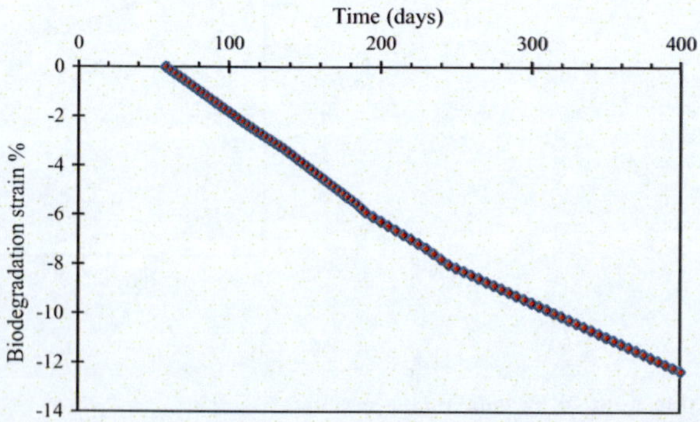

Fig. 11 Biodegradation-induced strain with time

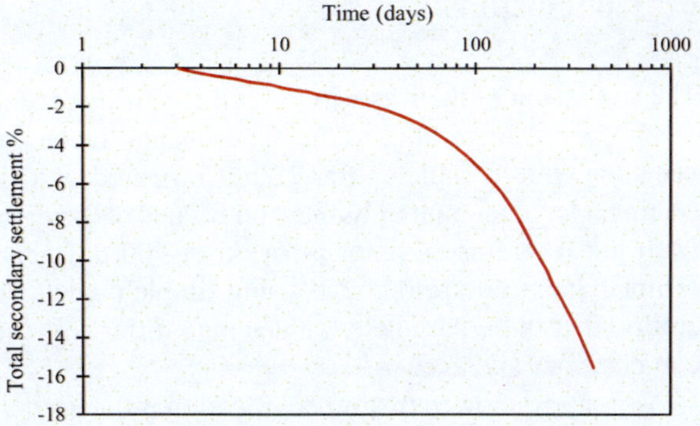

Fig. 12 Total secondary settlement

$$\varepsilon_{ST} = C'_{\alpha e}.\log\frac{t}{t_p} + \varepsilon_{bt}\left(1 - e^{-kt'}\right) \tag{7}$$

The results of total settlement indicate that the settlement mainly occurred by waste decomposition phase (Fig. 12). Great settlement performance was shown in the simulator during the decomposition phase rather than the creep phase. Therefore, it is concluded that waste decomposition plays a more important role than creep for landfill settlement.

Relationship between biodegradation strain and gas generation. The measured ε_b at time t' was divided by ε_{bt} to determine the degree of biodegradation-induced strain. Similarly, the gas generation degree was determined from the rate of the cumulative gas production (at the time t') to the total collected gas during the test period, as depicted in Fig. 13.

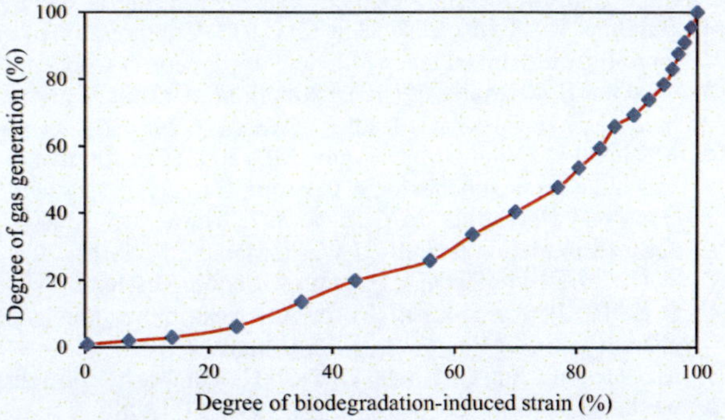

Fig. 13 Relationship between the degrees of biodegradation strain and gas generation

6 Conclusions

The settlement of the waste in the cell can be divided into immediate and secondary compression. According to the experimental results, the immediate settlement has been shown to complete after 72 h. Compared to a landfill of several decades and costing a large amount of money, the values of the parameters of this work extracted from the laboratory test for a period of 400 days can be employed in landfills to estimate long-term settlements using simple models. It should be taken into consideration that these parameter values may differ from site to site according to MSW compositions and age.

This study showed that biodegradation influenced the settlement of MSW significantly, which attributed to the high organic content present in the sample. The immediate settlement was found to be 20%; settlement owing to mechanical creep was 3.21%. Biodegradation-induced strain was found to be 12.5%. A relationship between gas generation and biodegradation rate (k) is useful for estimating biodegradation settlement. Biodegradation-induced settlement reduces with the waste degradable matter content. The relationship between biogas volume and settlement is approximately linear for the MSW investigated in this study.

References

1. Hossain MS, Gabr MA, Barlaz MA (2003) Relationship of compressibility parameters to municipal solid waste decomposition. J Geotech Geoenviron Eng 129(12):1151–1158
2. El-Fadel M, Khoury R (2000) Modeling settlement in MSW landfills: a critical review. J Crit Rev Environ Sci Tech 30:327–361
3. Bareither CA, Benson CH, Edil TB, Barlaz MA (2012) Abiotic and biotic compression of municipal solid waste. J Geotech Geoenviron Eng 138(8):877–888
4. Bareither CA, Benson CH, Edil TB (2013) Compression of municipal solid waste in bioreactor landfills: mechanical creep and biocompression. J Geotech Geoenviron Eng 139:1007–1021
5. Chakma S, Mathur S (2012) Estimation of primary and mechanical compression in MSW landfills. J Hazard Toxic Radioact Waste 16:298–303
6. Machado SL, Vilar OM, Carvalho MF (2008) Constitutive model for long term municipal solid waste mechanical behavior. J. Comput. Geotech. 35:775–790
7. Benson CH, Barlaz MA, Lane DT, Rawe JM (2007) Practice review of five bioreactor/recirculation landfills. J Waste Manage 27(1):13–29
8. Barlaz MA, Bareither CA, Hossain A, Saquing J, Mezzari I, Benson CH, Tolaymat TM, Yazdani R (2010) Performance of North American bioreactor landfills: II Chemical and biological characteristics. J Environ Eng 136(8):839–853
9. Marques AC, Filz GM, Vilar OM (2003) Composite compressibility model for municipal solid waste. J Geotech Geoenviron Eng 129(4):372–378
10. Chen Y, Ke H, Fredlund DG, Zhan L, Xie Y (2010) Secondary compression of municipal solid wastes and a compression model for predicting settlement of municipal solid waste landfills. J Geotech Geoenviron Eng 136(5):706–717
11. Ivanova LK, Richards DJ, Smallman DJ (2008) The long-term settlement of landfill waste. J Waste Resour Manage 161(WR3):121–133
12. Gourc JP, Staub MJ, Conte M (2010) Decoupling MSW settlement into mechanical and biochemical processes-modeling and validation on large-scale setups. J. Waste Manage 30:1556–1568

13. Chen YM, Xu XB, Zhan LT (2012) Analysis of solid–liquid–gas interactions in landfilled municipal solid waste by a bio-hydro-mechanical coupled model. J Sci China (Technol Sci) 55:81–89
14. Li YC, Liu HL, Cleall PJ, Ke H, Bian XC (2013) Influences of operational practices on municipal solid waste landfill storage capacity. Waste Manage Res 31:73–82
15. Xu X, Tony LT, Chen YM, Guo QG (2015) Parameter determination of a compression model for landfilled municipal solid waste: An experimental study. J Waste Manage Res 33(2):199–210
16. Sivakumar GL, Lakshmikanthan P (2015) Estimation of the components of municipal solid waste settlement. J Waste Manage Res 33(1):30–38
17. Lakshmikanthan P, Sughosh P, Babu GS (2018) Studies on characterization of mechanically biologically treated waste from Bangalore city. J Ind Geotech 48(2):293–304